T0073065

Mining Gold and Manufacturing Ignorance

Jock McCulloch • Pavla Miller

Mining Gold and Manufacturing Ignorance

Occupational Lung Disease and the Buying and Selling of Labour in Southern Africa

palgrave
macmillan

Jock McCulloch
Melbourne, VIC, Australia

Pavla Miller
School of Global, Urban and Social Studies (emerita), RMIT University
Melbourne, VIC, Australia

ISBN 978-981-19-8326-9 ISBN 978-981-19-8327-6 (eBook)
https://doi.org/10.1007/978-981-19-8327-6

This Palgrave Macmillan imprint is published by the registered company Springer Nature Singapore Pte Ltd.
The registered company address is: 152 Beach Road, #21-01/04 Gateway East, Singapore 189721, Singapore

To the memory of Jock McCulloch and the miners whose predicament he sought to bring to light.

Acknowledgements

Jock McCulloch finished research and a draft of this book before he died of mesothelioma in January 2018. As in his previous projects dealing with occupational health in southern Africa, he received generous support from a number of people and organisations. The funding for this project came from a Discovery Grant on 'Tuberculosis in Southern Africa: The History of a Pandemic', awarded by the Australian Research Council in 2013.

The colleagues who provided help, advice and a sounding board for the project included Tony Davies, David Rees, Jill Murray and Mary Ross at the National Institute of Occupational Health in Johannesburg, Jonny Meyers at the University of Cape Town and Geoff Tweedale in Manchester. Jock was fortunate to interview Oluf Martiny, who spent his career working as a medical officer with the Witwatersrand Native Labour Association in Johannesburg. He also owed thanks to Emile Schepers and his late father Gerrit Schepers in Grand Falls, Virginia, who candidly talked about his career in mine medicine in Johannesburg from the late 1940s. Jillian Smith was a research assistant for a major part of this long project, and her help with managing copies of archival documents has been invaluable.

Jock was also grateful for the friendly and competent assistance from the staff of the South African National Archives, the Archives at the

University of Johannesburg, the National Archives of Malawi, the National Archives of Swaziland, the National Archives of Botswana, the British National Archives, the Library of Parliament in Cape Town and the Adler Museum of Medicine in the Faculty of Health Sciences at the University of the Witwatersrand.

When Jock fell ill, I promised I would complete the manuscript and get it published. The task turned out to be more complex and time consuming than I initially envisaged. Several colleagues provided most valuable help and advice along the way. Jock's erstwhile collaborator Geoff Tweedale copyedited the whole original text. One of the initial readers of the manuscript, Thomas Cousins from the University of Oxford, provided invaluable suggestions for sharpening central arguments and setting the material in a wider scholarly context. David Rosner from New York University corrected some factual inaccuracies and helped make many points of the argument more precise. Rodney Ehrlich from the University of Cape Town made detailed comments on each chapter, pointed out inconsistencies in some of the arguments, asked many important and pertinent questions, and in the process provided me with a sound basic education in epidemiology. Catherine Meyburgh, one of the makers of the documentary *Dying for Gold* and an active member of the Justice for Miners campaign (JFM), helped clarify a number of issues regarding the JFM. Many other people have provided support, help and encouragement with this project. I hope that we have all done justice to Jock's legacy.

Contents

Abbreviations

AAC:	Anglo American Corporation
AASA:	Anglo American South Africa Limited
AVL:	Assisted voluntary labourers
AVS:	Assisted Voluntary Scheme
BCG:	Bacillus Calmette-Guerin: a vaccine to prevent tuberculosis
CCOD:	Compensation Commission for Occupational Diseases
Chamber:	Chamber of Mines
COIDA:	Compensation for Occupational Injuries and Diseases Act No. 130 of 1993
Council:	British Social Hygiene Council
DMS:	Director of Medical Services used in connection with colonial medical services
DNA:	Department of Native Affairs
DOTS:	Directly Observed Therapy Short-course
GMO:	Group Medical Officers Committee
GPC:	Gold Producers Committee
HTCs:	High Commission Territories (or British High Commission Territories [BCTs]): These comprised Botswana (Bechuanaland) Swaziland and Lesotho (Basutoland)
ILO:	International Labour Organisation
JFM:	Justice for Miners Campaign
MBOD:	Medical Bureau for Occupational Diseases

MRC: Medical Research Council
MPPC: Miners' Phthisis Prevention Committee, Miners' Phthisis Medical Bureau (Bureau) or (MPMB) established in 1916
MMOA: Mine Medical Officers' Association
MMR: Mass Miniature Radiography
MPMB: Miners' Phthisis Medical Bureau
MWU: Mine Workers Union
NHLS: National Health Laboratory Service
NCOH: National Institute for Occupational Health (note on changing names)
NTP: National Tuberculosis Programme
NAC: Native Advisory Council
NIOH: National Institute for Occupational Health
NRC: Native Recruiting Corporation
NUM: National Union of Mineworkers
ODMWA: Occupational Diseases in Mines and Works Act No. 78 of 1973. Passed in 1973, it superseded the previous Phthisis Acts, and still governs compensation in the Mines.
OFS: Orange Free State
PAS: Para-aminosalicylic acid (a chemotherapeutic agent used in the treatment of tuberculosis)
PDC: Pulmonary Disability Committee
PMF: Progressive massive fibrosis
PRU: Pneumoconiosis Research Unit
PRO: Public Records Office
SAIMR: South African Institute of Medical Research
SIMRAC: Safety in Mines Research Advisory Committee
SMB: Silicosis Medical Bureau
SANA: South African National Archives
SANTA: South African National Tuberculosis Association
SADC: Southern African Development Community: a regional organisation consisting of 14 Member Countries: Angola, Botswana, Congo (DR), Eswatini, Lesotho, Malawi, Mauritius, Mozambique, Namibia, Seychelles, South Africa, Tanzania, Zambia and Zimbabwe.

TEBA: The Employment Bureau of Africa (formed in 1977 from amalgamation of the NRC and the WNLA)
TRC: Tuberculosis Research Committee
WHO: World Health Organization
WNLA: Witwatersrand Native Labour Association

1

Introduction

In early August 2012, following failed wage negotiations and civil unrest, about three thousand striking platinum miners employed by Lonmin plc gathered on a hill close to the town of Marikana in the North West province of South Africa. Nine out of ten Lonmin miners were migrants, mainly from the Eastern Cape, who lived next to the mine in squalid shacks and informal settlements.[1] During a confrontation on 16 August, police opened fire, wounding 78 miners and killing 34 others. Many of those killed and injured were shot in the back.

As required under the Occupational Diseases in Mines and Works Act, the cardio-respiratory organs of the deceased men, who were deemed to have died suddenly on the mine, were sent to the National Institute for Occupational Health for examination.[2] The autopsy results were telling. Of the thirty-four, three miners had silicosis, with nodules in both lungs.

[1] For a study of the men's life conditions see D. van Wyk. 'A Review of Platinum Mining in the Bonjanala District of the North-West province. A Participatory Action Research Approach'. Johannesburg: The Bench Marks Foundation, 2012. Available at http://www.bench-marks.org.za/research/rustenburg.

[2] J.I. Phillips, G. Nelson, N. Vorajee, J. Murray, N. Ndlovu, J.C.A Davies. 'Marikana Autopsies Highlight Occupational Diseases Amongst Platinum Mine Workers'. Occupational Health Southern Africa, Vol. 20, No. 4, 2014, pp. 6–12.

J. McCulloch, P. Miller, *Mining Gold and Manufacturing Ignorance*,
https://doi.org/10.1007/978-981-19-8327-6_1

Another four had silicotic nodules in adjacent lymph glands. Evidence of previous pulmonary tuberculosis was found in two of the miners and a further two had emphysema. One contract worker from the Eastern Cape, who had not been diagnosed in life, had extensive silicosis, warranting first-degree certification. The other two mine workers in whom pulmonary silicosis was found at autopsy had also not been previously diagnosed. The deceased had a median age of thirty-three and more than half had previously been employed in gold mines. Five were migrant workers from neighbouring countries; like all such workers, they had limited previous access to health services. Lonmin had incomplete work histories for the deceased, and this compromised their families' claims for compensation.

In various ways, Marikana presents a microcosm of South Africa's mining history. Mining has always been dangerous work, much of it performed by migrant workers under the eye of brutal management and a violent state. For most of the twentieth century, South Africa's gold mines were the most profitable sector of the national economy. Gold was the single most important export, and gold mining the largest employer. Today, platinum is one of South Africa's leading exports. By 2010, the industry exceeded the value of gold mining and employed more labour.

Like previous generations of miners, the men who died at Marikana were migrant workers paid below subsistence wages. A number had occupational disease for which they had not been compensated. Most had worked at other types of mines before arriving at Marikana. Over time, the lungs of gold miners fill with dust. As well as leading to silicosis, this greatly increases the likelihood of developing active tuberculosis. Eventually the men fail the entry medical examination. As their health deteriorates, they move along the chain of employment from skilled work in larger and well-resourced mines to poorer paying jobs. Many now work on the platinum mines.

Against the background of continuity, there is one aspect of the Marikana tragedy which is radically different from the past. Under the various twentieth-century Miners' Compensation Acts, workers had no effective right to legal redress against employers. Prior to majority rule the concept of an individual miner suing an employer for lung disease was foreign to South African law. The compensation system established from 1911 was always assumed to preclude such claims and there is no evidence of a black miner ever bringing a case to court. The law also made no provision for class actions. This has now changed. In July 2019, the settlement of a massive class action in which former gold miners sought damages for their occupational lung disease was approved by the South African High Court. Under the terms of the settlement, all eligible workers suffering from silicosis and/or tuberculosis, who worked in the companies' mines from March 1965 to date, are eligible for a graduated schedule of compensation. Importantly, the settlement makes no distinction between citizens of South Africa and those of the adjoining states and allows damages to be claimed by the dependents of deceased miners.[3]

The Political Economy of Mining

When gold was first discovered on the Rand in the 1880s, most Africans lived in subsistence agricultural communities. Expropriation of land by white settlers, the imposition of taxation by colonial administrations and regulations prohibiting black ownership of land outside of 'native reserves' were all in part designed to drive indigenous peoples into wage labour and the money economy. Below subsistence wages entrenched Africans' reliance on paid labour. It was around this dynamic that the South African gold mining industry was built. Extracting the low-grade ore only seemed possible if wages of the huge mining workforce were kept very low. Since most locals could get better pay in other occupations, they preferred to find employment elsewhere. The fixed price of gold made it difficult for companies to respond to changing production costs. The resulting chronic shortages of labour—and the companies' increasing reliance on migrants—were compounded by the health risks associated with mining. It is probable that at some point in the mid-1920s, the South African industry accepted that it was impossible to engineer dust out of the mines and that silicosis and tuberculosis, especially among migrant workers,

[3] 'Media Release: Settlement of the silicosis and TB class action. South Africa—Gold mining companies', Johannesburg, 3rd May 2018; 'Media Release: Court approves settlement of the silicosis and TB class action', Johannesburg, 26th July 2019. http://goldminersilicosis.co.za/about-the-silicosis-litigation/updates/.

were inevitable.[4] Rather than aiming to eliminate silicosis, the mining industry began attempting to push back its onset.[5] When, in 1943, the Stratford Commission had the Government Actuary cost recommended compensation of miners with occupational lung disease, it became evident that the industry could not operate at profit if its legal obligations were met.[6] It was in the context of the ongoing threat to their viability that the mining houses entered debates about the character and incidence of occupational lung disease, set up clinics and hospitals, employed medical staff, collected (or avoided collecting) statistics, negotiated with public officials, funded research, contracted with insurances companies, influenced the flow of information and helped shape the public perception of risk.

While the mining houses grappled with shortages of labour in a dangerous industry, colonial administrations of the adjoining states faced chronic shortages of resources. Next to funding their own bureaucracies, they were responsible for keeping the peace, supporting local agriculture and industry, providing services such as hospitals and schools and ensuring the wellbeing of the population. The administrators' attempts to steer a path between their responsibilities and constraints spanned vigorous advocacy on behalf of black miners at one end and collusion between the gold mining companies, the South African state and the British Colonial Office in the buying and selling of labour at the other.

Mines, Occupational Lung Disease and the Development of Scientific Knowledge

The establishment of South Africa's gold mines, the consolidation of colonial empires and the intensification of global trade coincided with a critical period in the making of modern medical science. It also occurred

[4] Jock McCulloch. *South Africa's Gold Mines and the Politics of Silicosis*. James Currey, 2012, p. 52.

[5] Paul-André Rosental. 'Truncating a Disease. The Reduction of Silica Hazards to Silicosis at the 1930 International Labor Office Conference on Silicosis in Johannesburg'. *American Journal of Industrial Medicine*, vol. 58, Suppl. 1, November 2015, p. S9.

[6] *Report of the Miners' Phthisis Acts Commission, 1941–1943 (Stratford Commission)*. Pretoria: Government Printer, 1943, p. 9.

at a crucial juncture in the development of 'statistical thinking'.[7] Government attempts to control and to tax unruly populations, entrepreneurs' struggle to maximise profits, outbreaks of epidemics in new industrial and population centres, debates about risk and occupational health disease, contests about authority and knowledge between different groups of bureaucrats, professionals and scientists, all contributed to this process of intellectual innovation. Importantly, information, technologies, conceptual tools and procedures of governance were not simply disseminated from the metropole to colonial outposts, but rather circulated among different jurisdictions. At times, the colonies became laboratories for refining core administrative techniques, problematics and conceptual distinctions of Western scientific, social and political thought.[8]

What is today known as tuberculosis and silicosis are ancient diseases. In Western cultures, debates about their distinguishing features, diagnosis, causes and cures have stretched back hundreds of years. Spurred on by epidemic crises in several sectors of increasingly mechanised industries, the debates intensified during the nineteenth century. The precision of medical arguments was enhanced by unprecedented accumulations of mortality statistics, more systematic collection of morbidity data and new diagnostic methods and instruments, some based on developments in analytic chemistry. By the late nineteenth century, scientific advances made it possible to distinguish between diseases caused by the inhalation of dust and those due to bacterial infection. The terms pneumoconiosis (a collective name for a large group of chronic lung illnesses and systemic diseases caused by the inhalation of dust) and silicosis were first suggested

[7] There is an extensive literature on this topic. Key texts include Theodore Porter, *Trust in Numbers. The Pursuit of Objectivity in Science and Public Life* (1996); Mary Poovey, *A History of the Modern Fact: Problems of Knowledge in the Sciences of Wealth and Society* (1998); Ian Hacking, *The Taming of Chance*, (1990); and Alain Desrosières, *The Politics of Large Numbers: A History of Statistical Reasoning* (2002).

[8] Among key texts in the literature on imperial social formation and circulation of knowledge are works by Ann Laura Stoler, such as *Along the Archival Grain: Epistemic Anxieties and Colonial Common Sense* (2009); Helen Tilley, *Africa as a Living Laboratory* (2011); Nancy Rose Hunt, *A Nervous State: Violence, Remedies, and Reverie in Colonial Congo* (2016); Keith Breckenridge, *Biometric State: The Global Politics of Identification and Surveillance in South Africa, 1850 to the Present* (2014) and Douglas Hay and Paul Craven (eds), *Masters, Servants, and Magistrates in Britain and the Empire, 1562–1955* (2004).

by scholars in 1867 and 1871[9]; the bacteria that causes tuberculosis was discovered in 1882. In practice, it would be decades before tuberculosis and silicosis could be separately diagnosed in living subjects; in many regions, reluctance to collect relevant statistics hindered epidemiological research.

Importantly, the process of medical knowledge-making did not simply involve discovering and naming something—like a mountain range—that was clearly already there. Rather, debates about tuberculosis mobilised—and helped create—an array of derogatory stereotypes about the nature and habits of working people and colonial subjects. At the same time, the complexity of tuberculosis gave rise to a changing array of technical terms used to describe different forms and stages of the disease and to map its statistical incidence. Today, it is estimated that only between 5 and 10 per cent of people *infected* with TB bacteria (or those with latent TB) will develop the disease during their lifetime; most do so when their bodies come under additional stress. Since those with latent (rather than active) TB show no symptoms, few are diagnosed, not least since diagnosis is expensive and relies on advanced medical technology. Current World Health Organization statistics distinguish between TB 'incidence', 'prevalence' and 'mortality'. Incidence refers to the number of new and relapse cases of TB arising in a year and is derived from the rates of notification of TB cases in countries with well-funded health systems, and a range of other surveys and estimates elsewhere. 'Prevalence' refers to the total number of cases of TB at a given point in time and is subject to similar constraints regarding funding and technical capacity.[10] In effect, working out the true extent of TB is a complicated, expensive, technically demanding and highly negotiated process.

Similarly, the complex process of identifying silicosis involved drawing agreed boundaries around a highly complex *continuum* of interaction between workers' exposure to various forms of dust and pathogens and

[9] A. Baldasseroni and F. Carnevale. 'The Genesis and Development of the Scientific Concept of Pulmonary Silicosis during the Nineteenth Century'. In Paul-Andre Rosental, ed. *Silicosis: A World History*. Baltimore: Johns Hopkins University Press, 2017, p. 39.

[10] For an overview of the four main ways of deriving current TB statistics, see *Global Tuberculosis Report 2019*. Geneva: World Health Organization, 2019, pp. 26–34.

their lung tissue, while excluding the role of silica dust in other pathologies (such as auto-immune diseases and some forms of carcinoma).[11] The growing body of medical knowledge was never enough to determine these developments. Rather, professionals, political interests and economic constituencies all played important roles in interpreting the events in people's lives that came to be identified as industrial disease and disability. In turn, these changing interpretations became the basis of claims about causes and origins (or aetiology), latency, onset, incidence, prevalence, morbidity and mortality.[12] Indeed, some scholars argue that the social debates around silicosis set the agenda of national and international controversies about what constituted industrial disability and who should bear the financial and societal burden of dependency, family and community trauma and personal cost.[13] In this shifting terrain, the older term *miners' phthisis* was a useful but unstable category that sometimes described only tuberculosis but at other times included combinations of lung diseases such as silicosis, asbestosis and pneumonia. It is telling that black miners in Southern Africa continued to describe their chest problems as *sifuba* or *phthisis* well into the twenty-first century.[14]

[11] Paul-André Rosental, David Rosner and Paul D. Blanc. 'From Silicosis to Silica Hazards: An Experiment in Medicine, History, and the Social Sciences'. *American Journal of Industrial Medicine* Vol. 58, No. S1, 2015, p. S4.

[12] Gerald E. Markowitz and David Rosner. 'The Limits of Thresholds: Silica and the Politics of Science, 1935 to 1990'. *American Journal of Public Health*, Vol. 85, No. 2, 1995, p. 250. See also Paul-André Andre Rosental. 'Introduction: Why silicosis?' In *Silicosis: A World History*, ed., Rosental, 2017, p 3; and 'Truncating a Disease. The Reduction of Silica Hazards to Silicosis at the 1930 International Labor Office Conference on Silicosis in Johannesburg'. *American Journal of Industrial Medicine*, Vol. 58, No. S1, pp. S6–S14.

[13] Gerald E. Markowitz and David Rosner. 'Why is Silicosis so Important,' Chapter 1 in *Silicosis: A World History*, ed., Rosental, 2017, p. 15.

[14] Phthisis is a Greek word for consumption. Throughout the first half of the twentieth century, and until more precise and reliable modes of diagnosis became available in the 1950s, the term was used in South Africa for unspecific wasting lung disease. Rodney Ehrlich notes that black former gold miners attending the Occupational Medicine Clinic at Groote Schuur Hospital in Cape Town referred to silicosis and tuberculosis as 'phthisis' well into the twenty first century; he does not remember a black miner spontaneously offering the term 'silicosis'. Personal communication, July 2019.

Health Crises and the Public Perception of Risk

The viability of the South African gold mining industry was profoundly challenged by four health crises. The first three beset the gold mines in the period from 1900 to 1916. These crises overlapped in time, and each was affected by the dust levels underground, the conditions in the mine compounds and the industry's insatiable demand for labour. The initial crisis was over an acute form of silicosis. In the first decade of mining this disease took the lives of hundreds of rock drillers, many of them tin miners recruited from Cornwall.[15] The second crisis revolved around non-tuberculous infectious pneumonia, which killed thousands of migrant workers in the mine compounds. The death rates were particularly high among recruits from the Tropical north. The focus of the next crisis was tuberculosis, which became more obvious once mechanical ventilation and blasting regulations had reduced extreme dust levels. Those changes saw acute silicosis replaced by a chronic form of fibrosis. Although not in itself immediately fatal, this contributed to high rates of tuberculosis disease among black miners. The final health crisis began in the mid-1980s with the arrival of HIV/AIDS which, combined with silica dust exposure and tuberculosis, again led to the deaths of thousands of workers.

The histories of tuberculosis and the recent HIV/AIDS epidemic show there is no simple path for infection associated with oscillating migration. Tuberculosis, for example, moved in both directions between urban and rural settings. Once a disease arrived in a rural area its rate of transmission depended upon factors such as 'land hunger' or chronic shortage of land needed for subsistence, nutritional status, access to public health, crowded living conditions and generational poverty. The complex relationship between the social and material conditions in each labour-sending area and the disease process meant that the impact of migrant labour varied from region to region. The consequences were also uneven within a single setting, with women at times bearing much of the burden.[16]

---

[15] Elaine Katz. *The White Death: Silicosis on the Witwatersrand Gold Mines 1886–1910*. Johannesburg: Witwatersrand University Press, 1994.

[16] For an account of that process regarding HIV/AIDS see Simonne Horwitz. 'Migrancy and HIV/AIDS: A Historical Perspective'. *South African Historical Journal*, No. 45, November 2001, pp. 103–123. See also K. Zuma, E. Gouws, B. Williams and M. Lurie. 'Risk Factors for HIV infection among Women in Carletonville, South Africa: Migration, Demography and Sexually Transmitted Diseases', *International Journal of STD & AIDS*, Vol. 14, No. 12, December 2003, pp. 814–817.

Between the health crises in early twentieth century and the recent class action, the gold mines were widely believed to have resolved the problem of occupational lung disease. Indeed, for most of the twentieth century, the data published annually by the Miners' Phthisis Medical Bureau showed that the mines were safe. In the period from 1917–1920, the silicosis rate among white miners was reported to be 2.195 per cent, a low figure by international standards. In 1935 it had fallen to 0.885 per cent. The rate for black miners was even lower. In 1926–1927 it was 0.129 per cent and in 1934–1935 it had fallen to 0.122 per cent.[17] In 1946 the official silicosis rate among black miners increased somewhat to 0.178 per cent.[18] Advances in medical diagnosis technologies, such as the introduction of mass miniature radiography in the early 1950s, brought no rise in the reported disease rates. Over time conditions apparently improved still further, and by the early 1980s the Chamber insisted that silica exposure no longer caused disability or death. In 1985 Dr Schroder, from the Chamber's Air Pollution Division, claimed that: 'With the dust control in South African … Gold Mines, no one today dies from silicosis'.[19] As late as 1990, the South African specialists Drs Robert Cowie and Salmon Mabena wrote that: 'In many studies of silicosis under modern mining conditions, including a study of white South African gold miners, silicosis is shown to be a benign disorder associated with little dysfunction and no disability'.[20] In answering their critics, apartheid minority governments in Pretoria took great care to publicise the safety of the mines and the benefits they offered to migrant labour.

The dismantling of apartheid in the early 1990s was associated with a radical shift in the public perception of risk. The Commission into Safety and Health in the Mining Industry (Leon Commission), held under the

[17] 'Table 1: Incidence of Silicosis and Tuberculosis' in *The Prevention of Silicosis on the Mines of the Witwatersrand*. Pretoria: Government Printer, 1937, pp. 236; 242.

[18] *Report of the Silicosis Medical Bureau for the Year ended 31st March 1949*. Pretoria: Government Printer, 1950, pp. 6–7.

[19] H.H. Schroder. 'The Properties and Effects of Dust. Environmental Engineering in South African Mines'. The Mine Ventilation Society of South Africa 1982' quoted in Rina King. *Silicosis in South African Gold Mines, a study of risk of disease for black mineworkers*. TAG/WITS Sociology Research 1985, p. 12.

[20] Robert L. Cowie and Salmon K. Mabena. 'Silicosis, Chronic Airflow Limitation, and Chronic Bronchitis in South African Gold Miners'. *The American Review of Respiratory Diseases*, Vol. 143, No. 1, 1991, pp. 80–84.

first majority rule government in 1994, found that dust levels were hazardous and that they had probably been so since the 1940s.[21] More recent research conducted at University of Cape Town puts the silicosis rate in living retired miners at between 22 per cent and 30 per cent.[22] Jill Murray's post mortem data suggests that up to 60 per cent of miners will eventually develop what is a life-threatening and untreatable disease.[23] In addition to silicosis, the mines are also accused of playing a major role in the spread of tuberculosis, a well-known consequence of exposure to silica dust. Recent research shows that even subclinical amounts of dust in the lungs (that is dust not visible or identifiable in scans or X-rays) almost double the risk of contracting TB.[24] Not only is the burden of tuberculosis in Southern Africa currently among the highest in the world, the incidence of the disease *among miners* has been estimated to be as much as ten times higher than in the populations from which they originate.[25] As in the past, the majority of miners are migrant workers, whose exposure to silica dust and tuberculosis results in the transmission of disease to rural communities and across national borders.[26] There is evidence

[21] *Report of the Commission of Inquiry into Safety and Health in the Mining Industry*. Pretoria, Department of Minerals and Energy Affairs, 1995, pp. 51–53.

[22] T.W. Steen et al. 'Prevalence of Occupational Lung Diseases among Botswana Men Formerly Employed in the South African Mining Industry'. *Occupational and Environmental Medicine*, No. 54, 1997, pp. 19–26; and Jill Murray, Tony Davies and David Rees. 'Occupational Lung Disease in the South African Mining Industry: Research and Policy Implementation'. *Journal of Public Health Policy*, Vol. 32, Suppl 1, June 2011, pp. S65–S79.

[23] Jill Murray. Development of Radiological and Autopsy Silicosis in a Cohort of Gold Miners Followed up into Retirement. Paper presented at the Research Forum, National Institute for Occupational Health, Johannesburg, 26 May 2005.

[24] Eva Hnizdo and Jill Murray. 'Risk of Pulmonary Tuberculosis Relative to Silicosis and Exposure to Silica Dust in South African Gold Miners'. *Occupational and Environmental Medicine*, Vol. 55, No. 7, 1998, pp. 496–502; Rodney Ehrlich, Jill Murray and David Rees. 'Subradiological Silicosis'. *American Journal of Industrial Medicine*. Vol. 61, 2018, pp. 877–885.

[25] *Tuberculosis Strategic Plan for South Africa, 2007–2011*. Pretoria, South Africa: Ministry of Health; 2007, cited in David Stuckler, Sanjay Basu, Martin McKee and Mark, Lurie. 'Mining and Risk of Tuberculosis in Sub-Saharan Africa'. *American Journal of Public Health*, Vol. 101, No. 3, 2010, p. 524.

[26] For example, around 40 per cent of the adult male tuberculosis patients in Lesotho's hospitals work or have worked in South Africa mines. *The Mining Sector: Tuberculosis and Migrant Labour in Southern Africa*. AIDS and Rights Alliance for Southern Africa, July 2008, p. 2.

stretching back to the 1920s that South Africa's mines have been spreading tuberculosis to neighbouring states.[27]

By the late nineteenth century, statistics came to play a significant role in the official delineation of risk and in scientific debates more generally.[28] Despite widespread problems in identifying and recording the main causes of death, mortality rates were among the most robust statistical indicators of occupational health. Importantly, these figures were contingent on where workers died. In Southern Africa, that depended upon two things: the disease process and the industry's repatriation policies. The inferior health, accommodation and working conditions of black miners, and of migrants in particular, meant that unlike white workers they were highly susceptible to pneumonia. And even though all who worked on the mines were at risk of silicosis, the superior conditions enjoyed by white miners meant that fewer developed tuberculosis. Because pneumonia killed within a matter of days or weeks, sick men died on the mines. In contrast, silicosis and tuberculosis could take years to kill. Unlike whites, most black ex-miners who succumbed to these diseases had been repatriated to rural areas, and so their deaths were not recorded as mining deaths. Despite repeated requests over several decades, local authorities were not notified of repatriated miners, and no statistics were collected on their subsequent health. These problems were compounded by systematic disagreements about classifying the causes of death, and wide discrepancies in the way statistics were collected and statistical categories made up from place to place and year to year. The messiness and inadequacy of statistical data was the source of a chorus of contemporary complaints, and is reflected in frequent discrepancies in the figures cited in the following chapters.

The silicosis and tuberculosis crises were resolved politically by several commissions of enquiry and the Miners Phthisis Acts of 1912 and 1916, which made both diseases compensable. In contrast, there were no commissions of enquiry into pneumonia, no dedicated legislation—and no compensation. Instead, in 1913 the South African government imposed

---

[27] Stuckler et al. 'Mining and Risk of Tuberculosis', p. 529; McCulloch, *South Africa's Gold Mines*, 2012, pp. 85–105. See also R. M. Packard. *White Plague, Black Labour: Tuberculosis and the Political Economy of Health and Disease in South Africa*. Berkeley: University of California Press, 1989.

[28] See note 7.

a ban on recruitment from the Tropical north, and waited for the Chamber to implement a range of public health measures and develop a vaccine to prevent further fatalities.

The system of oscillating migration, whereby people moved periodically between their rural homes and mines or factories, has long been associated with the transmission of epidemic diseases.[29] As with the earlier epidemics, from the early 1980s the migrant labour system facilitated the spread of HIV/AIDS throughout Southern Africa. As in the past, in addition, the fear of epidemics infused wider debates about pressing social issues.[30] Between 1988 and 1992, for example, about 13,000 Malawi miners were repatriated from South Africa. The official reason given was that in the previous two years some 200 of them had tested HIV/AIDS positive. The Chamber of Mines then requested that the Malawi government screen all prospective migrant workers for the disease before they left for South Africa. The government refused, and the Chamber stopped recruiting. The gold mines were restructuring in response to falling profitability and it is likely that they used HIV/AIDS to justify previously planned retrenchments.[31]

In recent decades, a combination of social and biological factors created a perfect storm for the interaction between silicosis, tuberculosis and HIV. Risk factors such as migrancy and single-sex mine compounds increase HIV rates. Together, silica dust, silicosis and HIV have a multiplicative effect on the development of tuberculosis.[32] Silica is a biologically active dust and the biological processes which are initiated with exposure do not cease once it ceases. Silica exposure is associated with tuberculosis even in the absence of silicosis, and the increased risk is

[29] The diamond and gold rushes of the late nineteenth century, for example, were associated with the spread of syphilis, followed by the bubonic plague in 1901 and the Spanish flu in 1918. Howard Phillips. 'AIDS in the Context of South Africa's Epidemic History: Preliminary Historical Thoughts'. *The South African Historical Journal*, No. 45, Nov. 2001, pp. 11–26.

[30] John Illife. *A History of the African AIDS Epidemic*. Oxford: James Currey, 2006.

[31] Wiseman Chijere Chirwa. 'Aliens and AIDS in Southern Africa: The Malawi-South Africa Debate'. *African Affairs*, Vol. 97, No. 386, 1998, pp. 53–79.

[32] Jill Murray, Tony Davies and David Rees. 'Occupational Lung disease in the South African Mining Industry: Research and Policy Implementation'. *Journal of Public Health Policy*, Vol. 32, Suppl. 1, June 2011, pp. S65–S79; Rodney Ehrlich, Jill Murray and David Rees. 'Subradiological Silicosis'. *American Journal of Industrial Medicine*, Vol. 61, 2018, pp. 877–885.

life-long. In effect, silica retained in the lung and silicosis represent a type of acquired immune deficiency.[33] Failure to complete treatment has led to the emergence of multiple drug-resistant TB. By 2007, miners in sub-Saharan Africa were shown to have greater incidence of TB than any other working population in the world.[34] In its *Global Tuberculosis Control Report* for 2019, the WHO identified fourteen countries with the most severe burden of TB, multi-drug-resistant TB and TB and HIV combined. Of those fourteen, eight supplied labour to South Africa's mines.[35] Gold miners from these countries were the fittest members of their communities. They were employed in a technologically advanced industry with an admired record of medical surveillance dating back almost a century. From 1912, they were subject to compulsory medical examination before entering the industry. Once employed, they were much better fed than people in their home communities. They were also the only group of black people subject to case finding through regular medical inspections, which after 1955 featured the routine use of mass miniature X-rays. And yet, until majority rule, the extent of the burden of occupational lung disease among gold miners remained all but invisible.

Miners' Class Actions

The Leon Commission (1994) was highly critical of hazardous conditions on the mines and questioned the claim that they posed negligible health risk to workers. However, it brought no sustained reduction in dust exposures. On the contrary, the use of subcontracting underground labour arguably worsened working conditions, as did the use of

[33] Rodney Ehrlich. 'A century of miners' phthisis on the south African gold mines. Any end in sight?' Presentation at the 25th Anniversary meeting of the Collegium Ramazzini, held in Carpi, Italy on October 25–28, 2007, p. 15.

[34] Stuckler et al. 'Mining and Risk of Tuberculosis', 2010, p. 524, citing *Tuberculosis Strategic Plan for South Africa, 2007–2011*. Pretoria, South Africa: Ministry of Health, 2007.

[35] *Global tuberculosis report 2019*. Geneva: World Health Organization, 2019. Figure 2.5: 'Countries in the three high-burden country lists for TB, TB/HIV and MDR-TB being used by WHO during the period 2016–2020, and their areas of overlap', p. 24; and Table 2.4: 'The three high-burden country lists for TB, TB/HIV and MDR-TB being used by WHO during the period 2016–2020', p. 25.

production-related bonuses.[36] An increase in the average length of service for miners resulted in longer periods of exposure, while the introduction of full-calendar-year operations increased the annual hours worked underground.[37] The recorded prevalence of tuberculosis in gold miners increased from 806 cases per 100,000 in 1991 to 3821 in 2004, while HIV prevalence rose from less than 1 per cent to 27 per cent in 2000. The system for dealing with compensation for occupational lung disease was unable to cope with the extent of the crisis. The ANC government was reluctant to assume responsibility for laws enacted during minority rule, or for the citizens of neighbouring states. The South African Medical Bureau of Occupational Diseases (MBOD), which adjudicated compensation awards, had in effect ceased to function because of a lack of staff and the volume of claims. Massive backlogs meant that mineworkers were practically unable to access the statutory compensation to which they were entitled.[38] The problems were compounded by the difficulties the men faced in establishing their work histories. Most were only provided with a sheet of paper acknowledging their employment at a particular mine. The following year they might be re-recruited by the same employment agency, but often placed at a different mine, with their name transcribed in a different way.[39] The Employment Bureau of Africa (TEBA) electronic employment records are incomplete as are the records held by individual mines.[40] While the MBOD should have a complete record of all risk shifts worked by gold mineworkers, a substantial

[36] Murray et al. 'Occupational Lung disease in the South African Mining Industry', p. 70.

[37] Gillian Nelson. Living in the Shadow of a Dust Cloud: Occupational Respiratory Diseases in the South African Mining Industry, 1975 to 2009, PhD thesis, Faculty of Health Sciences, University of the Witwatersrand, Johannesburg, 2012, pp. 16–26. See also Anna Trapido. An Analysis of the Burden of Occupational Lung Disease in a Random Sample of Former Gold Mineworkers in the Libode District of the Eastern Cape. PhD thesis University of the Witwatersrand, 2000, pp. 27–28.

[38] Chris Bateman. 'Silicosis: 10,000 Gold Miners Getting Set to Sue'. *The South African Medical Journal*, Vol. 102, No. 6, 2012, pp. 338–340; Judgement 2019, 1/p37/#60; p. 42/#69.

[39] Judgement 2019, 2/p. 18/#26.

[40] Rodney Ehrlich, Alex Montgomery, Paula Akugizibwe, Gregg Gonsalves. 'Public Health Implications of Changing Patterns of Recruitment into the South African Mining Industry, 1973–2012: A database Analysis'. Public Health Vol. 18, No. 1, August 2017. As one critic observed: 'At face value, the mining companies appear to have little interest in maintaining complete and accurate records'. Nelson, Living in the Shadow of a Dust Cloud, 2012, p. 82.

breakdown of the MBOD's system after 1994 resulted in records either lost, destroyed or simply not kept.[41]

It was in this climate that, in 2013, Abrahams Kiewitz Attorneys, Richard Spoor Attorneys and the Legal Resources Centre separately commenced class actions against a number of companies on behalf of miners suffering from silicosis and/or tuberculosis. Against strenuous opposition from the mining companies, in 2016 the South African High Court amalgamated their separate actions into one. Importantly, the court also amended the common law regarding what is called 'the transmissibility of damages' so it was in line with South Africa's post-apartheid constitution, and in particular its clauses regarding human rights. In brief, the changes allow for the wives and other dependents of miners taking part in the class action to inherit damages awards in case the men die before the court process is completed.[42] In May 2018, the parties to the dispute reached an historic agreement to settle the Silicosis and Tuberculosis Class Actions. The settlement was confirmed by the South African High Court in July 2019. Its provisions not only represent a dramatic change in the legal rights of mine workers throughout the region but have considerable symbolic importance as part of South Africa's transformation from minority rule. Securing the rights of migrant workers has been a highly significant part of that transformation.[43]

The Manufacture of Ignorance

The making and dismantling of the complex edifice of racialised employment and management of risk in the South African gold mining industry is at the core of the chapters which follow. In particular, the book focuses on the discrepancy between official visibility of the extent of occupational lung disease during most of the twentieth century and since South Africa's transition to majority rule. A good shorthand description of the

[41] Judgement 2019, 1/p37/#60; p. 42/#69.

[42] Republic of South Africa, The High Court of South Africa, Gauteng Local Division, Consolidated Case Number 48226/12, 13 May 2016.

[43] Interview with Richard Spoor, Convention Centre, Cape Town, 23 October 2014.

construction and maintenance of this discrepancy is the manufacture of ignorance.

The contours of such manufactured ignorance are well captured in the core arguments made by the mining companies through the years of litigation before the class action was settled. In contesting the claims made against them, the companies not only challenged the appropriateness of class action, but also their responsibility and liability for past injuries. The companies' Defendant's Pleas emphasised the care the companies took to prevent occupational disease, to comply with legislation and state regulations, to monitor health, to inform labour of the risks of dust exposure and finally to provide appropriate treatment to miners who contracted silico-tuberculosis. At the same time, they drew attention to state failure to deal with a health crisis which was not of the companies' making, on the miners' own negligence and on their implicit acceptance of risk when they entered employment contracts.

The miners' litigation, in contrast, took as its starting point the extraordinarily high rates of occupational disease among retired miners revealed by recent medical research, now established to be up to a hundred times higher than the official rates published annually by the Bureau and its successors in the period from 1916 to 1990. Given this discrepancy, the miners' legal representatives asked the courts to identify when the mining houses and the Chamber became aware that there were high rates of uncompensated silicosis and tuberculosis? And when were they provided with reliable information that the mines were spreading tuberculosis to labour-sending communities? Did they, in effect, act to manufacture ignorance of workplace risks? On the face of it, these questions were difficult to answer. The documentary trail is fractured, and researchers and the miners' legal representatives had limited access to the extensive archives of the Chamber of Mines and Barlow Rand.[44] There is, however, a rich vein of compelling evidence from the 1920s of serious hazards which the mines failed to address. It is from this material that the story of manufacturing ignorance told in this book is woven.

[44] Originally incorporated in 1918, in 1971 Barlow purchased Rand Mines Ltd., South Africa's oldest mining house. Although it has since divested itself of many parts of the business, in the 1980s Barlow Rand was the largest company in South Africa, and an international parent company of businesses in Southern Africa, Britain, the United States and Europe.

The first strand of evidence concerns the knowledge of risk. The Commissions of 1902, 1912 and 1914, augmented by the 1914 Gorgas Report, provided sufficient evidence of occupational disease for the South African government, under pressure from London, to ban the recruitment of labour from the Tropical north. The decision to suspend the recruitment of miners from tropical regions was supported by research conducted in the 1920s and 1930s. Drs Watt, Mavrogordato and Macvicar all suggested that there were high rates of silicosis and tuberculosis, and that the mines' policy of repatriating sick men was spreading tuberculosis to rural communities within South Africa and through the migrant labour system to adjoining states. These findings were endorsed by the Stratford and Lansdown Commissions of 1943, and again by the Oosthuizen Committee Enquiry of 1954. The second strand of the story encompasses the Chamber's failure, over several decades, to inform the administrations of the British High Commission Territories, Nyasaland and of Portuguese East Africa of the risk to public health posed by men with tuberculosis returning from the mines. The third strand comprises the industry's failure, from 1922 until the 1990s, to carry out follow-up research into the health of miners once they left the industry. The final strand involves the ease with which the epidemic of occupational lung disease was identified in the early 1990s by independent researchers.

As it proudly claimed, the Chamber was the major funder of research into silicosis and pulmonary tuberculosis. During his annual address in June 1990, the Chamber's President, K.W. Maxwell, for example noted that: 'Some 60 per cent of the R65 million spent in 1989 by COMRO (the Chamber's research arm) was devoted to wide-ranging research projects directly related to safety'.[45] Between the 1970s and 1994, the industry spent almost R1.5 billion on research and development aimed at improving safety and productivity.[46] The scale of such investment makes the official invisibility of lung disease among miners all the more intriguing. Anna Trapido was a PhD candidate when she uncovered alarming rates of uncompensated silicosis and tuberculosis among former miners

[45] Addendum to Presidential Address, K.W. Maxwell, *The Chamber of Mines Annual Report*, 19th June 1990, p. 23.

[46] Submission of the Chamber of Mines to the (Leon) Commission of Inquiry into Health and Safety June 1994, p. 9.

at Libode in the Eastern Cape. The total outlay on Trapido's research—including her salary, the cost of reading X-rays, transport and incidental expenses—was at most R500,000. As the result of her findings, the Chamber paid out more than R5 million in compensation to Libode miners.[47]

The corporate strategy of shifting the costs of production onto labour, the state and civil society has been one of the defining features of modern industrial workplaces. In charting the histories of the asbestos and tobacco industries we have access to internal company correspondence. Released through legal discovery,[48] those documents reveal how, over decades, senior management and company medical officers in a dozen countries systematically suppressed evidence of hazardous work practices and the risks associated with asbestos and tobacco products. Studies such as those by Castleman, McCulloch and Tweedale, and Proctor and Schiebinger show that those industries also corrupted the science.[49] This process, which can be described as medical involution, or the creation of scientific ignorance, is very different from simply hiding evidence of hazards. The creation of ignorance, which Proctor and Schiebinger call *agnotology*, is subtler and more effective than suppression. Suppression simply leaves gaps in knowledge of risks; false knowledge, the outcome of cultural and political struggles, fills such gaps with misleading information.[50]

South Africa's mines were reputed to lead the world in the prevention, detection and compensation of occupational lung disease. The recent litigation—and this book—shows that this reputation was undeserved. Instead, they provide evidence of a quite different contribution to the

[47] J.C.A. Davies, communication with Jock McCulloch, 15th July 2009.

[48] This is part of the legal process, where parties to a dispute are able to demand from each other access to as much information about the case as possible so that all parties can go to trial without keeping secrets from each other.

[49] Barry Castleman. *Asbestos: Medical and Legal Aspects*, 4th edn. New Jersey: Aspen Law and Business Books, 1996; Jock McCulloch and Geoffrey Tweedale. *Defending the Indefensible: The Global Asbestos Industry*. Oxford: Oxford University Press, 2008; Robert N. Proctor. *Golden Holocaust: Origins of the Cigarette Catastrophe and the Case for Abolition*. Berkeley: University of California Press, 2012.

[50] Robert N. Proctor and Londa Schiebinger, eds. *Agnotology: The Making and Unmaking of Ignorance*. San Francisco: Stanford University Press, 2012.

world economy and scientific community. They suggest that the gold mines pioneered techniques, later adopted and refined by the asbestos and tobacco industries, for hiding the evidence of disease and denying employees and their families' compensation. The appropriation and corruption of medical knowledge in South Africa has been on such a scale that without it oscillating migration, especially from the British High Commission Territories, would probably have been politically unsustainable. Such manufactured ignorance had consequences far outside Southern Africa. The systematic downplaying of the risk of silicosis, for example, has arguably contributed to the current worldwide epidemic of occupational disease among stone benchtop manufactures and installers.[51] In all, rather than constituting an example of spectacular biomedical failure, the history of mining, medical knowledge and migrant labour can be read as a particular kind of corporate success.

The successful manufacture of ignorance has a number of possible explanations, ranging from incompetence to collusion between the mining industry and the state. In assessing the relative merits of such explanations, this book builds on the research of many other scholars. The literatures on gold mining and migrant labour are particularly rich, and this project draws upon the work of scholars such as Wilson, Harries, Delius, van Onselen, Moodie, Phimister and Crush. Until recently, surprisingly little has been published on the politics of occupational disease in South Africa and even less on oscillating migration and health.[52] The exceptions are major histories by Katz and Packard, and more recently

[51] See, for example, Michael Atkin. 'The Biggest Lung Disease Crisis since Asbestos: Our Love of Stone Kitchen Benchtops is Killing Workers'. ABC news, Updated 10 Oct 2018, 5:02pm, https://www.abc.net.au/news/2018-10-10/stone-cutting-for-kitchen-benchtops-sparks-silicosis-crisis/10357342.

[52] There are, however, several unpublished histories. See, for example, Harold Jack Simons. Migratory Labour, Migratory Microbes. Occupational Health in the South African Mining Industry: The Formative Years 1870–1956. Unpublished manuscript, 1960; Julie J. Baker. 'The Silent Crisis': Black labour, Disease, and the Economics and Politics of Health on the South African Gold Mines, 1902–1930. PhD thesis, Queen's University, Ont., 1989; and Matthew John Smith. Working in the Grave: The Development of a Health and Safety System on the Witwatersrand Gold Mines, 1900–1939. MA thesis, Rhodes University, 1993.

McCulloch's *South Africa's Gold Mines and the Politics of Silicosis.*[53] Katz's study of pneumonia and silicosis ends in 1916. Packard was writing a broader and more complex political economy of tuberculosis in South Africa, which takes the story up to the eve of majority rule. While Packard focuses on the complexities of tuberculosis infection among different communities *within* the borders of South Africa, this text deals with the interaction between the three main occupational lung diseases among miners and focuses on the dynamics of the migrant labour system. Second, while Packard fully acknowledges the political nature of statistics collection, presents evidence regarding the mines' attempts to minimise mortality and morbidity figures and mentions the synergy between silica dust and tuberculosis, he largely relies on the official rates of occupational injury. Given that the sheer extent of silicosis among miners was unknown at the time he was writing (when the apartheid regime was still in power), his otherwise damning social history of tuberculosis understates the significance of mines as a source of TB infection.[54]

McCulloch's *South Africa's Gold Mines and the Politics of Silicosis* looks at the story told in this book from the outside in. Focusing on South Africa, it situates the mines and mine medicine in a world context. It provides detailed evidence on the political economy of mining, the everyday dynamics of mine medicine, political developments and union struggles. The focus of the book is on the politics of silicosis. *Mining gold and manufacturing ignorance* looks at the same industry from a different perspective. Rather than silicosis, the main emphasis is on tuberculosis.

---

[53] Katz. *The White Death*; Packard. *White Plague, Black Labour*. See also Randall M. Packard. 'The Invention of the "Tropical Worker": Medical Research and the Quest for Central African Labor on the South African Gold Mines, 1903–36'. *The Journal of African History*, Vol. 34, No. 2, 1993, pp. 271–292; Randall M. Packard & David Coetzee. 'White Plague, Black Labour Revisited: TB and the Mining Industry'. In Jonathan Crush & James Wilmot eds. *Crossing Boundaries: Mine Migrancy in a Democratic South Africa*. Edited collection of papers presented at an international conference on 'Transforming Mine Migrancy in the 1990s', Institute for Democracy in South Africa, Cape Town, 1995; Jock McCulloch. *South Africa's Gold Mines and the Politics of Silicosis*. Oxford: James Currey, Oxford, 2012. See also Charles van Onselen. *The Night Trains: Moving Mozambican Miners to and from the Witwatersrand Mines, 1902–1955*. London: Hurst Publishers, 2018.

[54] Packard. *White Plague, Black Labour*, pp. 96; 215.

Rather than looking at the industry from the outside in, it charts how machinations about the buying and selling of mine labour behind closed doors helped engender an epidemic of tuberculosis throughout the Southern African region. This time, the main emphasis is not on South Africa itself, but on the labour-sending states: present-day Lesotho, Mozambique, eSwatini, Botswana, Zambia, Malawi and Zimbabwe. Finally, the underlying theoretical narrative of *The Politics of Silicosis* revolves around the political economy of mining, while *Mining Gold and Manufacturing Ignorance* is concerned with the contested making of medical knowledge, and medical involution in particular.

References

Baker, Julie J. 'The Silent Crisis': Black Labour, Disease, and the Economics and Politics of Health on the South African Gold Mines, 1902–1930. PhD thesis, Queen's University, Ontario, 1989.

Baldasseroni, A. and F. Carnevale. 'The Genesis and Development of the Scientific Concept of Pulmonary Silicosis during the Nineteenth Century'. In Paul-André Rosental, ed. *Silicosis: A World History*. Baltimore: Johns Hopkins University Press, 2017, pp. 30–63.

Bateman, Chris. 'Silicosis: 10,000 Gold Miners Getting Set to Sue'. *The South African Medical Journal*, Vol. 102, No. 6, 2012, pp. 338–340.

Breckenridge, Keith. *Biometric State: The Global Politics of Identification and Surveillance in South Africa, 1850 to the Present*. Cambridge: Cambridge University Press, 2014.

Castleman, Barry. *Asbestos: Medical and Legal Aspects*, 4th edn. New Jersey: Aspen Law and Business Books, 1996.

Chirwa, Wiseman Chijere. 'Aliens and AIDS in Southern Africa: The Malawi-South Africa Debate'. *African Affairs*, Vol. 97, No. 386, 1998, pp. 53–79.

Cowie, R.L. 'The Five Ages of Pulmonary Tuberculosis and the South African Goldminer', *The South African Medical Journal*, Vol. 76, 18th November 1989, pp. 566–567.

Desrosières, Alain. *The Politics of Large Numbers: A History of Statistical Reasoning*. Cambridge, MA: Harvard University Press, 2002.

Ehrlich, Rodney. 'A century of miners' phthisis on the south African gold mines. Any end in sight?' Presentation at the 25th Anniversary meeting of the Collegium Ramazzini, held in Carpi, Italy on October 25–28, 2007.

Ehrlich, Rodney, Alex Montgomery, Paula Akugizibwe and Gregg Gonsalves. 'Public health implications of changing patterns of recruitment into the South African mining industry, 1973–2012: A database analysis'. *BMC Public Health*, Vol. 18, No. 1, article no. 93, August 2017, pp. 1–12.

Ehrlich, Rodney, Jill Murray and David Rees. 'Subradiological silicosis'. *American Journal of Industrial Medicine*, Vol. 61, No. 11, November 2018, pp. 877–885.

Global Tuberculosis Report 2019. Geneva: World Health Organization, 2019.

Hacking, Ian. *The Taming of Chance*. Cambridge: Cambridge University Press, 1990.

Hay, Douglas and Paul Craven (eds). *Masters, Servants, and Magistrates in Britain and the Empire, 1562–1955*. Chapel Hill and London: The University of North Carolina Press, 2004.

Hnizdo Eva and Jill Murray. 'Risk of Pulmonary Tuberculosis Relative to Silicosis and Exposure to Silica Dust in South African Gold Miners.' *Occupational and Environmental Medicine*, Vol. 55, No. 7, 1998, pp. 496–502.

Horwitz, Simonne. 'Migrancy and HIV/AIDS: A Historical Perspective'. *South African Historical Journal*, Vol. 45, November 2001, pp. 103–123.

Hunt, Nancy Rose. *A Nervous State: Violence, Remedies, and Reverie in Colonial Congo*. Durham, NC: Duke University Press, 2016.

Illife, John. *A History of the African AIDS Epidemic*. Oxford: James Currey, 2006.

Katz, Elaine. *The White Death: Silicosis on the Witwatersrand Gold Mines 1886–1910*. Johannesburg: Witwatersrand University Press, 1994.

King, Rina. *Silicosis in South African Gold Mines: A Study of Risk of Disease for Black Mineworkers*. TAG/WITS Sociology Research, 1985.

Markowitz, Gerald E. and David Rosner. 'The Illusion of Medical Certainty: Silicosis and the Politics of Industrial Disability, 1930–1960.' *The Milbank Quarterly*, Vol. 67, Suppl. 2, Part 1. Disability Policy: Restoring Socioeconomic Independence, 1989, pp. 228–253.

Markowitz, Gerald E. and David Rosner. 'Why Is Silicosis So Important'. Chapter 1 in Paul-André Rosental Ed. *Silicosis: A World History*. Baltimore: John Hopkins University Press, 2017, pp. 14–29.

McCulloch, Jock. *South Africa's Gold Mines and the Politics of Silicosis*. Oxford: James Currey, 2012.

McCulloch, Jock and Geoffrey Tweedale. *Defending the Indefensible: The Global Asbestos Industry*. Oxford: Oxford University Press, 2008.

Murray, Jill, Tony Davies and David Rees. 'Occupational Lung Disease in the South African Mining Industry: Research and Policy Implementation'. *Journal of Public Health Policy*, Vol. 32, Suppl. 1, June 2011, pp. S65–S79.

Nelson, G. Living in the Shadow of a Dust Cloud: Occupational respiratory diseases in the South African mining industry, 1975 to 2009, PhD thesis, Faculty of Health Sciences, University of the Witwatersrand, Johannesburg, 2012.

Packard, Randall M. and Dr David Coetzee. 'White Plague, Black Labour Revisited: TB and the Mining Industry'. In Crush, Jonathan & Wilmot James, eds. *Crossing Boundaries: Mine Migrancy in a Democratic South Africa*, Edited collection of papers presented at an international conference on 'Transforming Mine Migrancy in the 1990s' held in Cape Town, June 1994. Institute for Democracy in South Africa, Cape Town, 1995.

Packard, R.M. 'The Invention of the "Tropical Worker": Medical Research and the Quest for Central African Labor on the South African Gold Mines, 1903–36'. *The Journal of African History*, Vol. 34, No. 2, 1993, pp. 271–292.

Packard, R.M. *White Plague, Black Labour: Tuberculosis and the Political Economy of Health and Disease in South Africa*. Berkeley: University of California Press, 1989.

Phillips, Howard. 'AIDS in the Context of South Africa's Epidemic History: Preliminary Historical Thoughts'. *The South African Historical Journal*, Vol. 45, November 2001, pp. 11–26.

Phillips J.I., G. Nelson, N. Vorajee, J. Murray, N. Ndlovu and J.C.A Davies. 'Marikana Autopsies Highlight Occupational Diseases Amongst Platinum Mine Workers'. *Occupational Health Southern Africa*, Vol. 20, No. 4, 2014, pp. 6–12.

Poovey, Mary. *A History of the Modern Fact: Problems of Knowledge in the Sciences of Wealth and Society*. Chicago: The University of Chicago Press, 1998.

Porter, Theodore M. *Trust in Numbers: The Pursuit of Objectivity in Science and Public Life*. Princeton, NJ: Princeton University Press, 1995.

Proctor, Robert N. and Londa Schiebinger, eds. *Agnotology: The Making and Unmaking of Ignorance*. San Francisco: Stanford University Press, 2012.

Proctor, Robert N. *Golden Holocaust: Origins of the Cigarette Catastrophe and the Case for Abolition*. Berkeley: University of California Press, 2012.

Rosental, Paul-André, David Rosner and Paul D. Blanc. 'From Silicosis to Silica Hazards: An Experiment in Medicine, History, and the Social Sciences'. *American Journal of Industrial Medicine*, Vol. 58, No. S1, 2015, pp. S3–S5.

Rosental, Paul-André. 'Truncating a Disease. The Reduction of Silica Hazards to Silicosis at the 1930 International Labor Office Conference on Silicosis in Johannesburg'. *American Journal of Industrial Medicine,* Vol. 58, Suppl. 1, November 2015, pp. S6–S14.

Simons, Harold Jack. Migratory Labour, Migratory Microbes. Occupational Health in the South African Mining Industry: The Formative Years 1870–1956. Unpublished manuscript, 1960.

Smith, Matthew John. Working in the Grave: The Development of a Health and Safety System on the Witwatersrand Gold Mines, 1900–1939. MA thesis, Rhodes University, 1993.

Steen, T.W.; K.M. Gyi; N.W. White; T. Gabosianelwe; S. Ludick; G.N. Mazonde; N. Mabongo; M. Ncube; N. Monare; R. Ehrlich and G. Schierhout. 'Prevalence of Occupational Lung Disease Among Botswana Men Formerly Employed in the South African Mining Industry'. *Occupational & Environmental Medicine*, Vol. 54, No. 1, 1997, pp. 19–26.

Stoler, Ann Laura. *Along the Archival Grain: Epistemic Anxieties and Colonial Common Sense.* Princeton: Princeton University Press, 2009.

Stuckler, David, Sanjay Basu, Martin McKee and Mark, Lurie. 'Mining and Risk of Tuberculosis in Sub-Saharan Africa'. *American Journal of Public Health*, Vol. 101, No. 3, March 2011, pp. 524–530.

Tilley, Helen. *Africa as a Living Laboratory: Empire, Development, and the Problem of Scientific Knowledge, 1870–1950.* Chicago: Chicago University Press, 2011.

Trapido, A. An Analysis of the Burden of Occupational Lung Disease in a Random Sample of Former Gold Mineworkers in the Libode District of the Eastern Cape. PhD thesis, University of the Witwatersrand, 2000.

van Onselen, Charles. *The Night Trains: Moving Mozambican Miners to and from the Witwatersrand Mines, 1902–1955.* London: Hurst Publishers, 2018.

van Wyk, D. 'A Review of Platinum Mining in the Bonjanala District of the North-West Province. A Participatory Action Research Approach.' Johannesburg: The Bench Marks Foundation, 2012. Available at http://www.bench-marks.org.za/research/rustenburg

Zuma, K., E. Gouws, B. Williams and M. Lurie. 'Risk Factors for HIV Infection among Women in Carletonville, South Africa: Migration, Demography and Sexually Transmitted Diseases'. *International Journal of STD & AIDS*, Vol. 14, No. 12, December 2003, pp. 814–817.

BY

Part I

The Gold Mines and Migrant Labour

2

A Most Modern Industry: The Migrant Labour System and Crisis Management

The first gold mines were opened on the East Rand in the late 1880s. The mines were concentrated on a single geological deposit around Johannesburg and were distinguished by their depth, scale and the size of their workforce. The deposits were low grade: it took three tons of ore to yield one ounce of gold.[1] The narrow seams made mechanisation difficult and so the mines relied on a massive workforce. Those features encouraged the dominance of a few mining houses which, because of their importance to employment, foreign exchange and state revenue, wielded great political influence. Although some small mines persisted, within twenty years ownership was concentrated into the hands of six giant mining and finance companies. By contrast, at the beginning of the twentieth century, the British coal mining industry consisted of over 1000 individual companies which competed against each other for share of a finite market. The South African mining houses provided the structure for the mobilisation of capital, and by pooling resources and expertise they

[1] *Report of the Commission of Inquiry into Safety and Health in the Mining Industry*. Pretoria, Department of Minerals and Energy Affairs, 1995, p. 27.

J. McCulloch, P. Miller, *Mining Gold and Manufacturing Ignorance*,
https://doi.org/10.1007/978-981-19-8327-6_2

lowered operating costs and spread risk across the industry as a whole.[2] Until 1940, the most influential mining group was Rand Mines Limited and its London parent, the Central Mining and Investment Corporation. After the Second World War, the development of the Free State goldfields saw Anglo American South Africa Ltd (AASA) become dominant.[3] The industry-wide restructuring of the mid-1990s produced a smaller number of focussed gold mining companies, leaving only AngloGold, AvGold, GoldFields, Rand Gold and the remnants of Johannesburg Consolidated Investments.

Mines, Miners and Profits

The gold mines drove South Africa's economic development. Gold had been central to the functioning of the world's financial markets until the end of the First World War, when countries (including South Africa) abandoned the gold standard. Even after that date, the official price of gold was set by overseas money markets (particularly in London and New York) and, with the exception of periods of major financial crises such as the Depression, was either stable or rising. In this way, gold mining protected the national income from the kinds of fluctuations associated with economies reliant upon agricultural exports. By 1930, the Lansdown Commission claimed, almost half of the population obtained its livelihood directly or indirectly from mining, which also accounted for almost half of government revenue. During the Second World War, the mines paid more in tax than in dividends, making the state the industry's major beneficiary.[4] Between 1961 and 1981, the South African gold mines were

---

[2] *Report of the Witwatersrand Mine Native Wages Commission on the Remuneration and Conditions of Employment of Natives on the Witwatersrand Gold Mines*. Pretoria: Government Printer 1943, p. 2.

[3] By 1976 Anglo's various interests were held through a network of over 150 holding and operating companies. See Duncan Innes. *Anglo American and the Rise of Modern South Africa*. New York: Monthly Review Press, 1984, p. 204.

[4] *Report of the Witwatersrand (Lansdown) Mine Native Wages Commission*, 1943, p. 2.

producing between 65 and 78 per cent of the world output and generating more than 40 per cent of the country's exports.[5]

Low-grade ore, its high silica content and the stable price of gold presented the mining industry with unique challenges. As a spokesman for the Chamber lamented in 1929: 'Nothing that the mines can do can alter the quality of their raw material or the price which they receive for their product; consequently, the cost of production is a factor of even more vital importance in the gold mining industry than in other industries.'[6] Dust in mines is created by drilling, blasting and moving ore. The damage caused to miners' lungs by exposure to free silica dust is irreversible and greatly increases the risk of pulmonary tuberculosis. The host ore in South African gold mines has, by world standards, very high silica content.[7] In 1913 a South African chemist, Dr John McCrae, found that only dust particles invisible to the naked eye can penetrate lung tissue and thereby cause silicosis.[8] McCrae's discovery overturned the idea that safety was proven by a lack of visible dust. Given the challenges it faced, the struggle to control and reduce labour costs dominated the industry policy. Expenditure on mine compounds, rations, health services and compensation were additional areas which could be controlled. The system which the mining houses gradually built up around these matters drove the industry for the rest of the twentieth century.

[5] Francis Wilson. 'Miners and Migrants: How the Mining Industry Has Shaped South Africa'. *Daedalus*, Vol. 130, No. 1, Winter 2001, p. 102.

[6] 'Statement of the Gold Producers Committee of the Transvaal Chamber of Mines' to the *Miners Phthisis Commission 1929–1930*, 10th December 1929, Johannesburg, Parliamentary Library, Cape Town, p. 34.

[7] A 1927 paper, for example, noted that on the Rand mines, the gold-bearing conglomerate contained about 86 per cent of free silica. W. Watkins-Pitchford. 'The Silicosis of the South African Gold Mines, and the Changes Produced in it by Legislative and Administrative Efforts'. *The Journal of Industrial Hygiene*. Vol. 9, No. 4, April 1927, p. 110.

[8] Dr John McCrae. *The Ash of Silicotic Lungs*. Johannesburg: The South African Institute for Medical Research, 1913.

The Chamber of Mines

The Chamber of Mines was established in 1886.[9] According to its constitution, the Chamber's role was to promote and protect members' interests, collect and circulate statistics, assist with technical and financial advice and represent members at industrial disputes and in the framing of industrial legalisation. The Chamber's other activities included health monitoring, the prevention of accidents and public relations. Formally, the organisation was comparable to a chamber of commerce. However, owing to the relatively stable price of gold and the concentration of the mines in a single region, the Chamber enabled the gold mines to achieve 'a far greater degree of co-operation and unified action … than can occur amongst members of Chambers of Commerce'.[10] Through the medium of the Chamber, the six mining houses spoke with one voice at the numerous Mines and Miners' Phthisis Commissions.[11] The Chamber was a sophisticated organisation which employed various techniques to counter its critics. These included representations to ministers, the placement of commissioned articles in the daily press, advertisements and submissions to public enquiries. One of the Chamber's most powerful tools in these public interventions was data on disease and mortality rates. The Chamber also acted as a secretariat to the industry's recruiting agencies.

In 1921, Sir Ernest Oppenheimer, the head of Anglo American, suggested the creation of a new committee to bypass what he now saw as the cumbersome Chamber of Mines Executive. Six months later, the Gold Producers' Committee (GPC) was formed. The GPC centralised decision making into the hands of the six large mining houses and a small group

[9] The Chamber has undergone several name changes during its history and its membership has been expanded to include various types of mining. For an official history see John Lang. *Bullion Johannesburg: Men, Mines and the Challenge of Conflict*. Johannesburg: Jonathan Ball Publishers, 1986. See also Innes, *Anglo American*, 1984, pp. 52–57.

[10] Statement of Evidence by the Transvaal Chamber of Mines in Industrial Legislation Commission of Enquiry 1948–1956. SANA NCR 543 1 & 2, p. 3.

[11] That was not the case with the British and USA asbestos corporations, which had a shared interest in resisting state regulation and compensation claims but were not always united in their political struggles for survival. See Jock McCulloch and Geoffrey Tweedale. *Defending the Indefensible: The Global Asbestos Industry*. Oxford: Oxford University Press, 2008.

of senior officials initially headed by William Gemmill and later by his son James.[12] The GPC was supported by numerous sub-committees. Perhaps the most important of these was the Group Medical Officers Committee (GMO), which reported on occupational health. The Gold Producers Committee influenced every aspect of mine legislation, from recruitment to safety and compensation.[13] Successive drafts of each new act were submitted by government to the GPC for review, passed onto the Chamber's legal advisor and usually a sub-committee for comment and then returned to the GPC for final evaluation.[14] The same process was followed with the Chamber's submissions to Parliamentary Select Committees and Commissions of Enquiry. There was voluminous correspondence between the GPC and the Departments of Mines, Health and Native Affairs. Curiously, today the South African National Archives hold only a handful of GPC documents.

The administration of the mines by industry and the state was so stable over time that a small number of individuals dominated policy making. These included Sir Robert Kotze, the Government Mining Engineer from 1910 to 1926; W. Watkins-Pitchford, Director of the South African Institute of Medical Research (SAIMR) from 1912 until 1926; H. M. Taberer, the General Superintendent of the Native Recruiting Corporation (NRC) until the early 1930s; and A. J. Orenstein, head of sanitation for Rand Mines Ltd., whose career stretched from 1914 until the mid-1960s.[15] Among these handful of powerful men, one stands out. William Gemmill was appointed as an Actuary in 1913, and soon rose to

[12] David Yudelman. *The Emergence of Modern South Africa: State, Capital, and the Incorporation of Organised Labor on the South African Gold Fields, 1902–1939*. Cape Town: David Philip, 1984, pp. 174–177.

[13] The TEBA Archives at the University of Johannesburg hold the papers from the Chambers two recruiting arms, WNLA and the NRC. They document the Gold Producers Committee's influence over state regulation.

[14] In 1955, for example, the Secretary for Mines initiated a review of the Silicosis legislation with a view to increasing compensation benefits. The GPC was the leading actor in that process, and it won major concessions for the Scheduled mines. See *Statement by the Transvaal and Orange Free State Chamber of Mines on the Revision and Consolidation of the Silicosis Act No. 17 of 1946, 12th July 1955*. Diseases and Epidemics. Tuberculosis. December 1954 to August 1955, WNLA 20L TEBA Archives, the University of Johannesburg.

[15] The career of A.J. Orenstein is the focus of Chap. 11.

become the Chamber's general manager and head of its recruiting agencies. William Gemmill played a major role in sustaining the long-term profitability of the large gold mines, not least by helping to sharply reduce the power of white organised labour, and playing a strategic role in decimating black union activity.[16] As a professional actuary Gemmill helped establish a preoccupation with quantifying production costs. An example is found in a briefing note from the late 1950s, which identifies the labour costs per ton of ore milled for several mines facing closure. The cost of foodstuffs was 1s 4s 2d and the cost of preparation 1s 4d; the Native Hospital came to 2s 7d; the maintenance of the compound 3s 2d; entertainment, education, welfare and recreation was 3d; clothing was 2d and the pneumoconiosis levies totalled 3d.[17] Perhaps his most important contribution, however, concerned the recruitment of Tropical labour, which Gemmill rightly saw as a way of diversifying the labour supply and keeping downward pressure on wages. Gemmill's dominating personality gave him a decisive influence on policy. He made regular tours of labour-sending areas, negotiated with the powerful farming lobbies in South Africa and Southern Rhodesia and handled negotiations with colonial administrations and the Colonial and Dominions Offices in London. In these forums, he represented the Chamber as the champion of free labour. In 1939, William Gemmill was exiled to the WNLAs office in Southern Rhodesia (now Zimbabwe) as punishment for having seduced his son James' fiancé. Despite that disgrace, Gemmill remained the Chamber's preferred representative in negotiating labour contracts with colonial governments, and in representing the industry at ILO meetings where the issues of occupational disease and migrant labour were discussed. While he remained useful to the industry, Gemmill never returned to

[16] Yudelman. *The Emergence of Modern South Africa*, 1984, Chs.5–6; V.L. Allen. *The History of the Black Mineworkers in South Africa: Vol.1, The Techniques of Resistance 1871–1948*. Keighley: The Moor Press, 1992.

[17] Draft Statement for a Memorandum from the Chamber to Round Table Conference of Vulnerable Mines: Native Labour Charges, from A.T. Milne, General Manager, Gold Producers' Committee to G.P.C. Members, 21st October 1957, p. 5. TEBA Archives, WNLA 14/3, Mass Miniature Radiography General, Dec. 1956 to Jan. 1958.

favour. His obituary in the Chamber's Annual Report for 1961 runs to just three lines.[18] Surprisingly, almost nothing has been written about him.[19]

In some respects, the gold mines were places of constant innovation. As shafts were sunk deeper, new technologies came on stream. Dust suppression technologies were improved and from 1922 the introduction of small jack hammers saw labour costs fall and drilling output double.[20] The frequent changes in technologies altered the nature and rhythms of work and the degree of risk. As the mines went deeper, the heat and humidity rose—as did the dangers of traumatic injury. By 1913, more than half the mines were working at depths of between 600 and 1200 metres; twenty years later depths of 1200–1500 metres were common. Some of the changes designed to minimise risk had unanticipated consequences. The introduction of water-fed drills from 1910 reduced the palpable dust, but the water sprays increased the humidity and with it the spread of tuberculosis.[21]

In other respects, however, the mines never changed. Despite their commercial success, the mines suffered from a chronic shortage of labour, which they attempted to resolve through an increasingly racialised work regime. The Chamber of Mines readily acknowledged that South Africa's low-grade deposits could not have been worked profitably in the USA, Canada or Australia because of the higher wages paid there.[22] White South African miners were unionised, militant and well paid. Black workers too conducted an impressive number of localised strikes, work

[18] See William Gemmill's Obituary in the Transvaal and Orange Free State Chamber of Mines 72nd Annual Report Year 1961, p. 7.

[19] For a notable exception see Allan H. Jeeves. 'William Gemmill and South African Expansion, 1920–1950'. Paper presented to The Making of Class history workshop, University of the Witwatersrand, 14th February 1987. http://wiredspace.wits.ac.za/bitstream/handle/10539/7824/HWS-198.pdf?sequence=1.

[20] Charles H. Feinstein. *An Economic History of South Africa: Conquest, Discrimination and Development Cambridge:* Cambridge University Press, 2005, p. 81.

[21] R.M. Packard. *White Plague, Black Labour: Tuberculosis and the Political Economy of Health and Disease in South Africa*. Berkeley: University of California Press, 1989, pp. 73–91.

[22] Transcript of discussion *Select Committee into the Miners Phthisis Commission Report*, Monday 15th May 1931, Parliamentary Library, Cape Town, p. 160.

stoppages, riots and go-slows, but faced overwhelming obstacles in forming associations and taking industrial action.[23] In this situation, the most effective way for employers to guarantee profitability was to blunt the militancy of white unions, tighten controls over black workers, keep paying them below-subsistence wages and maintain or increase their proportion of the workforce.

After an initial period of artisanal mining, with skilled and highly paid white miners assisted by unskilled non-white assistants (for several years including indentured Chinese labourers), the mining workforce was gradually reorganised. By the end of the first decade of the twentieth century, the last Chinese were deported, and blacks outnumbered whites by between seven and ten to one. Allowing for differences in the cost of living, the average wage paid to *white* miners on the Rand in 1913 was 1.56 times that paid to miners in Britain, 2.48 times to those in Germany and 1.23 times the pay in Australia.[24] Inflation during First World War, combined with lack of black labour and wage gains by white miners, saw output and profits fall. By 1918 the mines were short of 40,000 men and faced the challenge of expanding a low-paid black workforce for a hazardous industry under conditions of scarcity.[25] Migrant labour was a common feature of hard rock mining elsewhere in the world, including Australia and the USA. In terms of scale, longevity and the system's political foundations, however, South Africa's migrant labour system was probably unique.

[23] Alan H. Jeeves. *Migrant Labour in South Africa's Mining Economy: The Struggle for the Gold Mines' Labour Supply 1890–1920*. Kingston and Montreal: McGill-Queen's University Press, 1985, p.29; T. Dunbar Moodie and Vivienne Ndatshe. *Going for Gold: Men, Mines, and Migration*, Chs. 3 and 7; Allen. *The History of the Black Mineworkers Vol.1*, 1992, Chs. 14–20.

[24] Robert H. Davies. *Capital, State and White Labour in South Africa 1900–1960: An Historical Materialist Analysis of Class Formation and Class Relations*. New Jersey: Humanities Press, 1979, p. 73.

[25] Frederick A. Johnstone. *Class, Race and Gold: A study of class relations and racial discrimination in South Africa*. London: Routledge & Kegan Paul, 1976, p. 96.

Recruiting and Oscillating Migration

Before the 1870s, white farmers in the Western Cape recruited migrant workers to meet their seasonal needs. After the discovery of diamonds in 1874, more than 10,000 migrants, mostly from Swaziland and the Transkei, were employed in Kimberley's diamond mines. A pattern of oscillating migration was soon established, whereby men continued to live in rural areas but left their families for months at a time to earn money, whether in mines or on white farms. In addition, many black women were employed as domestics in white households. The Colonial Office viewed migrant labour as a temporary phase of economic and social development during which it was necessary for the state to protect workers from exploitation while simultaneously reducing the need for migration.[26] That did not happen. When gold was discovered on the Rand in 1886, what had begun as a transitional phenomenon common to industrialising regions became a permanent feature of the South African economy. In contrast to other employers, however, the gold mines recruited large and increasing numbers of non-South African men from regions where agricultural wages were low and there was little or no secondary industry.[27]

During the twentieth century, mining recruits came from three sources: from within South Africa's borders including the so-called Bantustans under apartheid; from the High Commission Territories (HCTs) of Bechuanaland (Botswana), Basutoland (Lesotho) and Swaziland (eSwatini) and from other colonies (or ex-colonies) such as Portuguese East Africa (Mozambique), Nyasaland (Malawi), Northern Rhodesia (Zambia) and Southern Rhodesia (Zimbabwe). 'East Coasters' were men from the East Coast of Mozambique, South of latitude 22° South. 'Tropical' referred to men recruited from North of latitude 22° South, or north of Beit Bridge, which separated South Rhodesia from South Africa. In 1914, the Tropical areas were listed as part of the German South-West African Protectorate (now Namibia); the Portuguese colonies of Beira and

[26] Letter from D. Watson, Colonial Office London to Mr. R. Turner, undated September 1952. PRO CO 859 306 Migrant Labour Conference.

[27] Innes. *Anglo American*, 1984, pp. 59–61; 68.

Chinde, Mozambique, Nyasa and Zambesia, Quelimane and Tete; and the British Nyasaland Protectorate, North-East Rhodesia, and part of South Rhodesia including North-West Rhodesia[28] [see map]. There were numerous shifts over time between those sites as the industry pursued its underlying strategy of keeping wages down and ensuring a sufficient supply of labour.[29]

South Africa & Bechuanaland Protectorate. "Handy Reference Atlas of the World"; by John Bartholomew, published by John Walker & Co. Ltd., London, 1949

[28] SANA NTS 189 335/F473 Secretary for Native Affairs, 23rd March 1914, Mortality amongst Tropical Natives. See also Randall M. Packard. 'The Invention of the "Tropical Worker": Medical Research and the Quest for Central African Labor on the South African Gold Mines, 1903–36". *The Journal of African History*, Vol. 34, No. 2, 1993, pp. 271–292. Packard argues that the category was as much political as geographic or medical and was invented by the mining industry to unify workers from these regions under one 'orientalist' category in order to shift emphasis from working conditions on the mines to the cultural and biological deficiencies of the recruits themselves. The account here adds more complexity to the story.

[29] See J.S. Harington, N. D. McGlashan and E. Z. Chelkowska. 'A Century of Migrant Labour in the Gold Mines of South Africa". *The Journal of the South African Institute of Mining and Metallurgy*, Vol. 104, No. 2, March 2004, pp. 65–71.

Soon after large-scale gold mining began in 1886, managers became aware that workers preferred those mines which paid the best wages.[30] The challenge for the industry was how to reduce competition between individual mines and keep wages low in a market where labour was scarce. The Chamber's response was to centralise recruiting. In 1896, the Rand Native Labour Association, subsequently the Witwatersrand Native Labour Association (WNLA), was formed to coordinate the supplies of labour from Portuguese East Africa, Northern and Southern Rhodesia and Nyasaland and, initially, Bechuanaland.[31] In October 1912 the Chamber of Mines created the Native Recruiting Corporation (NRC). The NRC operated within South Africa and the High Commission Territories of Basutoland, the southern half of the Bechuanaland Protectorate and Swaziland. William Gemmill headed both organisations and reported on their behalf to the Gold Producers Committee. The initiatives were successful, and the cost of recruiting fell.[32]

Two factors helped consolidate the oscillating migration system. In South Africa itself, stringently enforced pass laws, enacted by the state at the behest of the Chamber of Mines, limited the number of non-white residents in urban areas.[33] Until 1969, the gold mines were by law not allowed to provide family housing for more than 3 per cent of the black labour force, and none at all to 'foreign natives'.[34] Outside South Africa, mine wages enabled men to pay increasingly onerous government taxes, and avoid employment for much lower wages—or no pay at all.[35] In addition, the capitation fees paid by the WNLA became a major source

[30] Feinstein. *An Economic History of South Africa*, p. 69.

[31] *Report of the Witwatersrand Mine Native Wages Commission on the Remuneration and Conditions of Employment of Natives on the Witwatersrand Gold Mines*. Pretoria: Government Printer 1943, p. 2.

[32] Johnstone. *Class, Race and Gold*, p. 31.

[33] Francis Wilson. 'Miners and Migrants: How the Mining Industry Has Shaped South Africa', *Daedalus*. Vol. 130, No. 1, Winter 2001, p. 104. Those laws bound labourers to an employer and were often reinforced by Masters and Servants Acts which made a breach of contract a criminal offence. Feinstein. *An Economic History of South Africa*, 2005, pp. 55–56. See also Martin Chanock. *The Making of the South African Legal Culture, 1902–1936: Fear, Favour and Prejudice*. Cambridge: Cambridge University Press, 2001.

[34] Francis Wilson. *Migrant Labour in South Africa: Report to the South African Council of Churches*. Johannesburg: The South African Council of Churches and SPRO-CAS, 1972, p. 9.

[35] Charles van Onselen. *The Night Trains: Moving Mozambican Miners to and from the Witwatersrand Mines, 1902–1955*. London: Hurst Publishers, 2018, pp. 47–51.

of revenue for neighbouring colonial administrations. Bechuanaland (now Botswana), the location of pioneering research which helped inspire the miners' class action, is a case in point.[36] Proclaimed a British Protectorate in 1885, Bechuanaland covered a vast land mass.[37] Large numbers of men first began moving south in 1870 to work in the Kimberley diamond mines. The discovery of gold on the Witwatersrand in the late 1880s created a new and ever-increasing demand for labour. The seasonal flow of workers corresponded to the agricultural cycle. Men generally left home just after ploughing and planting in January or February, and returned in October or November, before the planting season began again. To help defray the costs of colonial administration, in 1899 a Hut Tax at the rate of one pound per annum was imposed on all adult males, with substantial penalties for non-payment. By 1919, Hut and other Native Taxes became the largest components of recurrent revenue.[38] Incomes from agriculture were rarely enough to support a family, let alone pay money to government.[39] There were few opportunities in Bechuanaland for wage labour, which in any case offered far lower pay than the Johannesburg gold mines.

The WNLA began signing up Tswana workers in May 1901. Between January 1903 and May 1905, it recruited 6743 miners. Some men found unfit for mine work were passed on to white farms in the Transvaal, a practice which was to become common.[40] Initially, the lack of local doctors meant that prospective miners were not medically examined before being shipped south. As a result, labour agents lost money when a man was rejected at Johannesburg. From 1910, most recruits were examined

[36] T.W. Steen; K.M. Gyi; N.W. White; T. Gabosianelwe; S. Ludick; G.N. Mazonde; N. Mabongo; M. Ncube; N. Monare; R. Ehrlich and G. Schierhout. 'Prevalence of Occupational Lung Disease Among Botswana Men Formerly Employed in The South African Mining Industry. *Occupational & Environmental Medicine*, Vol. 54, No. 1, January 1997, pp. 19–26.

[37] David Torrance. 'Britain, South Africa, and the High Commission Territories: An Old Controversy Revisited'. *The Historical Journal*, Vol. 41, No. 3, 1998, p. 753.

[38] Dile Delarey Kote. Recruitment of Mine Labour in Botswana 1899–1945, B.A. (History) thesis, University of Botswana, Lesotho and Swaziland, 1976, pp. 20–21.

[39] David Massey. 'A Case of Colonial Collaboration: The Hut Tax and Migrant Labour". *Botswana Notes and Records*, Vol. 10, 1978, pp. 99–112.

[40] Pauline Cuzon. The History of TEBA in Botswana, unpublished manuscript, July 1985, pp. 11; 14. BNA BNB 10,313.

by local government or mission medical officers. The WNLA paid a fee of 2s 6d per head for the first 500 recruits examined, and 1s per head thereafter. Through capitation fees paid by the WNLA, sending labour to the mines became a major source of revenue for neighbouring colonial administrations and local missions.[41] In 1940, for example, the Portuguese East Africa (Mozambique) government agreed to the maximum number of recruits for employment on South Africa's mines being raised to 100,000. In return, Mozambique received £1 14s 6d per recruit per annum from the mines. An annual fee of 10s was also paid by each recruit.[42] Despite the effectiveness of the Chamber's recruiting agencies, many prospective miners found their own way to Johannesburg. To gain control over this movement of labour, in 1928 the Chamber created the Assisted Voluntary Scheme (AVS) under which the NRC provided migrant workers advances or rail fares and food and allowed them to sign on at the mine of their choice after arriving on the Rand.[43] By 1937 around half of South African miners were AVS voluntaries.[44]

A black miners' typical day started at about 4:30 a.m. when men were wakened in the compounds, the company-provided cramped and primitive barrack-like hostels, and proceeded to the shafts. The day ended between 3 and 4 p.m., when the miners returned to the surface. Sunday was a rest day. Mining was gruelling, exhausting and dangerous work. It was common for miners to walk three to five kilometres underground before they reached the work face. The transportation of the large number of men took time, and some miners would wait an hour or more on the surface before they entered the lifts. On a cold winter's morning that

[41] Medical examinations were not compulsory until the Proclamation (No.21) of 1935. I. Schapera. 'Migrant Labour and Tribal Life: A Study of Conditions in the Bechuanaland Protectorate', Report Presented to the B.P. Administration, Public Records Office, DO 119 1268 Schapera, 1944, p. 148.

[42] *Report of the Witwatersrand Mine Native Wages Commission on the Remuneration and Conditions of Employment of Natives on the Witwatersrand Gold Mines* Pretoria: Government Printer 1943, pp. 3–4. See also Patrick Harries. *Work, Culture and Identity: Migrant Laborers in Mozambique and South Africa C 1890–1910.* Portsmouth: Heinemann, 1994.

[43] Letter from W. Gemmill, General Manager, Native Recruiting Corporation Ltd, Johannesburg, to the Director of Native Labour, Johannesburg, 5th November 1927. SANA, 1932, NTS 2047 64/280 Gold Mines Voluntary Native Labour. See also Alan Jeeves. 'Migrant Labour and South African Expansion, 1920–1950'. *South African Historical Journal*, No 18, 1986, pp. 73–92.

[44] William Gemmill and Mr. Wellbeloved, Chamber of Mines. Evidence presented before the Native Labour Committee, Johannesburg 13th December 1937, pp. 2–3. SANA K 356 Native Labour Commission 1937.

was often done in the open.[45] The mortality rate from pneumonia, particularly among men from the Tropical north, was very high. Most deaths were in new recruits who had no prior exposure to the pneumococcus bacteria and contracted acute infections from which they died, often within days. There was also a close association between deaths from pneumonia and meningitis, as those infections are caused by the same organism. One obvious remedy was to improve conditions in the compounds, but most mining houses resisted change.[46]

The depth and scale of the mines and their racialised labour regimes made them particularly dangerous. According to official records, between 1903 and 1920 a total of 88,252 African miners died from accidents or disease. In addition, an unknown number perished in rural areas from disease or injury sustained on the mines.[47] After 1920 the number of recorded deaths fell, and over the following sixty years the annual toll from mine accidents fluctuated around 800. This was still a huge count: British coal mines had a far larger workforce but the annual number of deaths from accidents seldom exceeded 200. The dangers of mining—and the resultant chronic shortages of mine labour—were compounded by occupational lung disease.

In the 1920s, white wages were cut and the reliance on skilled white miners further reduced, but the colour bar remained. The workforce in the gold mines now consisted of unionised white labour in supervisory roles, with the bulk of the manual work being done by black migrants. In 1918 there were 22,764 white miners and 179,628 blacks; 38,327 and 303,087 respectively by 1937.[48] When, in the early 1930s, rise in the gold price led to an increased demand for labour, the mines broadened the scope and breadth of their recruiting. By 1936, there were over 300,000 men from Portuguese East Africa, Nyasaland and the HCTs on

[45] *Tuberculosis in South African Natives with Special Reference to the Disease Amongst the Mine Labourers on the Witwatersrand.* South African Institute for Medical Research, Johannesburg, March 1932, pp. 71–72.

[46] Cartwright, A.P. *Doctors of the Mines: A History of the Work of Mine Medical Officers.* Cape Town: Purnell and Sons, 1971, pp. 19–20.

[47] Between 1906 and 1920 just over 6600 European miners died. H.J. Simons. Migratory Labour, Migratory Microbes. Occupational Health in the South African Mining Industry: The Formative Years 1870–1956. Unpublished manuscript, 1960, p. 2.

[48] Yudelman. *The Emergence of Modern South Africa*, pp. 54; 191.

the mines, while the numbers of black South Africans fell.[49] In general, the contracts in South Africa and the Protectorates were for less than a year, whilst those in Portuguese East Africa and Tropical areas varied between twelve and eighteen months. What Jonathan Crush and his colleagues called a *labour empire* eventually extended from Cape Town to present-day Angola in the west and the borders of Tanganyika in the northeast.[50] The empire included 154 recruiting stations and a regular air service from the more distant corners of Nyasaland (Malawi).[51]

The entry medical system set up and operated by WNLA and the NRC had far-reaching consequences. The ecology of the labour system involved the constant movement of black and sometimes white miners through a system in which the gold mines were at the top and the coal and asbestos mines at the base. At each step down that ladder, the pay and work conditions worsened, as did access to medical care and compensation. The most important issue in the strict guidelines for labour recruitment concerned diseased lungs: those who had been repatriated for tuberculosis or silicosis were not to be recruited again. Particular attention was to be given to recruits with flabby muscles suggestive of wasting: 'Special care should be taken in the examination of natives with long underground mining histories. It has been the experience that the majority of such natives develop tuberculosis and it is useless to send such men to the mines as they will invariably be rejected by the Mine Medical Officer.'[52] Similarly, men over the age of forty were not to be employed. Under these regulations, it was common for large numbers of men to fail the entry medical. During 1937, for example, around a quarter of recruits offering themselves to the NRC were rejected. Although unfit for mining, a

[49] Letter from District Commissioner's Office, Molepolole, to The Government Secretary, Mafeking, 20th January 1948, Subject: Sentences imposed on deserters from the mines. Botswana National Archives, S114/7/3, 1945–54 Labour Native Recruitment.

[50] Crush, J., A. Jeeves and D. Yudelman. *South Africa's Labor Empire: A History of Black Migrancy to the Gold Mines*. Boulder and Cape Town: Westview Press, 1991.

[51] Memo, A.T. Milne, General Manager, to G.P.C. Members, 27th September 1957, Memorandum to the Round Table Conference on Vulnerable Mines: Recruiting and Medical Examination of Mine Native Labourers, p. 2. TEBA Archives, WNLA 14/3, Mass Miniature Radiography General, Dec. 1956 to Jan. 1958.

[52] Suggestions Relating to the Medical Examination of Native Mine labourers (As Amended for Tropical Natives), WNLA Circular, Johannesburg August 1935. Emigrant Labour 1935–1936 M2/3/3. Malawi National Archives.

proportion of those men were considered suitable for farm labour, and at some NRC offices farmers' organisations were given access to rejects.[53] Soon, that practice became common, and the NRC and the WNLA began acting as *de facto* agents for farmers in the North-Eastern Transvaal and Southern Rhodesia.

Although the labour system was racialised, there is little reference to race in the miners' phthisis or silicosis legislation or in the mines acts. Instead, those laws are framed in terms of two categories: miners (whites) and native labourers (blacks). White miners were designated as skilled and native labourers as unskilled. They did different jobs for different rates of pay and for different contract periods. Whites were permanent employees and worked in supervisory roles. Blacks did the bulk of the physical labour, some of it highly complex, from drilling and blasting to loading skips, tasks which involved the highest dust exposures. Whites and blacks worked under different compensation regimes; they were subject to different forms of medical surveillance and they received different medical care: there were sanatoria for white silicotics and repatriations for blacks. The nomenclature of miners and native labourers pervaded both official discourse and South African science; it also influenced how the industry was viewed from the outside. The minutes from the ILO's 1930 Silicosis Conference in Johannesburg, for example, suggest that the overseas delegates were oblivious to racial segregation on the mines.[54]

The Chamber's Case for the Benefits of Migrant Labour

Throughout the twentieth century, the Chamber employed several arguments in support of oscillating migration. These included the economic benefits to labour-sending communities and the right of African men to

[53] Statement by the Native Recruiting Corporation Ltd. before the Native Labour Committee at Pietersburg undated December 1937, p. 2. SANA. K 356 Native Labour Commission 1937.

[54] So too were the editors of the International Labour Office's 1937 review of the South African compensation system. See *Workmen's Compensation for Silicosis in the Union of South Africa, Great Britain and Germany*. International Labour Office. Studies and Reports. Series F (Industrial Hygiene) No.16. London, 1937.

freely enter into contracts. The Chamber also claimed that the migrant labour system protected rural communities from the spread of tuberculosis. The populations of tradition-bound African societies, it noted, were highly vulnerable to infectious disease, and this made it important to avoid mistakes such as the ones caused by the speed of early nineteenth-century urbanisation in Europe. Oscillating migration overcame that threat by easing the transition between traditional rural and modern urban life. There was also another important health benefit. According to the Chamber, the low compensation rates for silicosis in black miners were due to breaks in employment. The Chamber also had a ready explanation for its refusal to raise below-subsistence wages. The subsistence of miners' families, it claimed, was covered by farming. Men only went to the mines to satisfy special needs such as the purchase of a gun, cattle or the payment of lobola or bride price. Once those needs were met, they returned home. Because miners had no incentive to remain in permanent work, higher wages would only reduce the labour supply.[55]

Under Gemmill's leadership, the Chamber may have been the first employers' association to produce its own propaganda films. One such film from 1920, titled 'With the WNLA in Portuguese East Africa', ran for ten minutes and was shown at cinemas throughout South Africa.[56] It portrays the orderly flow of labour to the gold mines and attacks the ban on the recruitment of workers from the Tropical north. The Chamber also founded the South African Industrial News Service to distribute favourable news items.[57] A monograph published by the Chamber in 1947 contains many of the arguments rehearsed before Commissions and Select Committee Enquiries. The pamphlet, almost certainly written by William Gemmill, extolls the benefits of migrant labour. The Chamber's recruiting arms, WNLA and the NRC, enabled migrant workers to proceed to the mines with a minimum of hardship. Recruiting officers explained the conditions of service, conducted medicals and advanced travelling expenses. On the mines, the WNLA helped men

[55] See L.G. Irvine and D. Macaulay. 'The Life-history of the Native Mine Labourer in the Transvaal". *Journal of Hygiene*, Vol. 6, No. 2, 1906, p. 160.

[56] 'With the WNLA in Portuguese East Africa', Cinematograph Pictures: WNLA Recruiting Operations August 1920–January 1935 WNLA 211 TEBA Archives, University of Johannesburg.

[57] Yudelman. *The Emergence of Modern South Africa*, 1984, pp. 195–196.

maintain contact with their families, remitted savings and dealt with grievances.[58] The excellent system of medical care included monthly inspections and an exit examination. There were 37 well-equipped hospitals, with over 7000 beds and 60 full-time mine medical officers. Miners were well fed and usually left the mines in far better health than when they arrived: 'There can be no doubt that the tribal Native working on the good mines of South Africa is much better off than his antitype in Asiatic countries and even many peasant communities in parts of Europe'.[59]

Rather than leading to immorality and ill health as some critics claimed, migrant work protected miners and their families from disease and the sudden erosion of traditional life. The settlement of large number of blacks into the cities would pose a health problem. 'Primitive people seem to lack the power—inherited by European races—to resist tubercular infection'. Continuous employment on the mines would also produce higher rates of silicosis. In the rural areas, the Chamber claimed, tuberculosis was not a serious hazard, but in the towns a breakdown may follow an abrupt transition to city life. 'There can be no doubt that not only would this [permanent employment] lead to a serious increase in the incidence of silicosis, but that the addition of tuberculosis would become almost inevitable.' The mines used X-rays 'so that tuberculosis or possible silicosis will be detected before clinical signs or symptoms become apparent, and treatment may be begun at the most propitious time'.[60]

When speaking outside South Africa, Gemmill's justifications of the migrant labour system emphasised the role of oscillating migration in preserving indigenous cultures. Following the ILO's Sydney Conference in 1951, for example, Gemmill wrote an overview of the WNLA's operations in Southern Africa. As he explained it, migrant labourers were strongly attached to their tribal lives and their aim was to consolidate a traditional way of life by earning the necessary money. In responding to those complex needs, the WNLA had devised what it called *the controlled*

[58] *The Native Workers on the Witwatersrand Gold Mines*. Transvaal Chamber of Mines Publication P.R.D. No. 7, 1947, p. 5.

[59] *The Native Workers on the Witwatersrand*, pp. 13–14; 19.

[60] *The Native Workers on the Witwatersrand*, pp. 14; 16–17.

migrant system. Under this regime, men were engaged on the mutually agreed condition that they must return to their homes at the end of a stipulated period. The wages paid were usually much higher than in their countries of origin, and the conditions of engagement conformed with those laid down by the ILO. The system had the approval of the South African and Portuguese governments as well as the British Colonial Office. Migrant labour, Gemmill noted, preserved family life, while the educative influence of men on returning to their homes promoted social progress. 'If the standard of living of the African peoples is to be raised, their productive capacity and efficiency must increase, and nothing is more conducive to this than employment on the Gold Mines.' Government restrictions on the free movement of labour took many forms, including quotas on the numbers the WNLA could engage. Such restrictions, Gemmill claimed, involved the exploitation of the African for the benefit of local low-wage-paying employers.[61]

Development of the Free State Mines After the Second World War

The development of the Free State mines after the Second World War changed the industry once more. Deeper and more mechanised than those around Johannesburg, they represented the industry's largest capital investment. By 1955, over R400 million had been spent. Within a decade, the twenty-two new mines accounted for over half of South Africa's total output and almost four fifths of the industry's working profit.[62] The average ore grade was appreciably higher than on the older

[61] William Gemmill, Statement Presented to the International Labour Office Dealing with the Restrictions Placed on the Spontaneous Flow of African Migrant Labour, pp. 3–5. PRO CO859/197/2, International Labour Organization Representations by Mr Gemmill (WNLA) Regarding Restrictions on Migrant Labour in Africa, 1951.

[62] Feinstein, *An Economic History of South Africa*, p. 166.

mines. In the period during 1970, gold production rose sharply while state revenue from mining trebled.[63]

The Free State mines were vast and complex workplaces which used millions of litres of water and massive amounts of energy. The President Steyn Mine at Welkom consisted of three distinct shaft areas, each of which was the size of a separate mine. Each working area comprised at least 18 levels, some extending to a depth of over 3 kilometres. The underground operations included hundreds of stope panels situated in an area spanning approximately 32 square kilometres and linked by more than 240 kilometres of tunnels.[64] Typically, during a 24-hour period, more than 14,500 men worked underground.[65] When Jan Prinsloo began as a mine overseer at President Steyn, it was the deepest shaft on earth.[66] Prinsloo was responsible for 500 to 600 men per shift but if another overseer was ill, he was in charge of double that number. He is a large man and he would walk constantly between one work area and another, most often bent double. Prinsloo recalls that dust was a problem despite Anglo American spending millions on dust prevention and management. He also recalls that AASA was a good employer. It created a highly regulated work environment, with overseers and managers meeting regularly to monitor the six monthly, yearly and two-year work plans which controlled all aspects of mine work.

In the immediate post-war period, the demand for labour became intense. Recruits to the mines were attracted by the continuity of employment, the provision of rations, accommodation and medical care, none of which were available on the diamond, coal or asbestos mines.[67] In addition, the highly efficient recruiting system provided them with a cash advance of £3, which enabled recruits to provide for their families until they could remit wages. The cost to the mines' recruiting agencies of

[63] See Innes, *Anglo American and the Rise of Modern South Africa*, p. 154.

[64] A stope is the actual site underground where the rock containing the gold is extracted.

[65] High Court of South Africa (Gauteng Local Division, Johannesburg), Case No. 08108/2013, Bongani Nkala et al. and Harmony Gold Mining Company Limited et al., 28th May 2014, pp. 37–49.

[66] Interview with Jan Prinsloo, Welkom, 21st October 2005.

[67] *Report of the Witwatersrand Mine Native Wages Commission on the Remuneration and Conditions of Employment of Natives on the Witwatersrand Gold Mines*. Pretoria: Government Printer 1943, p. 22.

recruiting and repatriating miners varied according to the distance from the mines. During the mid-1950s, the average costs per recruit were as follows: East Coasters £9 11s 4d, Tropicals £22 17s 4d and Union and High Commission Territory recruits £7 17s 4d. The average for all recruits was £10s 18s 11d.[68] While the outlays were high so were the returns. In January 1948, the District Commissioner's Office at Molepolole estimated that the cash value of each recruit to the industry was between £350 and £400.[69] This was a massive return: the average miner's pay for a twelve-month contact was £30. More broadly, recruiting outside of South Africa enabled the industry to resolve the paradox of a highly profitable industry based on scarce labour and low wages. Feinstein estimates that for five decades from the early 1930s, a doubling of black miners' wages would have cut total industry profits by over 60 per cent.[70] In 1921, black miners, who constituted the bulk of the workforce, received around one fifteenth of the wages of whites; by 1969, the differential had grown to one twentieth. The mining houses became so successful in controlling wages that in real terms the pay of black miners did not rise between 1911 and 1970.[71] During that period real wages of white workers increased by 70 per cent.[72]

The political environment in which the new mines operated was unstable. In August 1946, the African Mine Workers' Union launched a national strike with almost 100,000 miners demanding a minimum daily

---

[68] Draft Statement for a Memorandum from the Chamber to Round Table Conference of Vulnerable Mines: Native Labour Charges, from A.T. Milne, General Manager, Gold Producers' Committee to G.P.C. Members, 21st October 1957, pp. 3–4. TEBA Archives, WNLA 14/3, Mass Miniature Radiography General, Dec. 1956 to Jan. 1958.

[69] Letter from District Commissioner's Office, Molepolole, to The Government Secretary, Mafeking, 20th January 1948, Subject: Sentences imposed on deserters from the mines. Botswana National Archives, S114/7/3, 1945–54 Labour Native Recruitment.

[70] Feinstein, *An Economic History of South Africa*, p. 111.

[71] Francis Wilson. *Labour in the South African Gold Mines 1911–1969*. Cambridge: Cambridge University Press, 1972, pp. 45–46. Wilson argues that in real terms black wages were probably lower in 1969 than they had been in 1911.

[72] Anna Trapido. An Analysis of the Burden of Occupational Lung Disease in a Random Sample of Former Gold Mineworkers in the Libode District of the Eastern Cape. PhD thesis University of the Witwatersrand, 2000, p. 15.

wage of 10s. The brutal smashing of the strike and more extensive criminalisation of black union activity signalled the beginning of a new phase of labour regime.[73] By 1960, Malawi and Mozambique were supplying the bulk of mine labour. During that period the major mining houses, including Anglo American, diversified into asbestos, chrome and platinum. By the early 1970s, almost 80 per cent of miners were migrants from Malawi, Mozambique, Angola, Botswana, Lesotho and eSwatini.[74] As the workforce grew to well over 400,000 in the late 1970s, the racial imbalance was maintained. In 1986, stimulated by expansion following the remarkable rise in the gold price in the 1970s, employment in the gold mines reached an all-time peak of 534,000 persons. Thereafter, the numbers gradually declined, and by 2010 the workforce had shrunk to 160,000.

As regional and international opposition to apartheid strengthened, the Chamber became concerned about the mines' dependence on labour from the surrounding Front Line states. For the first time in its history, it sought to increase local recruitment by raising wages to a level which competed with South Africa's secondary industries.[75] In 1973, the year that the United Nations General Assembly denounced apartheid, the pay of African miners rose by 36 per cent. This was followed by rises of 61 per cent and 68 per cent in the following two years.[76] Most of the new recruits were from the former Transkei in the Eastern Cape. In 1966, 39.1 per cent of the black miners came from within the borders of South Africa;

---

[73] See Allen, *The History of the Black Mineworkers in South Africa, Vol.1*, Chs. 17–20.

[74] Jonathan Crush and James Wilmot eds. *Crossing Boundaries: Mine Migrancy in a Democratic South Africa*. Edited collection of papers presented at an international conference on 'Transforming Mine Migrancy in the 1990s' held in Cape Town, June 1994. Institute for Democracy in South Africa, Cape Town, 1995. In 1972, around 840,000 foreign migrants worked in the economies of Rhodesia and South Africa. Wilson, *Migrant Labour in South Africa*, p. 114.

[75] Wilson, 'Miners and Migrants', p. 102; Francis Wilson. 'Historical Roots of Inequality in South Africa". *Economic History of Developing Regions*, Vol. 26, No. 1 2011, Routledge, p. 7. The Frontline States were a loose coalition of African countries formed in 1970 to co-ordinate their responses to apartheid and formulate a uniform policy towards white minority rule in South Africa and Rhodesia. These were Angola, Botswana, Lesotho, Mozambique, Swaziland, Tanzania, Zambia and, from 1980, Zimbabwe.

[76] Feinstein, *An Economic History of South Africa*, p. 206.

by 1979, this grew to 54 per cent of a larger workforce.[77] During the same period, the Chamber initiated a policy of stabilisation, which saw an increase in the length of each contract, and in the average age of the miners.

Industrial Strife in a Dying Industry

For most of the twentieth century, unionism in South Africa was racialised. The radicalism (though not the racism) of the white South African Mine Workers Union was blunted, particularly after the Rand Revolt of 1922, as it and its successors were gradually co-opted into mine and government bureaucracy. Black miners, in contrast, faced the concerted opposition of state, industry and white supervisors in their attempts to organise; after a brief period of growth in the 1940s, the African Mine Workers Union all but ceased to exist. In 1982, a new black mineworkers union was established.[78] From its inception in 1982, the National Union of Mineworkers (NUM) fought to improve wages and protect its members against unfair dismissal. The NUM grew from 6000 members in 1982 to 187,000 in 1987. Following a breakdown in wage negotiations in August of that year, the Union embarked on a three-week strike. Its defeat forced the NUM into a series of agreements with the Chamber. These tied wages to profitability, further reduced the size of the workforce and led to a significant reduction in real wages.[79] As apartheid was being dismantled, the NUM's efforts to save jobs and protect labour rights were fought against the collapse of influx controls and the reincorporation of the homelands into the South African polity.[80] Paradoxically, stabilisation and subcontracting spanned the transition to majority rule when the new

---

[77] J.S. Harington, N.D. McGlashan and E.Z. Chelkowska. 'A Century of Migrant Labour in the Gold Mines of South Africa'. *The Journal of the South African Institute of Mining and Metallurgy*, Vol. 104, No. 2, March 2004, Figure 2, p. 67.

[78] V.L. Allen. *The History of the Black Mineworkers in South Africa. Vol.3: The Rise and Struggles of the National Union of Mineworkers 1982–1994*. Keighley: The Moor Press, 2003.

[79] Trapido, An Analysis of the Burden of Occupational Lung Disease, 2000, pp. 28–30.

[80] For discussion of that transformation process see Crush and Wilmot eds., *Crossing Boundaries*, 1995.

ANC government was seeking to improve workers' power in their dealings with employers.[81]

A spike in the gold price above $800 in 1980 did not last. By 1990, the mines were facing falling incomes and rising overheads, with labour accounting for over half of production costs.[82] Those problems were compounded by competition from open cut mines in the USSR, Canada and Australia.[83] The industry responded by retrenchments and closures resulting in the loss of 250,000 jobs, a development that was felt particularly hard in the Eastern Cape. Employers also introduced subcontracting and outsourcing. The major area for subcontracting was underground. Most of the new subcontractors were retrenched workers who were re-employed without the health benefits, compensation for injury, death or severance pay they were previously entitled to.[84] South Africa's gold output continued to fall as did the mines' contribution to national tax revenue, which dropped from 26 per cent in 1981 to 2 per cent in 1992.[85] At their peak in 1986, the mines employed well over half a million workers, by 1999, less than half that number, and by 2022, around 95,000. The recently settled class action against the gold mining companies was in effect fought against a dying industry.

References

Baker, Julie J. 'The Silent Crisis': Black Labour, Disease, and the Economics and Politics of Health on the South African Gold Mines, 1902–1930. PhD thesis, Queen's University, Ontario 1989.

Cartwright, A.P. *Doctors of the Mines: A History of the Work of Mine Medical Officers*. Cape Town: Purnell and Sons, 1971.

---

[81] Jonathan Crush, Theresa Ulicki, Teke Tseabne and Elizabeth Jansen Van Veuren. 'Undermining Labour: The Rise of Sub-contracting in South African Gold Mines". *Journal of Southern African Studies*, Vol. 27, No. 1, 2001, pp. 5–31.

[82] Submission of the Chamber of Mines to the Leon Commission, p. 26.

[83] Feinstein, *An Economic History of South Africa*, p. 207.

[84] Crush et al., 'Undermining Labour'.

[85] Nicoli Nattrass. 'South African Gold Mining: The 1980s and Beyond'. In Crush and Wilmot eds. *Crossing Boundaries*, p. 167.

Coetzee, A.M. 'The Saga of Pneumonia on the Mines'. *Mining Survey*, 1979, pp. 9–14.

Crush, J., A. Jeeves and D. Yudelman. *South Africa's Labor Empire: A History of Black Migrancy to the Gold Mines.* Boulder and Cape Town: Westview Press, 1991.

Ehrlich, Rodney, Jill Murray and David Rees. 'Subradiological silicosis'. *American Journal of Industrial Medicine.* Vol. 61, No. 11, November 2018, pp. 877–885.

Harington, J. S.; N. D. McGlashan and E. Z. Chelkowska. 'A Century of Migrant Labour in the Gold Mines of South Africa'. *The Journal of the South African Institute of Mining and Metallurgy*, Vol. 104, No. 2, March 2004, pp. 65–71.

Katz, Elaine. *The White Death: Silicosis on the Witwatersrand Gold Mines 1886–1910.* Johannesburg: Witwatersrand University Press, 1994.

Koornhof, H.J.; S.A. Madhi, C. Feldman, A. von Gottberg and K.P. Klugman. 'A Century of South African Battles Against the Pneumococcus—"the Captain of Death"'. *South African Journal of Epidemiological Infection*, Vol. 24, No. 4, 2009, pp. 7–19.

McCulloch, Jock. 'Hiding a Pandemic: Dr G.W.H. Schepers and the Politics of Silicosis in South Africa'. *The Journal of Southern African Studies*, Vol. 35, No. 4, December 2009, pp. 835–848.

McCulloch, Jock. *South Africa's Gold Mines and the Politics of Silicosis.* Oxford: James Currey, 2012.

Murray, Jill, Tony Davies and David Rees. 'Occupational Lung Disease in the South African Mining Industry: Research and Policy Implementation'. *Journal of Public Health Policy*, Vol. 32, Suppl. 1, June 2011, pp. S65–79.

Orenstein, A.J. 'Vaccine Prophylaxis in Pneumonia'. *Journal of the Medical Association of South Africa*, 1931, No. 5, pp. 339–346.

Packard, R.M. 'The Invention of the "Tropical Worker": Medical Research and the Quest for Central African Labor on the South African Gold Mines, 1903–36'. *The Journal of African History*, Vol. 34, No. 2, 1993, pp. 271–292.

Packard, R.M. *White Plague, Black Labour: Tuberculosis and the Political Economy of Health and Disease in South Africa.* Berkeley: University of California Press, 1989.

Simons, Harold Jack. Migratory Labour, Migratory Microbes. Occupational Health in the South African Mining Industry: The Formative Years 1870–1956. Unpublished manuscript, 1960.

van Onselen, Charles. *The Night Trains: Moving Mozambican Miners to and from the Witwatersrand Mines, 1902–1955*. London: Hurst Publishers, 2018.
Watkins-Pitchford, W. 'The Silicosis of the South African Gold Mines, and the Changes Produced in it by Legislative and Administrative Efforts'. *The Journal of Industrial Hygiene,* Vol. 9, No. 4, April 1927, pp. 109–139.
Wright, Almroth; W. Parry Morgan and R.W. Dodgson. 'Observations on Prophylactic Inoculation Against Pneumococcus Infections'. *The Lancet*, Vol. 183, No. 4715, 10th January 1914, pp. 87–95.

3

Mapping and Resolving a Health Crisis: 1902–1929

When deep mining began in Southern Africa in the late 1880s, the size of the workforce increased rapidly. Recruits from South Africa, British Protectorates and from Portuguese East Africa were crowded into squalid compounds and fed on a diet of maize meal. As on all hard rock mines, there was lung disease and deaths from rock falls, but the Johannesburg mines were particularly dangerous. In 1903, the deaths of 3762 black miners from disease provoked scrutiny from the South African and British Parliaments.[1] The Colonial Office was in regular correspondence with the Governor-General in Cape Town, who monitored the death rates among Nyasa recruits.[2] In addition, from the 1920s the ILO took a keen interest in the health of migrant workers. The Chamber had good reason to be concerned. Occupational disease was a serious threat to the labour supply, and through the compensation system it came to represent a major cost of production.

[1] See Elaine Katz. *The White Death: Silicosis on the Witwatersrand Gold Mines 1886–1910.* Johannesburg: Witwatersrand University Press, 1994, pp. 93–122.

[2] See, for example, Letter from Lewis Harcourt, British MP, Downing Street, London, to Viscount Gladstone, Governor-General of South Africa, 25th November 1911. *SANA, GG 1538 50–141, 1911.*

J. McCulloch, P. Miller, *Mining Gold and Manufacturing Ignorance*,
https://doi.org/10.1007/978-981-19-8327-6_3

While the different components of fatal lung disease on the mines were not always easy to distinguish, the three major ones were eventually identified as pneumonia, silicosis and tuberculosis. Silicosis was caused and tuberculosis associated with silica dust exposure while pneumonia was a problem of susceptibility and hygiene, especially for workers from the Tropical north. In addition, pneumonia was a complicating factor in tuberculosis. As one senior health official noted: 'A large number of natives die of tuberculosis following a pneumonic attack and amongst tropical natives especially tuberculosis runs a very acute course'.[3] All miners were at some risk of silicosis and tuberculosis, although the lower dust exposures and superior general health of white miners meant their risk, especially of contracting tuberculosis, was relatively low. In contrast, young migrant workers were the most susceptible to pneumonia. Black miners who contracted silicosis and tuberculosis were repatriated, and usually died out of sight in rural areas. The thousands of miners who caught pneumonia died quickly, and their deaths were recorded as mine deaths. These divergent dynamics of chronic and acute lung disease shaped responses to what soon came to be perceived as a profound health crisis.

Men coming from the lowveld, or low-lying regions with subtropical climate, suffered greatly when they reached the much colder altitudes of Johannesburg. From the first years of mining, medical officers noted the high incidence of acute pneumonia among recruits from the Tropical north. Some recruits contracted the disease on the long train journey to the mines in crowded open carriages.[4] Inadequate or wet clothing and the exposure of exhausted men emerging from the hot and humid mines into the cold surface air in winter produced surges in the mortality rate.[5] The

[3] Appendix 10, Letter from R.N. Kotze, Chair of the Miners Phthisis Prevention Committee to the Minister of Mines, Johannesburg, 4th July 1914, *in General Report of the Miners Phthisis Prevention Committee Johannesburg 15th March 1916* Pretoria, The Government Printing & Stationery Office 1916, p. 142.

[4] Charles van Onselen. *The Night Trains: Moving Mozambican Miners to and from the Witwatersrand Mines, 1902–1955*. London: Hurst Publishers, 2018, Ch. 5, pp. 70–90.

[5] Harold Jack Simons. Migratory Labour Migratory Microbes. Occupational Health in the South African Mining Industry: The Formative Years 1870–1956. Unpublished manuscript, 1960, p. 8.

pneumococcus, discovered in 1881 almost simultaneously by Sternberg and Pasteur, are normally found in the throat and nasal passages and cause no symptoms. The body's defences against the disease can, however, be broken down by environmental factors. The incidence of respiratory infections increases when people from different communities are suddenly brought into close contact.

Most deaths were in new recruits who had no prior exposure to the pneumococcus bacteria and contracted acute infections from which they died, often within days. Before chemotherapy, pneumonia ran a course of anything up to three weeks before reaching a crisis. This was followed by a minimum of two to three weeks' convalescence before a miner could resume underground work. In some cases, convalescence lasted for months.[6] The contributing factors were overcrowded, filthy compounds and poor nutrition. There was also a close association between deaths from pneumonia and meningitis, as those infections are caused by the same organism.[7] It was common knowledge that when large numbers of new recruits arrived on a mine, pneumonia deaths followed. It was also known that the incidence on the so-called popular mines was lower. The popular mines were those where the work and living conditions were best, and they drew a larger proportion of seasoned men who were more discerning about where they worked. The less popular mines had a larger number of novices.[8] The conditions on Kimberly's diamond mines were also hazardous, with pneumonia being the most serious disease. As on the Rand, the predisposing factors were fatigue, undernourishment, overcrowding and exposure to cold and wet.[9]

---

[6] Simons, Migratory Labour, pp. 78–81.

[7] For an overview of the industry's response to pneumonia see A.P. Cartwright. *Doctors of the Mines: A History of the Work of Mine Medical Officers*. Cape Town: Purnell and Sons, 1971, pp. 15–28; 109–124.

[8] See A.M. Coetzee. 'The Saga of Pneumonia on the Mines'. *Mining Survey*, 1979, pp. 9–14.

[9] *Report Tuberculosis Commission, 1914*. U.G. 34-'14, Cape Town: Government Printer, 1914. Union of South Africa, pp. 153–162.

Defining the Problem

The gold mines closed briefly during the South African War (1899–1902). When they reopened, many of the Cornish rock drillers who had left for Britain never returned. According to the Government Mining Engineer for the Transvaal, in the interim more than 200 men had died of silicosis. His report added to adverse publicity about the dangers of gold mining in the British press, and questions were asked in the House of Commons. There was much concern at the Colonial Office and in February 1902 a party of British MPs visited Johannesburg. Sir Gilbert Parker, who led that delegation, subsequently wrote to the Chamber: 'I am convinced that the importation of natives from the Zambezi valley and from Central Africa ought to be stopped, the percentage of deaths amongst the natives on mines being seriously high chiefly because of the mortality amongst these particular natives'.[10]

In response, the Commissioner for Native Affairs, Godfrey Lagden, met with the Chamber to discuss what was now widely understood to amount to a crisis. A committee was appointed that included Drs L.G. Irvine, D. MacAuley and Andrew Watt. Their report identified a serious problem with an annual mortality rate of 54.5 per 1000 and noted that: 'Native workers are particularly susceptible to this disease [pneumonia] and the conditions of mining work favour its incidence'.[11] Men exposed to cold and damp when coming off shifts were particularly vulnerable, as were those with scurvy.[12] Despite that warning, recruitment in the north continued—as did the deaths. On the Simmer and Jack mines in March 1904, the official annual mortality rate for Tropicals reached 360 per 1000.[13] In response to the deaths, the British Colonial Office several times briefly suspended the recruitment of miners from its

---

[10] Quoted in Cartwright, *Doctors of the Mines*, p. 16.

[11] Cartwright, *Doctors of the Mines*, pp. 16–18.

[12] Mine rations were deficient in protein and vegetables. In 1906 the Department of Native Affairs gazetted a standard diet that included meat, vegetables and mealie meal (maize). Further improvements during the 1920s saw a decline in deaths from malnutrition. See Simons, Migratory Labour, p. 8.

[13] Cartwright, *Doctors of the Mines*, p. 19.

own tropical areas.[14] The increasing death toll affected recruitment, and the mines struggled to attract labour. In August 1902, the Rand mines were employing 37,000 men, or less than half the number required for full production.[15]

Understandings and proposed solutions of the health crisis besetting the mines were closely linked to the composition of the mine workforce. This consisted of several distinct groups: British miners, many of them tin miners from Cornwall; white and black South Africans; and recruits from outside South Africa, including men from the High Commission Territories (HCTs) and Nyasaland. The presence of men from territories administered by Britain drew the Colonial Office into the controversy. On 7 November 1902, the High Commissioner for South Africa, Lord Milner, appointed a commission (known as the Weldon Commission) to investigate the causes and extent of what he described as miners' phthisis.[16]

Like the commissions which were to follow, Weldon was designed to protect the health and interests of white labour. Its report, submitted in 1903, devotes a single page, consisting of case notes taken at the Lancaster West Gold Mine at Krugersdorp, specifically to black miners.[17] Focussing on conditions underground, the Commission found that what witnesses variously described as 'miners' phthisis', 'chest disease', 'Miners' Lung' or silicosis was a major problem for all classes of mine labour but especially amongst rock drillers, who had an average working life of just seven years. In fact, it was probably closer to four, as the dust in Johannesburg mines had particularly high silica content.[18] Weldon correctly identified the cause of silicosis as the fine dust generated by pneumatic drills and the use of gelignite for blasting, a finding that was verified by the Medical Commission in 1912.[19] He recommended more stringent blasting regulations, wet drill-

[14] SANA NTS 189 335/F473 Mortality amongst Tropical Natives.

[15] Cartwright, *Doctors of the Mines*, p. 7.

[16] Within six days, the British Home Office in London announced the establishment of the Haldane Commission to study silicosis among Cornish miners. That decision was overdue, as the death rate among tin miners had long been a matter of public concern. Katz, *The White Death*, p. 25.

[17] 'Medical Report on Five Cases of 'Silicosis' among Natives', Appendix F in *Report of the Miners' Phthisis Commission 1902–1903*. Pretoria: Government Printer, 1903.

[18] The silica content of 85 per cent was far higher than on US mines. Katz, *The White Death*, p. 4.

[19] *Report of the Miners' Phthisis Commission 1902–1903*. Pretoria: Government Printer, 1903, pp. xxi–xxii, and *Report of a Commission into Miners' Phthisis and Pulmonary Tuberculosis (Medical Commission)*. Cape Town: Government Printer, 1912, p. 7.

ing, mechanical ventilation and water sprays for laying dust. The Commissioner made little reference to the toll that the different components of chronic and acute lung disease were taking on migrant labour, and his report led to only a small reduction in risk. Not satisfied with the outcome, the British scientists J.S. Haldane and Sir Thomas Oliver continued to pressure the British government to improve conditions on the mines.

Under pressure from the white Mine Workers Union, a second Commission into Miners' Phthisis and Pulmonary Tuberculosis, usually known as the Medical Commission because most of its members were physicians, was established in 1911. The Commission's Report, tabled in 1912, is notable for its attempts to distinguish between different constituent elements of phthisis, and its subtle description of the disease process. It was also the first Commission of its kind to use X-rays as a diagnostic tool, putting South African science at the forefront of medical research into occupational lung disease. The Commission helped the term 'silicosis' enter the occupational disease lexicon by pronouncing it to be a separate disease, noting that 'All true cases of miners' phthisis are thus primarily cases of silicosis; silicosis is the feature common to them all'.[20] According to the Medical Commission, silicosis is an insidious, life-threatening disease caused by exposure to silica dust. The severity of silicosis and the speed with which it develops will depend upon the volume of dust in a miner's lungs. However, there is often little symmetry between the disease process and disability. Furthermore, there is a strong synergy between silicosis and tuberculosis, so that a miner with dusted lungs is very likely to contract pulmonary tuberculosis. That synergy was most pronounced among the migrant workers who were drawn to the gold mines from impoverished rural communities. Where dust levels are low the disease will take some years to impair working capacity, but as it advances there will be shortness of breath and a persistent cough. [21]

The Commission noted that Miners' phthisis, and silicosis in particular, is extremely difficult to diagnose in its early stages, and found that even at post-mortem it may be impossible to arrive at a correct diagnosis.

[20] *Report of a Commission into Miners' Phthisis and Pulmonary Tuberculosis (Medical Commission)* Cape Town: Government Printer, 1912, p. 10.

[21] *Report of a Commission into Miners' Phthisis and Pulmonary Tuberculosis 1912*, pp. 5–117. See also McCulloch, *South Africa's Gold Mines*, pp. 1–13.

Consequently, the statistics on the prevalence of silicosis were unreliable. To make an accurate diagnosis requires an X-ray, a careful clinical examination and the patient's work and medical histories. To prevent silicosis and its associated heightened risk of tuberculosis, it is necessary to prevent the release of dust in the workplace.[22] Dr Andrew Watt, an experienced mine medical officer who assisted the Commission, was convinced that it was impossible to reduce dust to a level at which disease would not occur.[23] That view is consistent with studies published since 1994.[24]

Regulating the Recruitment of Tropical Labour

Before a blanket ban on all Tropical recruiting was imposed in 1913, the British colonial government several times prohibited recruitment of mine workers from the tropical regions of its African colonies. Formal recruiting from Nyasaland commenced in 1903 when, as an experiment, the colonial government granted the WNLA a quota of 1000 men. In February of the following year the quota was increased to 5000, but recruitment was suspended during the winter because of deaths from pneumonia. In January 1906, the Commissioner for Native Affairs in Pretoria recommended that if the fatalities rose above 100 per 1000, recruitment from the north should be suspended. Twelve months later, the annual death rate for Nyasa rose to 166 and WNLA recruitment was halted again. Despite the ban, many men continued to go south as voluntary labour without a medical examination in what the British authorities designated as 'unprotected emigration'.[25]

The mines' demand for labour continued to rise, and recruiters were encouraged to push men through the system. As a result, many seriously

[22] *Report of a Commission*, 1912, pp. 7; 9; 11.

[23] McCulloch, *South Africa's Gold Mines*, pp. 79–80.

[24] See Jill Murray, Tony Davies and David Rees. 'Occupational Lung Disease in the South African Mining Industry: Research and Policy Implementation'. *Journal of Public Health Policy*, Vol. 32 Suppl 1, June 2011, pp. S65–79; and Rodney Ehrlich, Jill Murray and David Rees. 'Subradiological silicosis'. *American Journal of Industrial Medicine*, Vol. 61, No. 11, 2018, pp. 877–885.

[25] The history of recruiting from Nyasaland is found in Memo: Introduction of Tropical Natives for labour in the Transvaal Mines Department of Native Affairs 1911, undated, unsigned. SANA NTS 189 335/F473.

ill men passed the medical examinations at the point of embarkation and at arrival at the WNLA Compound and were sent underground. A sample of post-mortems of twenty recruits who arrived at the WNLA Compound between January 1908 and mid-April 1908 is representative of the fate of thousands of young men. All were migrant workers shipped from Ressano Garcia in Portuguese East Africa (Mozambique), all died at the WNLA hospital, and all were subject to a post-mortem performed by a mine medical officer.[26]

The surviving records are distressingly brief. In some instances, such as that of case No. 66864 who died of meningitis, the notes run to only eight or ten words. The men have a serial number but no name. The man's age, his date of arrival at Johannesburg, the date of hospitalisation if relevant and the date of death are recorded. In most cases the cause of death is pneumonia, meningitis or tuberculosis. There are notes on external appearance, which is usually 'poorly nourished', and in a few instances 'emaciated'. Of those who died from pneumonia, case No. 71347 is typical. He arrived from Ressano Garcia on 13 February 1908 and was admitted to hospital on 8 March, having served less than a month underground. He died on 21 March from pneumonia. He is described as 'poorly nourished'. Case 70651 was aged 28 and he arrived from Ressano Garcia on 14 January 1908. He was admitted to the WNLA hospital on that same day. He died five weeks later from general tuberculosis. His condition is described as emaciated. The record of No. 66553, who was aged 23, is almost identical. He arrived on 13 January 1908 and died at the WNLA hospital from general tuberculosis on 28 March, having worked underground for only a few weeks.

In February 1910, the Colonial Office finally agreed to the resumption of WNLA recruiting from Nyasaland. By 1912, the Rand mines employed an estimated 23,000 Nyasa. The number of deaths rose sharply and within a year the Secretary of State in London threatened to ban *all* recruiting from British Central Africa.[27] When Dr G.D. Maynard of the Johannesburg Municipal Council carried out a study of mine deaths, he

[26] The original case notes are held at the National Institute of Occupational Health in Johannesburg by Professor Jill Murray.

[27] Memo: Introduction of Tropical Natives for labour in the Transvaal Mines Department of Native Affairs 1911, undated, unsigned. SANA NTS 189 335/F473 Mortality Tropical Natives.

found that the recruits were much more likely to die during the initial period of service. The mortality rates from pneumonia, for example, were at least three times higher during the first six months.[28] In contrast, phthisis (in this instance some combination of tuberculosis and silicosis) became more common with longer periods underground. The territorial origins of labour were also important, with the most vulnerable recruits coming from Nyasaland. The Department of Native Affairs was greatly concerned that if Britain limited recruitment, the Portuguese may do the same. It warned the Chamber: 'Unless a decided improvement can be effected at an early date the government will have no alternative to the measure of entirely prohibiting the introduction of tropical Natives'.[29] The WNLA agreed that the mines had to be made safe but denied responsibility. The industry was doing everything it could to 'bring the death rate to a normal figure'. The problem, it believed, was too much fresh air: the excessive ventilation in the compounds forced on the mines by the Coloured Labourers Health Regulations, it argued, had resulted in many fatalities.[30] Native Labour Bureau was confident the Chamber would soon find a remedy.[31]

In August 1911, Prime Minister Botha met with officials from the Colonial Office in London to discuss the mine deaths. The officials warned Botha that if the deaths from lung disease continued, Britain would prohibit recruiting from North of 22°. Botha asked that no sudden decision be made as it would damage the industry, and promised that his government would rush through legislation to enforce change. He

[28] Letter from Dr G.D. Maynard, the Municipal Council of Johannesburg to Dr Charles Porter, Medical Officer of Health to the Town Clerk, 5th January 1911. SANA NTS 189 335/F473 Mortality Tropical Natives.

[29] Letter from the Department of Native Affairs, Pretoria to the Chairman, Chamber of Mines, Johannesburg 12th June 1911. SANA NTS 189 335/F473 Mortality Tropical Natives.

[30] Letter from the Secretary WNLA to the Minister for Native Affairs, Pretoria 19th June 1911. SANA NTS 189 335/F473 Mortality Tropical Natives. The medical orthodoxy at the time was that ventilation and abundant fresh air reduced the likelihood of infection. The problem was that the compounds were not heated, and vigorous ventilation in winter made the men very cold. As they huddled together for protection against the cold, their susceptibility to cross infection was likely increased. See Packard, *White Plague, Black Labour*, p. 78.

[31] Letter from the Director of the Native Labour Bureau, Johannesburg to the Secretary for Native Affairs 21st June 1911. SANA NTS 189 335/F473 Mortality Tropical Natives.

also promised to appoint 'a strong Commission on Tuberculosis'.[32] On his return to South Africa, Botha was reassured by the Chamber that the pneumonia problem would soon be resolved.[33] The new Tuberculosis Commission, whose brief included identification of the extent and causes of mortality in black goldminers and their susceptibility to pneumonia, was appointed in February 1912, just as the Medical Commission was preparing to submit its report.[34] The same year, the government and the WNLA founded the South African Institute of Medical Research (SAIMR). The WNLA provided £40,000 for a building and equipment and shared the annual maintenance costs with the government.[35] The Institute's brief was to find a remedy for the high mortality rates from lung disease on the mines. One of its first projects was research on a pneumococcus vaccine.

Science to the Rescue?

In Western medical science, the first decade of the twentieth century was the golden era of vaccination, and it is here that the mining industry looked for a solution to the most statistically visible component of mine deaths. The search for a pneumonia vaccine began in 1903, when the Chamber awarded the Transvaal Government Bacteriologist, Dr Pakes, a research grant of £1000. Pakes submitted only one interim report before he resigned amid accusations that he had embezzled the grant.[36]

[32] Memorandum of Conversation between Lewis Harcourt, Colonial Office, and General Botha, 8th August 1911, London. SANA, GG 1538 50–141, 1911.

[33] Minute from Louis Botha, Prime Minister, Cape Town, 8th February 1912. Subject: Native Mortality on Rand Mines. SANA, GG 1538 50–141, 1911.

[34] *Report Tuberculosis Commission, 1914*. U.G. 34-'14, Cape Town: Government Printer, 1914. Union of South Africa, p. 1. The Tuberculosis Commission is discussed in Chap. 6.

[35] See *Memo: The Incidents and Correspondence which led to the formation of the South African Institute of Medical Research N. O'K Webber Johannesburg 23 July 1914*. The South African Institute for Medical Research 17/9/1912 to 24/7/1918. WNLA 144 TEBA Archives. See Chap. 6 for further discussion of the Institute.

[36] Julie J. Baker. 'The Silent Crisis': Black Labour, Disease, and the Economics and Politics of Health on the South African Gold Mines, 1902–1930. PhD Thesis, Queen's University, Ontario, 1989, pp. 96–97.

In early 1911, the Government Bacteriologist, Dr J.C. Mitchell, was commissioned to further explore the development of a vaccine.[37] Mitchell found that pneumonia among black miners was infectious, and that the organism involved was a common form of the pneumococcus. The incidence of the disease, he believed, was governed by racial susceptibility and environment. The high altitude and cold weather on the Rand lowered resistance to infection, and even after acclimatisation the incidence among Tropicals was far higher than for other groups. Mitchell identified several environmental factors which the mines should address. The compounds needed to be properly heated and well ventilated, and rations should include an abundance of fresh vegetables and lime juice. Mitchell also noted that since dust was a predisposing factor, the work conditions underground were important. Variations in the temperature between work places and where miners waited to be hauled to the surface were also hazardous, and Mitchell wanted miners to be issued with warm clothing and provided with warm change houses. The Government Bacteriologist concluded that the differences in the death rates on individual mines were probably due to variations in conditions, with exposure to infected bedding and clothing a particular threat. 'It is therefore important', Mitchell wrote, 'that the disinfection of infected places, bedding, clothing, etc., should be carried out with the greatest rigour. The receiving compounds should be thoroughly cleansed and disinfected with a Formalin autoclave after the departure of one gang and before the arrival of the next.'[38] Most mine managers ignored Mitchell's advice on improving living and working conditions. Instead, the industry commissioned further vaccine research.

While Mitchell was drafting his report, Sir Julius Wernher, Chairman of the Central Mining and Investment Corporation in London, received a visit from the prominent bacteriologist Sir Almroth Wright. Wright had read about Mitchell's work in the British press, and he offered to develop

[37] A summary of Mitchell's report is found in Letter from the Secretary, the WNLA, to the Acting Under-Secretary for the Interior, Department of the Interior, Pretoria, 3rd November 1911. TEBA Archive Medical Research Dr Mitchell's Investigations: Pneumonia WNLA 133/3.

[38] Letter from the Secretary, the WNLA to the Acting Under Secretary for the Interior, Department of the Interior, Pretoria 3rd November 1911. Medical Research Dr Mitchell's Investigations: Pneumonia WNLA 133/3 TEBA Archive.

a vaccine for the WNLA.[39] Following brief negotiations, Wright was engaged for a period of six months at the extravagant fee of £6000.[40] Wernher was concerned about a ban on recruitment, and he expected Wright to quickly produce a useful report. Wright arrived in South Africa in September 1911 and immediately began a series of mass inoculations. The programme was poorly designed. Neither the dosages nor the number of injections administered, the locales of the trials or the characteristics of the subjects were standardised in the initial experiments. Wright left Johannesburg abruptly in February 1912 and the trials were continued by his assistant Parry Morgan.[41]

After repeated delays, in December 1913 Wright finally tendered his report. According to him, 'there was a reduction in the incidence and death rates of pneumonia in the inoculated' in every one of his trials.[42] Dr Maynard, a statistician with the South African Institute for Medical Research who reviewed Wright's data, found that vaccination had no significant impact on the mortality rates. Dr A. J. Orenstein, the powerful Superintendent of Sanitation for Rand Mines, was also highly critical of the programme and did not believe the results 'justified the adoption of vaccinations'.[43] A subsequent review by the WNLA concluded that Wright's pneumonia investigations had produced no positive results.[44]

---

[39] Dr Samuel Evans, 'Re Pneumonia Research Work 1910–15', Enclosed with Confidential Circular, No. 145/28, from W. Gemmill, General Manager, GPC, Johannesburg, 28th August 1928. Subject: Research Work-Pneumonia, p. 3. TEBA Archives 133/1, Medical Research: Sir Almroth Wright's investigations concerning Pneumonia among Mine Natives.

[40] Letter from T.J. Milner, The Central Mining & Investment Corporation, Limited, to Sir Almroth Wright MD FRS, London, 29th July 1911. TEBA Archives, 133/1, Medical Research: Sir Almroth Wright's investigations concerning Pneumonia among Mine Natives.

[41] Letter from Dr W. Parry Morgan, to the Chairman of the Bacteriological Research Committee, Witwatersrand Native Labour Association, Ltd., Johannesburg, 2nd September 1912. TEBA Archives, 133/1, Medical Research: Sir Almroth Wright's investigations concerning Pneumonia among Mine Natives.

[42] Almroth Wright, with W. Parry Morgan and R.W. Dodgson. 'Observations on Prophylactic Inoculation Against Pneumococcus Infections'. *The Lancet*, Vol. 183, Issue. 4715, 1914, pp. 87–95.

[43] See A.J. Orenstein. 'Vaccine Prophylaxis in Pneumonia'. *Journal of the Medical Association of South Africa*, No. 5, 1931, pp. 339–346.

[44] Dr Samuel Evans, 'Re Pneumonia Research Work 1910–15', Enclosed with Confidential Circular, No. 145/28, from W. Gemmill, General Manager, GPC, Johannesburg, 28th August 1928. Subject: Research Work- Pneumonia, p. 3. TEBA Archives, 133/1, Medical Research: Sir Almroth Wright's investigations concerning Pneumonia among Mine Natives.

Wright's vaccine failed because it contained only four strains of pneumococci. By identifying the numerous strains present and including several more of them in a new preparation, Spencer Lister, a mine medical officer who had assisted Wright, succeeded in producing a more effective vaccine. Despite the scepticism of Maynard and Orenstein, prophylactic vaccination became universal on the Rand in 1917. The programme was gradually discontinued from 1927.

The 1913 Ban on Tropical Recruiting

In their efforts to monitor the health of migrant workers, the South African governments, like the imperial authorities in London, relied upon the data issued by the Chamber. That data was unreliable. A memo from Prime Minister Botha from 12 May 1913 makes clear his frustration that the mortality rates for Tropical labour collated over the previous three years 'had not reflected the true position'. The figures were confined to those men who died on individual mines and excluded the large number of deaths at the WNLA compound in Johannesburg, through which nine out of ten black miners passed. In 1910 the annual death rate on the mines was 75 per 1000, but that rose to 97 when the WNLA deaths were included. Three years later that gap had widened appreciably. In January 1913 the death rate on the mines was 45 but with the WNLA deaths that rose to 115. The figures for February were 64 and 117 and for March 72 and 119 per thousand respectively.[45] The revised data (which excluded the significant number of men who died on the way home or soon after repatriation) showed that the death rate had risen in an alarming manner and Botha decided to prohibit further recruitment from the Tropical north.

The bans on Tropical recruitment were the first and only state interventions designed specifically to protect the health of migrant labour. When the general ban was announced by Prime Minister Botha on 13 May 1913, the Chamber immediately began a multi-pronged campaign

[45] Memo from Prime Minister Botha, Cape Town to the Governor General 12th May 1913. Draw Attention to the Misleading Nature of Statistics. SANA GG 1541 50/301 Mortality Among Tropical Natives.

to have it rescinded.[46] Within a week, the Johannesburg press published several optimistic reports. *The Star* noted that much progress had been made in improving mine conditions and in developing a vaccine.[47] In addition, the Chamber commissioned Major W.C. Gorgas, famous for reducing deaths among workers on the Panama Canal, to investigate the causes of high mortality rates from pneumonia and to find a remedy. Gorgas arrived in Johannesburg in December 1913 and submitted his report the following year.[48] While Gorgas's key recommendations on reorganising the mines' workforce were rejected by the Chamber, the largest and most profitable mining house, Rand Mines Ltd, appointed Gorgas's assistant to oversee sanitation and to reorganise the company's health services, a task which A.J. Orenstein undertook with marked success.[49]

Between 1912 and 1920, the official annual mortality rate on the mines fell dramatically, from 12.5 per 1000 in 1912 to 3.4 in 1920. The fall was almost certainly due to improved compounds, better diets and the provision of change rooms to protect miners from chills rather than vaccinations.[50] The ban on recruitment of Tropical labour imposed in 1913 was also significant in the sustained decline in the mortality rate from pneumonia which followed. Even so, between 1933 and 1938 pneumonia accounted for between 29 and 37 per cent of deaths from disease on the mines.[51] Importantly, while pneumonia probably remained the major

[46] The controversy is discussed by Randall M. Packard in 'The Invention of the "Tropical Worker": Medical Research and the Quest for Central African Labor on the South African Gold Mines, 1903–36', *The Journal of African History*, Vol. 34, No. 2, 1993, pp. 271–292. Packard puts major emphasis on the racial politics of discursive construction of the issues and downplays the role the ban played in protecting the health of migrant workers.

[47] 'Tropical Natives'. *The Star*, Johannesburg, 19th May 1913.

[48] Gorgas, W.C. *Recommendation as to Sanitation Concerning Employees of the Mines on the Rand made to the Transvaal Chamber of Mines*. Johannesburg, 1914. For further discussion of Gorgas's work, see Chap. 6. For an overview of Orenstein's contribution, see Chap. 11.

[49] Cartwright, *Doctors of the Mines*, pp. 38–39.

[50] Deaths from pneumonia continued, albeit as a reduced rate, until the introduction of sulphonamide drugs in 1938, which produced a dramatic decline in mortality. See H.J. Koornhof, S.A. Madhi, C. Feldman, A von Gottberg and K.P. Klugman. 'A Century of South African Battles Against the Pneumococcus—"the Captain of Death"'. *South African Journal of Epidemiological Infection*, Vol. 24, No. 4, 2009, pp. 7–19.

[51] Dr A. Millar, 'Discussion, Pneumonia', *Proceedings of the Transvaal Mine Medical Officers' Association*, Vol. 19, No. 214, 1940, pp. 229–30.

cause of death among *black* miners, it was never the subject of a commission of inquiry, nor was it ever declared an occupational disease. In this case as in many others, the keenly contested politics of medical classification, rather than the actual burden of mortality, proved decisive.[52] And while Wright's attempts to develop a vaccine produced no positive results, his Johannesburg studies did bolster the idea that racial inferiority, rather than squalid compounds, poor diets and hazardous work environments were behind the high mortality rates among black miners.[53]

Negotiating Occupational Disease: Laws, Medicals and Repatriations

While Almroth Wright and his successors worked on a vaccine and the mining houses gradually improved conditions in the compounds and mines, the health crisis was resolved politically by a series of Mines and Miners' Phthisis Acts. Those acts, passed between 1911 and 1925, established a system of compulsory medical examinations to ensure the fitness of recruits for underground work, and to prevent the spread of tuberculosis. In principle, medical surveillance combined with a compensation system subjected the industry to external review. South Africa was the first state to compensate for silicosis (1911) and tuberculosis (1916) as occupational diseases, and it was the first to introduce medical certification for hard rock miners. The Miners' Phthisis Act No. 44 of 1916 made tuberculosis a compensable disease in gold miners and barred any person with tuberculosis from underground work. The 1916 Act also created the Miners' Phthisis Medical Bureau (Bureau), which in theory was responsible for the conduct of all mine medicals. The Bureau also compiled the official disease data. Although the Act appeared to bring medical reviews under the Bureau's and therefore the state's authority, it in fact decentralised the system. White miners were examined at the Bureau. In contrast,

[52] As McCulloch and Rosental put, occupational disease is a 'negotiated disease'. Jock McCulloch and Paul-André Rosental, with Joe Melling. 'Johannesburg and Beyond: Silicosis as a Transnational and Imperial Disease, 1900–1940'. In Rosental, Paul-André, ed. *Silicosis: A World History*. Baltimore: Johns Hopkins University Press, 2017, p. 75.

[53] See also Packard, 'The Invention of the "Tropical Worker"'.

all pre-employment, periodic and exit examinations of black miners were conducted at individual mines by mine medical officers, or at the WNLA compound. William Gemmill explained how the system worked: 'Every Mine Medical officer is by law an officer of the Miners' Phthisis Medical Bureau, and is subject to the instructions of the Bureau in all Miners' Phthisis matters affecting natives employed by the Mines'.[54] In theory he was correct, but as with all such legislation there was a gulf between what was prescribed and what actually happened in the workplace.[55] The crushing workloads of Bureau interns meant that in practice, the medical examinations of black miners were never supervised or reviewed by the Bureau.[56]

It was easy for government to justify outsourcing medical examinations to employers. By 1915, there were over one hundred thousand migrant workers on the Rand, all of whom were subject to compulsory examination. There was also a high labour turnover, which increased the medical workload even further. The cost for the state in running that system would have been prohibitive, and no doubt would have been a source of conflict between the government, the mines and the white electorate. For the mining industry there was another important consideration. Within months of the passage of the 1911 Act, compensation costs began rising, and the Chamber was keen to minimise the number of claims. The mines were willing to bear the cost of medical examinations so long as they also controlled compensation referrals to the Bureau.[57]

Under the Miners' Phthisis Act No. 40 of 1919, when a miner who had worked underground for a period exceeding one month left employment, he was to be given a final examination by the Bureau or by a designated medical practitioner. The Act's most important innovation was to lay down the criteria for diagnosing tuberculosis. A miner was to be compensated only where the tubercle bacillus was present in his sputum or if

[54] William Gemmill, General Manager Tropical Areas, note for Mr. K. Lambert Hall, Secretary, Nyasaland, Northern and Southern Rhodesia Inter-territorial Conference, Salisbury, 20th December 1940. Malawi National Archives. Emigrant Labour Governors Survey M2/3/19,

[55] Simons, Migratory Labour, pp. 27–43.

[56] Jock McCulloch. 'Hiding a Pandemic: Dr G.W.H. Schepers and the Politics of Silicosis in South Africa'. *The Journal of Southern African Studies*, Vol. 35, No. 4, 2009, pp. 835–848.

[57] For a discussion of the compensation system see McCulloch, *South Africa's Gold Mines*, pp. 33–54.

he was suffering from a serious impairment. There was no pathology testing at the WNLA in 1919; indeed, there were no such facilities at the WNLA hospital as late as the early 1950s.[58] Nor did the Act offer guidelines as to what constituted serious impairment. Despite these constraints, the sophistication of the system increased with each subsequent Act. In terms of scale, longevity and the degree to which it was racialised, the South African medical system was unique. The burden of proof regarding occupational lung disease which was gradually built into it became one of the principal targets of the recent miners' class action.

Repatriations and Occupational Disease

The first health crisis on the mines was resolved by the legislation and by the creation of two scientific bodies, the South African Institute of Medical Research and the Miners' Phthisis Medical Bureau. There was, however, a third element which was arguably more important in quelling criticism of the mines' mortality rates: the repatriation of sick and dying miners.[59] The extent of the practice varied between different mines and over time but was always substantial. Following complaints from the Department of Native Affairs, the Commissioner appointed in 1913 to investigate the grievances of black mine workers found that the annual repatriation rates from individual mines varied from 2 per cent to more than 13 per cent and reflected a divergence in the practice of individual medical officers.[60] Between 1916 and 1920, according to the official data, a total of 4121 men were repatriated because of tuberculosis. A further 964 died from that disease at individual mines before they could be sent home.[61] In 1915, the Rand Mines Group reported that 29.66 per 1000 or 1720 of its 57,990 black miners were repatriated. In 1926, WNLA and the NRC together repatriated 5087 black miners. Two years later the

[58] McCulloch, 'Hiding a Pandemic'.

[59] For more detailed discussion of technologies and repatriations, see Chap. 13.

[60] *Report of the Native Grievances Inquiry 1913–1914*. Cape Town: Government Printer, 1914, p. 25.

[61] Discussion *in Proceedings of the Transvaal Mine Medical Officers' Association*. Vol. II, No. 1, May 1922, p. 7.

combined total was 6924.[62] Neither the Chamber nor the government accepted responsibility for repatriations, and it was left to mine doctors to decide if a patient was 'fit to travel'. Men with pneumonia were acutely ill and far less likely to be sent home than those with silicosis or tuberculosis. Those miners who survived the train ride to the regional depots did not necessarily complete the journey home.

The industry had good reason to repatriate sick men. Hospital wards were overcrowded and there was insufficient staff to provide long-term care. Writing in 1915, the WNLA's senior medical officer, Dr G. Turner, was blunt about the financial advantages. 'This system [repatriations] has saved the Mines large sums of money. … Take the case of a native with scurvy. If he stays on the Reef he costs the Mine about 2s 6d per day for two, possibly three, months. … Whereas if returned to his kraal, this expense is eliminated.' Repatriations had the further advantage of lowering the mines' official mortality rates. Perhaps the most telling comment on repatriations comes from the 1914 report of the Tuberculosis Commission. It found that the repatriation rate was rising much faster than the mortality rate on the mines was falling, suggesting that repatriations were masking the actual number of deaths.[63] In other words, the increasingly efficient export of dying men to their home communities made it seem, in statistical returns, that fewer miners succumbed to occupational injury and disease.

The issue of repatriations resurfaced again and again over the following decades. Prior to the passing of the Miners' Phthisis Act of 1925, for example, there was intense debate about mine medicals, tuberculosis and repatriations. In October 1924, the Chairman of the Miners' Phthisis Medical Board, A.B. Du Toit, aired his concerns to the Minister of Mines. For some years, Du Toit had a 'presentment' that black miners were not getting 'a square deal' in regard to compensation, and that if ever he had an opportunity to investigate the situation he would do so. The new Act gave him that opportunity. He was particularly critical of the conduct of mine medicals. Du Toit was also concerned about the mines' repatriation

---

[62] Cited by Baker, 'The Silent Crisis', pp. 126–129.

[63] *Report Tuberculosis Commission, 1914*. U.G. 34-'14, Cape Town: Government Printer, 1914. Union of South Africa, p. 204.

policies. Tuberculosis was rampant in the Native Territories and the position was made worse by the repatriation of infected miners. 'Nothing appears to have been done in this matter and I am informed that the condition of affairs in some of the kraals is ghastly.' He went on: 'The further one goes back into the records the larger the annual number becomes of tuberculotic natives discharged by the Mines. In 1915 there were 1200. In this way it is computed that during the past 20 years an *ARMY OF 20,000 OR MORE TUBERCULOUS NATIVES* have been discharged by the mines back to their kraals to spread the disease.' Mine medical officers were not much concerned about the health of workers *leaving* the mines. As a result, it was common for men to be discharged from one mine as free from silicosis or tuberculosis and found later by the WNLA or another mine to be ill. Those men were missing out on compensation and would die in their villages. Du Toit warned the minister: 'The important question of health as it affects millions of people cannot be brushed aside for all time. The day will come when the European will have to answer for it.'[64] In a lengthy response, the Acting Director of Native Labour addressed each of du Toit's allegations. Recruits were examined in batches of 200 to 700 but the WNLA medical officers were expert, and the medicals were thorough. It was common knowledge that tuberculosis was a serious menace and he admitted that some men died on their way home.[65] This was not surprising given the fact that, as a later commentator put it, 'it appears that the criterion for deciding whether [sick mine workers] are fit for repatriation is fitness to travel, the measurement of which is ability to stand'.[66]

The deaths of East-Coast miners while *en route* from Johannesburg to Ressano Garcia were a constant source of bad publicity for the mining industry. In February 1925, the Chief Medical Officer of WNLA, Dr

---

[64] Letter from A.B. Du Toit, Chairman of the Miners' Phthisis Medical Board to the Minister of Mines and Industries, 21st October 1924, pp. 4–6 Subject: Confidential: Compensation to Native labourers who Contract Silicosis and Tuberculosis. SANA Miners Phthisis NTS 6720 33/315 Vol. 1. Emphasis in the original.

[65] Letter from the Acting Director of Native Labour to the Secretary for Native Affairs, 4th December 1924, subject: Confidential Compensation for Native Labourers. SANA Miners Phthisis NTS 6720 33/315 Vol. 1.

[66] *Report of the Departmental Committee of Enquiry into the Relationship Between Silicosis and Pulmonary Disability and the Relationship Between Pneumoconiosis and Tuberculosis (Oosthuizen Committee)*. Pretoria: Government Printer, part 2, 1954, Ch. 8, p. 133.

A.I. Girdwood, wrote a report on the practice. Convalescent men from mine hospitals were received at the WNLA compound in Johannesburg on Mondays and Tuesdays for repatriation. Those fit to travel were dispatched each Tuesday morning from the Booysens Railway Station. One train went to Ressano Garcia and the other to the Cape Province. A WNLA medical officer saw the train off, but there was no doctor on board. Each train had special coaches for those unable to walk and a white conductor. The seriously ill were given a stretcher while the others were in berths in the third-class carriages. There were no more than six patients in each compartment, so the men could lie down at night. During the journey they were fed and cared for by an attendant.[67]

In addition to the repatriation of injured men, around twenty miners suffering from tuberculosis were repatriated each week. 'It is this class of case', wrote Dr Girdwood, 'that is so liable to die on the train. They differ from the ordinary cases inasmuch as they are suffering from an incurable disease.' It was common for miners to wait for weeks at the WNLA hospital for the Bureau to process their compensation claims. Consequently, many were seriously ill by the time they were shipped out. 'If one considers the pathology of the lungs in these cases, large, ragged, breaking-down cavities full of pus, which might at any moment ulcerate through the blood vessel and cause a fatal haemorrhage, it should not be a matter of surprise that cases do die in the train, but that so few do.' Some miners believed a traditional healer would be able to cure them, and all sick miners wanted to see their families before they died. Dr Girdwood wrote: 'If we attempted to detain all natives of this class who were not absolutely fit to travel, apart from the fact that we would very soon be overcrowded, there would be continual dissatisfaction and complaints, not only up here but in the Territories'.[68] For all these reasons, repatriations became a permanent feature of mine medicine.

[67] For a detailed account see Charles Van Onselen. *The Night Trains: Moving Mozambican Miners to and from the Witwatersrand Mines, 1902–1955*. London: Hurst Publishers, 2018.

[68] Letter from A.I. Girdwood, Chief Medical Officer, to General Manager WNLA, Johannesburg, 17th February 1925. TEBA Archives. WNLA 20L Diseases and Epidemics Tuberculosis February 1923 to December 1930. For a harrowing account of these train journeys, see Van Onselen, *The Night Trains*, Ch.7: 'The Down Passage: Returning the Living Dead', pp. 114–132.

In 1929, the Director of Native Labour, Major Herbert Cook and the WNLAs Chief Medical Officer, Dr Arthur Girdwood, appeared before the Miners' Phthisis Commission. Under the Act, Cook was responsible for the payment of compensation to black miners. He told the Commission that the time was coming when his department must know exactly what happened to those men who had received an award and therefore what was being achieved by compensation. Cook had without success canvassed local magistrates, and he believed that the life expectancy of compensated miners was less than three years. Cook was also concerned that neither his department nor the mines provided practical instruction on health care to men who were repatriated with compensable disease.[69]

Dr Arthur Girdwood was asked about the WNLA's policies aimed at preventing the spread of tuberculosis. Girdwood reassured the Commission: 'We have been doing everything possible to try and discover ways and means of eliminating tuberculosis in the mines and prevent white miners becoming infected'. The black miners who spread disease were those in a very bad way and Girdwood rejected the suggestion that the mines should educate its workforce about how to reduce the risk of spreading tuberculosis once they returned home.[70] By 1929, there had been six Miners' Phthisis Commissions, a Commission into Tuberculosis and Pneumonia and several Parliamentary Select Committees. Cook and Girdwood were key figures within the migrant labour system, yet it is obvious from the Commission transcripts that neither had any knowledge about the post-mining health of the more than 300,000 migrant workers who served each year. Such lack of clarity persists. Today, there is no certainty about the number of former miners eligible for compensation for their occupational lung disease under the recent class action settlement, with estimated numbers ranging from seventeen thousand to half a million.[71]

[69] Evidence of Major Herbert Sutton Cook, Director of Native Labour, before the Miners' Phthisis Commission, 1929–30, pp. 1443–1448. SANA, K162, Miners' Phthisis Commission, 1929–30, Minutes Vol. 2.

[70] Evidence of Dr Arthur Inglis Girdwood, Chief Medical Officer of the Witwatersrand Native Labour Association, before the Miners' Phthisis Commission, 1929–30, pp. 1466–1468. SANA, K162, Miners' Phthisis Commission, 1929–30, Minutes Vol. 2.

[71] Republic of South Africa, The High Court of South Africa, 2019, 1/35–36 #57, #58.

Conclusion

The first health crisis on South Africa's gold mines lasted roughly from 1902 to 1916. Resolving it was a messy process. For the mining houses, the key issue was access to labour. There was also the perception of risk, which was important to the imperial authorities in London and later to the ILO. Under pressure from critics at home and abroad, the Chamber set up the South African Institute of Medical Research and the Miners' Phthisis Medical Bureau and agreed to the world's most progressive legislation for hard rock miners, under which silicosis (1911) and tuberculosis (1916) became compensable occupational diseases. Pneumonia was managed by the State and the Chamber in a very different way. White miners did not die from pneumonia, and their union, the MWU, took no interest in that issue. There were no commissions of enquiry, no dedicated legislation and no compensation schemes. Despite the huge death toll it caused among black miners, pneumonia was never declared an occupational disease.

Men who succumbed to pneumonia died on the mines, while those with tuberculosis and silicosis tended to die in rural areas. Pneumonia was also different because of the elevated death rate among recruits from Portuguese East Africa and the British colonial territories of Botswana, Nyasaland and Northern Rhodesia, a factor which attracted scrutiny from imperial authorities. Threatened with ban on the recruitment of tropical labour, the Chamber employed the international experts Sir Almroth Wright from Britain and William Gorgas from the USA to find a way of reducing the death rate. That money was not well spent. Wright failed to develop an effective vaccine, and the Chamber largely ignored Gorgas' excellent advice on how to prevent contagious disease. The fact remains that pneumonia was the only disease that led to a ban on Tropical recruitment, the only state intervention designed specifically to protect the health of migrant labour.

The first health crisis became a model for how the industry would respond to occupational disease. The mining companies gained a decisive role in the scientific formulation of the relevant issues, played a key role in framing the legislation and externalised the principal health costs onto labour-sending communities.

References

Jeeves, Alan. 'Migrant Labour and South African Expansion, 1920–1950'. *South African Historical Journal*, No. 18, 1986, pp. 73–92.

Allen, V.L. *The History of the Black Mineworkers in South Africa: Vol. 1, The Techniques of Resistance 1871–1948*. Keighley: The Moor Press, 1992.

Allen, V.L. *The History of the Black Mineworkers in South Africa: Vol. 3: The rise and struggles of the National Union of Mineworkers 1982–1994*. Keighley: The Moor Press, 2003.

Cartwright, A. P. *Doctors of the Mines: A History of the Work of Mine Medical Officers*. Cape Town: Purnell and Sons, 1971.

Chanock, M. *The Making of the South African Legal Culture, 1902–1936: Fear, Favour and Prejudice.* Cambridge: Cambridge University Press, 2001.

Crush, J. and J. Wilmot eds. *Crossing Boundaries: Mine Migrancy in a Democratic South Africa*. Edited collection of papers presented at an international conference on 'Transforming Mine Migrancy in the 1990s' held in Cape Town, June 1994. Institute for Democracy in South Africa, Cape Town, 1995.

Crush, J.; T. Ulicki; T. Tseabne and E. J. Van Veuren. 'Undermining Labour: The Rise of Sub-contracting in South African Gold Mines'. *Journal of Southern African Studies.* Vol. 27, No. 1, March 2001, pp. 5–31.

Cuzon, P. The History of TEBA in Botswana, unpublished manuscript, July 1985.

Davies, R. H. *Capital, State and White Labour in South Africa 1900–1960: An Historical Materialist Analysis of Class Formation and Class Relations*. New Jersey: Humanities Press, 1979.

Feinstein, Charles H. *An Economic History of South Africa: Conquest, Discrimination and Development.* Cambridge: Cambridge University Press, 2005.

Harington, J. S.; N. D. McGlashan and E. Z. Chelkowska. 'A Century of Migrant Labour in the Gold Mines of South Africa'. *The Journal of the South African Institute of Mining and Metallurgy*, Vol. 104, No. 2, March 2004, pp. 65–71.

Harries, P. *Work, Culture and Identity: Migrant Laborers in Mozambique and South Africa C 1890–1910*. Portsmouth: Heinemann, 1994.

Innes, Duncan. *Anglo-American and the Rise of Modern South Africa.* London: Monthly Review Press, 1984.

Irvine, L. G. and D. Macaulay. 'The Life-history of the Native Mine Labourer in the Transvaal'. *Journal of Hygiene*, Vol. 6, No. 2, April 1906, pp. 149–174.

Jeeves, A. H. 'William Gemmill and South African Expansion, 1920–1950', paper presented to The Making of Class history workshop, University of the Witwatersrand, 14th February 1987. http://wiredspace.wits.ac.za/bitstream/handle/10539/7824/HWS-198.pdf?sequence=1.

Jeeves, Alan H. *Migrant Labour in South Africa's Mining Economy: The struggle for the gold mines' labour supply 1890–1920*. Kingston and Montreal: McGill-Queen's University Press, 1985.

Johnstone, F. A. *Class, Race and Gold: A Study of Class Relations and Racial Discrimination in South Africa*. London: Routledge & Kegan Paul, 1976.

Kote, D. D. Recruitment of Mine Labour in Botswana 1899–1945, B.A. (History) thesis, University of Botswana, Lesotho and Swaziland, 1976.

Lang, J. *Bullion Johannesburg: Men, Mines and the Challenge of Conflict*. Johannesburg: Jonathan Ball Publishers, 1986.

Massey, D. 'A Case of Colonial Collaboration: The Hut Tax and Migrant Labour'. *Botswana Notes and Records*, Vol. 10, 1978, pp. 95–98.

McCrae, J. *The Ash of Silicotic Lungs*. Johannesburg: The South African Institute for Medical Research, 1918.

McCulloch, Jock and Geoffrey Tweedale. *Defending the Indefensible: The Global Asbestos Industry*. Oxford: Oxford University Press, 2008.

McCulloch, Jock and Paul-André Rosental, with Joe Melling. 'Johannesburg and Beyond: Silicosis as a Transnational and Imperial Disease, 1900–1940'. In Rosental, Paul-André, ed. *Silicosis: A World History*. Baltimore: Johns Hopkins University Press, 2017, pp. 64–104.

Murray, Jill, Tony Davies and David Rees. 'Occupational Lung Disease in the South African Mining Industry: Research and Policy Implementation'. *Journal of Public Health Policy*, Vol. 32 Suppl. 1, June 2011, pp. S65–79.

Moodie, T. Dunbar and Vivienne Ndatshe. *Going for Gold: Men, Mines, and Migration*. Berkeley: University of California Press, 1994.

Nattrass, N. 'South African Gold Mining: The 1980s and Beyond'. In Jonathan Crush and James Wilmot eds. *Crossing Boundaries: Mine Migrancy in a Democratic South Africa*. Edited collection of papers presented at an international conference on 'Transforming Mine Migrancy in the 1990s' held in Cape Town, June 1994. Institute for Democracy in South Africa, Cape Town, 1995.

Packard, R. M. 'The Invention of the "Tropical Worker": Medical Research and the Quest for Central African Labor on the South African Gold Mines, 1903–36'. *The Journal of African History*, Vol. 34, No. 2, 1993, pp. 271–292.

Packard, R. M. *White Plague, Black Labour: Tuberculosis and the Political Economy of Health and Disease in South Africa.* Berkeley: University of California Press, 1989.

Simons, H.J. Migratory Labour, Migratory Microbes. Occupational Health in the South African Mining Industry: The Formative Years 1870–1956. Unpublished manuscript, 1960.

Steen, T. W.; K. M. Gyi; N. W. White; T. Gabosianelwe; S. Ludick; G. N. Mazonde; N. Mabongo; M. Ncube; N. Monare; R. Ehrlich and G. Schierhout. 'Prevalence of Occupational Lung Disease Among Botswana Men Formerly Employed in The South African Mining Industry'. *Occupational & Environmental Medicine*, Vol. 54, No. 1, 1997, pp. 19–26.

Torrance, D. 'Britain, South Africa, and the High Commission Territories: An old controversy revisited'. *The Historical Journal.* Vol. 41, No. 3, 1998, pp. 751–772.

Trapido, A. An Analysis of the Burden of Occupational Lung Disease in a Random Sample of Former Gold Mineworkers in the Libode District of the Eastern Cape. PhD thesis, University of the Witwatersrand, 2000.

van Onselen, Charles. *The Night Trains: Moving Mozambican Miners to and from the Witwatersrand Mines, 1902–1955.* London: Hurst Publishers, 2018.

Wilson, Francis. 'Miners and Migrants: How the Mining Industry Has Shaped South Africa'. *Daedalus,* Vol. 130, No. 1, Winter 2001, pp. 99–121.

Wilson, Francis. 'Historical Roots of Inequality in South Africa'. *Economic History of Developing Regions*, Vol. 26. No. 1, 2011, pp. 1–15.

Wilson, Francis. *Labour in the South African Gold Mines 1911–1969.* Cambridge: Cambridge University Press, 1972a.

Wilson, Francis. *Migrant Labour in South Africa: Report to the South African Council of Churches.* Johannesburg: The South African Council of Churches and SPRO-CAS, 1972b.

Yudelman, D. *The Emergence of Modern South Africa: State, Capital, and the Incorporation of Organised Labor on the South African Gold Fields, 1902–1939.* Cape Town: David Philip, 1984.

Part II

Science Solves Health Crises

4

Identifying Risk and Compensating Tuberculosis: 1916–1957

By the 1830s, Scottish physicians had confirmed the everyday knowledge of hard rock miners, cutlery grinders and stone masons that men exposed to silica dust were likely to die from tuberculosis by their mid-thirties. From 1916, what was by then a medical orthodoxy was embodied in the South African legislation. Dr Andrew Watt, who appeared before the Parliamentary Select Committee in that year, was strongly opposed to silicotics returning underground as they would quickly develop tuberculosis. He pointed out that following the South African War, many former miners came back from the veldt in good health, having lived an open-air life. Those who went back underground soon fell ill.[1] To Watt it was unthinkable that a man with tuberculosis should continue to work in a mine.

South Africa was the first state to compensate for silicosis and tuberculosis as occupational diseases. Miners, stonemasons and foundry workers in the UK faced similar risks. From the final decades of the nineteenth century, there was a succession of British commissions into tuberculosis. To prevent the spread of disease, by 1910 public health authorities had

[1] Dr Watt, evidence before *Miners' Phthisis Working of Acts Parliamentary Select Committee 1916* SC 10-15 1915 Third Report AN 4923, p. 628.

J. McCulloch, P. Miller, *Mining Gold and Manufacturing Ignorance*,
https://doi.org/10.1007/978-981-19-8327-6_4

implemented three key policies: a state directed system of case identification, the compulsory isolation of infective patients and the education of patients and their families about disease prevention.[2] However, tuberculosis was not compensable in hard rock miners,[3] and it was not until 1918 that silicosis became recognised as an occupational disease for specific classes of British workers. The International Labour Office Convention No. 42 of 1948 included silicosis (with or without pulmonary tuberculosis) in its Schedule of Occupational Diseases, providing it was proven to be an essential factor in causing incapacity or death.[4] Under the National Insurance (Industrial Injuries) Act of 1951, tuberculosis became a prescribed occupational disease for British health workers or research and laboratory staff coming into contact with patients or infective material.[5]

The response by South Africa's gold mines was markedly different. In contrast to the UK, where compensation was awarded for demonstrable disability, under South African law workers were entitled to compensation even in the absence of impairment. The South African decision to make tuberculosis compensable was remarkable. The legislative recognition of an occupational disease usually involves the play of various social, political and economic factors which are far removed from actual workplaces or medical discovery.[6] To qualify as an occupational disease, an injury typically has to meet a number of tests. Is the disease specifically associated with the person's employment? Does the disease occur more frequently in a particular industry than in the general population? Are the risks peculiar to that industry, or do they occur in other industries or

[2] *Report Tuberculosis Commission, 1914*, pp. 16–17.

[3] On the UK experience see P. Weindling (ed.) *The Social History of Occupational Health*. London: Croom Helm, 1985 and N.J. Wikeley. *Compensation for Industrial Disease*. Aldershot: Dartmouth Publishing, 1993. See also Arthur McIvor. 'Germs at Work: Establishing Tuberculosis as an Occupational Disease in Britain, C.1900–1951'. *Social History of Medicine*, Vol. 24, No. 4, 2012, pp. 812–829.

[4] Note on the Status of Lung Diseases in Workmen's Compensation, p. 3. *PRO CO859/259 Lung Diseases Workmens' Compensation 1952*.

[5] See Arthur McIvor. 'Germs at Work: Establishing Tuberculosis as an Occupational Disease in Britain, C.1900–1951'. *Social History of Medicine*, Vol. 24, No. 4, 2012, pp. 812–829.

[6] See, for example, Allard E. Dembe. *Occupation and Disease: How Social Factors Affect the Conception of Work-Related Disorders*. New Haven: Yale University Press, 1996.

in workers' daily lives? The 1906 British Committee report on the Workmen's Compensation Act, for example, excluded *tuberculosis phthisis* as an occupational disease because it also occurred outside the workplace.[7]

The South African legislation was driven by two key concerns of the white Mine Workers Union (MWU), formed in 1902 as the Transvaal Miners' Association. The first was ensuring that its members were compensated for silicosis. Many miners whose doctors diagnosed serious lung disease were denied compensation by the companies.[8] The plight of the widows of white miners was raised at a number of Commissions and Select Committee hearings, and there is a voluminous correspondence from women appealing for assistance to the Prime Minister's Office.[9] Until 1956, the South African Acts applied only to gold mines, thereby excluding the thousands of men and women who worked on coal and asbestos mines. The legislation also excluded workers exposed to silica dust in other industries.[10]

The second concern related to the high tuberculosis rate among black miners, which the MWU believed put white miners and their families at risk. There is little surviving correspondence about the 1916 Act, but we do know it was designed to protect white families from infection. We also know that once passed, the MWU fought to ensure the Act was never repealed. White supervisors were in close contact with black miners in confined spaces. We also know from the surviving records that at any time, large numbers of infected men were employed on the mines. It was in fact common for black miners to die underground from haemorrhages. There were also cultural factors at play which worked in favour of the MWU. Infectious diseases exercised a powerful influence on the origins and development of urban segregation in Southern Africa. Between 1900 and 1904, the threat of bubonic plague, cholera and smallpox saw the hasty removal of African urban populations to locations in Cape Town

[7] Dr W. Watkins-Pitchford. 'The Industrial Diseases of South Africa', *South African Medical Record*, Vol. 11, No. 3, 14th February 1914, p. 34. See also Beris Penrose. 'Medical Monitoring & Silicosis in Metal Miners: 1910–1940'. *Labour History Review*, Vol. 69, No. 3, 2004, pp. 285–303.

[8] *Report of the Miners' Phthisis Commission of Enquiry 1929–30* (Young Commission). Union of South Africa. Pretoria: The Government Printer, 1930, Part 1, p. 17; Part 2, p. 83.

[9] See Prime Minister's Office Correspondence 1/1/275 148/15/15 SANA.

[10] See McCulloch, 'Compensation', in *South Africa's Gold Mines*, pp. 33–54.

and Port Elizabeth. What Maynard Swanson has called 'the sanitation syndrome' equated black urban settlement with a threat to the health and security of white communities.[11] The powerful metaphor linking black urban settlement with a danger to the health of white families became a routine part of public health discourse; the resultant panic played a role in the passing of the 1916 Act.

Using a linguistic sleight of hand, the Miners' Phthisis Acts were woven around the binary pairing of miners (whites) and native labourers (blacks).[12] In the early twentieth-century period of reconstruction after the South African War, that device enabled legislators to racialise the labour laws without mentioning race. The technologies used in medical examinations and the methods of paying compensation were also racialised. White miners were examined by specialists at the Bureau and had access to private physicians. The screening of whites involved a clinical examination, a chest X-ray and the taking of medical and work histories. Black miners were examined by mine medical officers and only that small number referred to the Bureau for further examination were X-rayed. In practice, access to compensation for 90 per cent of the workforce fell under the authority of mine medical officers rather than the state. There is a reference to this anomaly in a 1949 report by the Northern Rhodesian Commission of Inquiry into Silicosis. It recommended that the State Medical Bureau assume the sole responsibility for certifying compensable disease and performing all entry and exit examinations. As it was, mine medical officers were part of both the examining and the compensation system.[13] That arrangement, with its inbuilt conflict of interests, remained at the heart of South African mine medicine for decades. As a result, epidemiology, as it is usually understood, played little part in producing the official disease data. The reported silicosis and

[11] See Maynard W. Swanson. 'The Sanitation Syndrome: Bubonic Plague and Urban Native Policy in the Cape Colony'. *Journal of African History*, Vol. 18, No. 3, 1977, pp. 387–410.

[12] See 'Delict and Compensation' in Martin Chanock. *The Making of the South African Legal Culture, 1902–1936: Fear, Favour and Prejudice*. Cambridge: Cambridge University Press, 2001, pp. 189–196.

[13] *Northern Rhodesia Report of the Commission on Silicosis Legislation* Lusaka: Government Printer, 1949, p. 11.

tuberculosis rates were in effect the compensation rates, and both were controlled by the mine doctors.

In the event, the mines complied with the Miners' Phthisis Acts by setting up the world's first system of case finding based on entry, periodic and exit medicals of a working population. However, the vast majority of those black miners identified with tuberculosis were then repatriated to their home communities without provision for their isolation or care. There were no notifications to local health authorities, no supplementary feeding programmes for children and no education programmes for miners or their families on how to prevent the spread of infection.[14]

A Looming Crisis of Tuberculosis Infection

From the beginning of the twentieth century, there were numerous warnings about a looming tuberculosis crisis on the mines. Those concerns were expressed by the Departments of Public Health and Native Affairs, the SAIMR, the Bureau and the MPPC. They were reiterated by the international authorities William Gorgas and Lyle Cummins, who acted as consultants for the Chamber. The common themes were the known hazard of exposure to silica dust and the spread of infection from the mines to rural areas. The constant counter-theme was the Chamber's insistence that tuberculosis was not contracted because of dust exposure and cramped, arduous and unsanitary working and living conditions, but was brought to the mines by migrant workers.

In 1905, the *Transvaal Medical Journal* published an editorial warning of an imminent public health disaster. Tuberculosis was easily transferred from individual to individual and the medical profession agreed on the urgent need for action. The South African native was 'tremendously susceptible' to infection, and many deaths recorded as enteric fever were in fact due to tuberculosis. According to the editor, the main area for concern was the mines: 'That the disease in its infective phases is developed by the boys, in the majority of cases, after their arrival on the Rand is

[14] See Chap. 13 for detailed discussion of the repatriation regime.

undoubted'. In the absence of isolation care by the state, miners should be repatriated as soon as possible. White miners must be protected and the risk of repatriations to the families of those infected was less crucial than that involved in allowing them to remain on the mines. The editorial concluded: 'That Tuberculosis occurs amongst mine boys in epidemic form is clearly proved: is it any wonder that so many cases of Miners' Phthisis contract a tuberculous secondary infection?'[15]

The decision to compensate tuberculosis as an occupational disease had its origins in a Parliamentary Select Committee enquiry of 1916. Dr George Albert Turner, the medical officer for the WNLA, appeared as an expert witness. Each year, Turner examined thousands of East Coasters (miners from Portuguese East Africa). He also conducted post-mortems of miners who died at the WNLA Compound. Turner told the Committee that the high mortality rate was due to the employment of Tropicals who were susceptible to pulmonary tuberculosis and pneumonia. He suspected that most men contracted tuberculosis on the mines.[16] Turner's evidence was supported by Dr L.G. Irvine, an experienced mine medical officer, who later became Director of the Bureau. Irvine was sure that East Coast recruits were more prone to tuberculosis than were men from the Cape, who had some acquired immunity. White miners showed silicosis in all its stages, whereas the typical black patient had early or intermediate stage silicosis with a mass of tuberculosis infection on top.[17]

The Select Committee found an urgent need to reduce the prevalence of tuberculosis. It recommended that infected miners be excluded from underground work, and for that reason receive compensation.[18] Such awards would be for the loss of earnings rather than for a workplace

[15] Editorial, 'The Tuberculosis Peril'. *The Transvaal Medical Journal*, Vol. 1, No. 2, Johannesburg, 1st September 1905, p. 26.

[16] Evidence of Dr George Albert Turner, Monday 17th January 1916 in 1916 Miners' Phthisis Working of Acts Parliamentary Select Committee 1916 SC 10-15 1915 Third Report AN 4923, pp. 580–589.

[17] Evidence of Dr L.G. Irvine, Monday 17th January 1916 in *1916 Miners' Phthisis Working of Acts Parliamentary Select Committee 1916 SC 10-15 1915 Third Report AN 4923 15th December 2015*, pp. 606–614.

[18] *Third and Final Report of the Select Committee on Working of Miners' Phthisis Acts*. Union of South Africa Select Committee 10-15. AN 492 April 1916, pp. xxiv–xxvi.

injury. In the case of white miners, that recommendation went against the evidence of number of expert witnesses. Dr Andrew Watt told the Committee that he had examined hundreds of white miners with tuberculosis but had not seen a single case in which a man's wife or children had been infected.[19] Dr Francis Aitken, the superintendent of the Springkell Sanatorium, agreed that it was rare for tuberculosis to spread from a white miner to his wife and children. Infection was influenced by living conditions: if conditions were good, the rate of infection was likely to be low.[20]

The Select Committee's Report received wide support. Three months later, the Miners Phthisis Act No. 44 of 1916 was passed. The Act made it an offence for any person to work underground knowing he had tuberculosis. His employer was also held to be guilty. During the first reading the Minister of Mines, Mr Malan, noted that the work conditions had improved to such an extent that the number of awards for miners' phthisis was falling, as was the severity of the disease. Despite those improvements, men with ordinary tuberculosis remained a danger, especially to miners suffering from silicosis. Malan explained that the aim of the Act was to prevent men with tuberculosis from entering the industry, and to identify and exclude those who had become infected. To that end, there were to be six monthly examinations of white miners, 'while blacks were also to be examined with regard to tuberculosis'. The legislation gained unanimous support. Dr Davies (Yeoville) noted that men with silicosis were very much at risk from tuberculosis. Mr Sampson (Siemert) warned the House that tuberculosis was spreading throughout South Africa and that: 'It should be made worth the while of the industry to take every possible means to eliminate dust from the mines'.[21]

[19] Evidence of Dr Andrew Watt, Monday 17th January 1916 in 1916 Miners' Phthisis Working of Acts Parliamentary Select Committee 1916 SC 10-15 1915 Third Report AN 4923, p. 617.

[20] Miners' Phthisis Working of Acts Parliamentary Select Committee 1916, SC 10-15 1915 Third Report, AN 4923, pp. 617–618; 693.

[21] See Debates of the House of Assembly of the Union of SA Vol 1, First Session, Second Parliament, 19th November 1915 to 17th June 1916. Reproduced in *The Cape Times*, Tuesday 6th June 1916, pp. 386–387.

The Chamber and Tuberculosis

The Act of 1916 represented a defeat for the Chamber. Over the following decades, each time the legislation was reviewed, the Chamber challenged the legitimacy of designating tuberculosis as an occupational disease. Chamber officials pointed out, quite correctly, that in the UK and the USA tuberculosis was not compensable in hard rock miners. In 1921, the Chamber's Legal Advisor, George Barry, noted the relative infrequency of pure tuberculosis in (white) miners and that 'contrary to preconceived ideas, employment in the gold mines at the present time does not specially increase the liability in our miners to ordinary Pulmonary Tuberculosis'. Barry concluded: 'I think a strong case can be made for the exclusion of compensation for pure pulmonary tuberculosis from the Act'.[22]

In their submissions to the 1931 Select Committee, George Barry and Frans Unger, representing the Chamber, argued that since tuberculosis was widespread in labour-sending communities, it should be removed from the Act. During the Second World War, the Gold Producers' Committee (GPC) reviewed the findings of the recently tabled Miners' Phthisis (Stratford) Commission Report. The GPC argued that Stratford had presented no evidence why tuberculosis should be classed as an industrial disease. Tuberculosis was endemic in the Native Territories, and with the change of environment and occupation latent tuberculosis was liable to become active. The illness was not due to the inhalation of dust and there was no justification for the gold mines compensating tuberculosis. The medical treatment of infected blacks should be the responsibility of the state.[23]

In his review of the proposed Silicosis Act of 1946, the Chamber's legal advisor, George Barry, again noted that 'grossly exaggerated' statements had been made in the press about the mines spreading tuberculosis. 'It is

[22] Report by George Barry, Legal Advisor to The Chamber on Miners' Phthisis Commissions (of 1919) and the Draft Act, December 1921 Miners Phthisis' Act 1919 WNLA 144 B July 1919 to January 1922, p. 5.

[23] Gold Producers' Committee Comments upon Report of Miners' Phthisis Commission, February 1944, p. 15. SANA, K105 Miners' Phthisis 1941–48, Vol. 1 Correspondence, The Stratford Commission.

common knowledge that tuberculosis is endemic in native territories and that latent tuberculosis which is liable to flare up at any time is difficult of detection at a medical examination.' According to Barry, the incidence of tuberculosis on the mines compared more than favourably with other industries, and the mines did far more than any other industry to combat the spread of disease.[24]

In response to a sudden increase in the number of awards in 1954, an enquiry chaired by Professor Sarel Oosthuizen was asked to review the compensation system.[25] In its submissions, the Chamber again challenged the status of tuberculosis as a compensable disease. 'It is the view of the Chamber that there is nothing specific in the nature of employment underground or elsewhere on a mine which is particularly conducive to the spread of tuberculosis.'[26] Oosthuizen agreed that tuberculosis was not in a strict sense an occupational disease. However, his report was highly critical of mine medicals and the industry's repatriation policies.[27] Oosthuizen noted that at recruitment, miners were a physical elite far less vulnerable to infection than the populations from which they were drawn. Therefore, the tuberculosis which did occur among miners must be related to their employment. Oosthuizen remarked that the mining industry had never recognised tuberculosis as an occupational disease peculiar to mining. At various times the industry admitted that men with silicosis were more susceptible to tuberculosis, but it objected to any suggestion that mining be regarded as causative factor in the contraction or spread of tuberculosis.[28]

---

[24] George Barry, Legal Advisor, Chamber of Mines, Notes on Silicosis Bill-Natives, Cape Town, 14th May 1945. SANA, K105, Miners' Phthisis 1941–48 [Stratford] Correspondence, Volume 4.

[25] See Report of the Departmental Committee of Enquiry into the Relationship Between Silicosis and Pulmonary Disability and the Relationship Between Pneumoconiosis and Tuberculosis August 1954. SANA F 33/671 Treasury. See also Jock McCulloch. 'Hiding a Pandemic: Dr G.W.H. Schepers and the Politics of Silicosis in South Africa'. *The Journal of Southern African Studies*, Vol. 35, No. 4, 2009, pp. 835–848.

[26] Memorandum from the Legal Adviser, B.T. Tindall, dated 2nd May 1955, p. 8. TEBA Archive, WNLA 14/3, Mass Miniature Radiography, General File, March 1954 to January 1956.

[27] Minutes of a Meeting of the Sub-Committee of Group Medical Officers held on 21st June 1955, in the Chamber of Mines Building, p. 2. *TEBA Archive, WNLA 14/3, Mass Miniature Radiography, General File, March 1954 to January 1956.*

[28] For a review of the Chamber's view of tuberculosis as an occupational disease see Report of the Departmental Committee of Enquiry into the Relationship Between Silicosis and Pulmonary Disability and the Relationship Between Pneumoconiosis and Tuberculosis. Part 2 The Relationship Between Pneumoconiosis and Tuberculosis, 1954 pp. 57–67. SANA, F33/671 Treasury: 29.

Oosthuizen found the medical examination of black miners inadequate, and he recommended that they be examined radiologically every three months and at the end of their contracts. He also strongly disapproved of the repatriation of tubercular miners and recommended that suspected cases be placed under observation in a mine hospital until they were no longer infectious. Oosthuizen recommended that miners not be discharged unless authorised by the Bureau and endorsed by the Native Affairs Department. The local authority of the area to which a miner was returning should be notified so that his case could be followed up. Indeed, Oosthuizen proposed that it should be an offence for a mine owner to discharge an employee suffering from active tuberculosis.[29]

The Group Medical Officers Committee (GMO) met to review the Oosthuizen Report. As was so often the case, Dr A.J. Orenstein served as chair. The Committee opposed periodic X-rays as both unjustified and expensive. It also opposed Oosthuizen's recommendation that it be an offence for a mine owner to discharge an employee suffering from active tuberculosis.[30] Two weeks later, the Committee met again. Those present included Orenstein, again in the Chair, and the Chamber's Assistant Legal Adviser, who was usually present when tuberculosis was discussed. In principle, the Committee agreed that miners certified by the Bureau with tuberculosis should be admitted to hospital for observation or treatment.[31] There were, however, several practical difficulties with such a policy. In some cases, treatment was ineffective, and it would be unfair to detain a man indefinitely who wanted to return home. Given these problems, the Committee rejected Oosthuizen's recommendation that miners be treated until they were no longer infectious.

The industry adopted the same position when negotiating labour contracts. During talks with the government of Nyasaland, for example, William Gemmill emphasised a distinction between miners' phthisis

[29] Memorandum from the Legal Adviser, B.T. Tindall, 2nd May 1955, pp. 12–13; 16. TEBA Archive, WNLA 14/3, Mass Miniature Radiography, General File, March 1954 to January 1956.

[30] Minutes of a Meeting of the Sub-Committee of Group Medical Officers 21st June 1955, in the Chamber of Mines Building, pp. 2–3. TEBA Archive, WNLA 14/3, Mass Miniature Radiography, General File, March 1954 to January 1956.

[31] Minutes of a Meeting of the Sub-Committee of Group Medical Officers 4th July 1955, in the Chamber of Mines Building, p1. TEBA Archive, WNLA 14/3, Mass Miniature Radiography, General File, March 1954 to January 1956.

(which in this instance he used as a synonym for silicosis) and tuberculosis.[32] Tuberculosis, he argued, had existed in the rural areas long before blacks went into the mines. While the mines did all they could to ensure that miners' phthisis was compensated, they took a very different view towards tuberculosis, which black miners contracted while in their villages. According to Gemmill, it had always been the industry's contention that pure tuberculosis was not an occupational disease.[33]

Negotiating Diagnoses

The Chamber's opposition to tuberculosis compensation was one element in a wider conflict over the Miners' Phthisis Acts. That conflict was due in part to legitimate problems with diagnosis, and in part due to flaws in the conduct of mine medicals. Silicosis was not easy to diagnose then, and remains difficult to diagnose today, with vastly improved medical technology.[34] It was also difficult for mine doctors relying upon a stethoscope to diagnose tuberculosis in a fibroid lung. Many cases initially showed little, if any, evidence of infection. As one mine medical officer commented, making two hundred daily stethoscopic examinations with the object of discovering early tuberculosis was a wasted effort.[35]

Weighing was the other tool for diagnosis. In 1916, the Miners Phthisis Medical Bureau made weighing compulsory and the mines were required to weigh men every thirty days. Miners who had lost five pounds or more between two weighings or six pounds over three consecutive weighings

[32] Mr. Frans Unger and Mr. George Ernest Barry representing the Transvaal Chamber of Mines, transcript of evidence before Select Committee of Inquiry into the Miners' Phthisis Commission Report, AN 756-1931 SC12-31, Parliamentary Library, Cape Town, pp. 85–87.

[33] William Gemmill, General Manager Tropical Areas, Note for Mr. K. Lambert Hall, Secretary, Nyasaland, Northern and Southern Rhodesia Inter-Territorial Conference, Salisbury, 20th December 1940, Emigrant Labour Governors Survey M2/3/19, Malawi National Archives.

[34] Rodney Ehrlich, Jill Murray and David Rees. 'Subradiological Silicosis'. *American Journal of Industrial Medicine*, Vol. 61, 2018, pp. 877–885.

[35] Dr L.F. Dangerfield. 'Pulmonary Tuberculosis in South Africa and the Problem of the Native Mine Labourer'. In *Proceedings of the Transvaal Mine Medical Officers' Association*, March 1943, Vol. 22, No. 249, p. 173.

were to receive a more detailed examination.[36] Although weight loss was usually present in active tuberculosis, it was unreliable as an indicator of early disease. Weight loss was caused by a number of factors, and even in advanced tuberculosis it was not always present.[37] Dr W. Watkins-Pitchford thought that weighing was a clumsy tool. 'Experience has shown … that, for one reason or another, this system [of weighing] does not lead to the detection of more than about 50 per cent of the cases of simple tuberculosis and of tuberculosis with silicosis which are finally discovered.' Watkins-Pitchford concluded, 'In order to mitigate the mischief of which the overlooked native "carrier" of tuberculous infection is the source, I have lately proposed utilising large-scale X-ray examinations'.[38] In addition, medical records were not kept, so that when a man returned to the mines, as most did, there was no file on his previous service.[39]

There were major differences between white and black communities regarding tuberculosis. During minority rule, the mortality rates in whites were among the lowest in the world, while those among blacks were among the highest.[40] The Bureau data suggested that the differences between white and black miners were just as great. Silicosis was fourteen times more common in whites while tuberculosis was twice as common in blacks. During 1939, a total of 1164 black miners received compensation under the Miners' Phthisis Act 1925. Of those, 377 were for ante-primary silicosis, 78 for primary silicosis, 150 for silicosis and tuberculosis and 559 for tuberculosis.[41] The Chamber claimed that the low rate of silicosis in blacks proved that migrant labour protected black miners from lung disease. It also proved that dust exposure played no role in tuberculosis.

[36] 'Miners' Phthisis Act No. 35 (1925) Condensed précis of regulations and procedure (Natives and Non-Europeans) for mine medical officers', *Proceedings of the Transvaal Mine Medical Officers' Association*, Vol. 6 No. 4, October 1924, p. 4.

[37] Dr N.R.A. MacColl. 'The Early Diagnosis of Tuberculosis'. In *Proceedings of the Transvaal Mine Medical Officers' Association, May 1940*, Vol. 20, No. 218, pp. 1–2.

[38] Watkins-Pitchford, 'The Silicosis of the South African Gold Mines', pp. 128–129.

[39] E.H. Cluver. 'The Progress and Present Status of Industrial Hygiene in the Union of South Africa'. *Journal of Industrial Hygiene*, Vol. 11, No. 6, June 1929, p. 204.

[40] See B.A. Dormer and F.J. Wiles. 'Tuberculosis in the Bantu'. *The South African Medical Journal*, Vol. 20, No. 10, 25th May 1946, p. 264.

[41] Annual Report, 1939. Witwatersrand Native Labour Association, p. 3. BNA S305/9 Witwatersrand Native Labour Association.

Because blacks were excluded from supervisory roles and whites did not do the actual drilling and blasting, white miners had relatively lower dust exposures. When diagnosed, they were given free specialised treatment at the Springkell Sanatorium, established in 1911 by the Chamber and the government for their care, or they could enter one of the provincial hospitals. Suitable sheltered employment was found on the mines or elsewhere for men able to work. Others were assisted with loans, allotments and other benefits so they could start small businesses.[42] Black miners had higher rates of exposure to silica dust, lived in crowded and unhygienic compounds and received a fraction of the pay of whites. Once they left the mines, their nutrition was poor and their access to biomedical care minimal.

Changing Patterns of Disease

In the first decade of mining on the Rand, the dust levels were very high. The life expectancy of drillers, most of them white immigrants from Europe, was as little as six years. After 1910, the introduction of mechanical ventilation, blasting regulations and watering down reduced the palpable dust and with it the severity of silicosis. As a consequence, the form of miners' phthisis in which the lungs were a solid mass of fibrosis became rare, and a typical X-ray showed only patches of scarring.[43] In 1912, Dr Andrew Watt invented the term *tuberculo-silicosis* in acknowledgment of the shift in disease forms.[44] As the threat of acute silicosis receded, tuberculosis became more prominent. From that point, the divide in the health profiles of white and black miners became more pronounced. Few white silicotics contracted tuberculosis, and those with simple silicosis were given surface jobs, pensions and retraining. By 1920 the life expectancy of a white miner with silicosis was as much as fifteen years from diagnosis. The small number who developed tuberculosis went to the Springkell

[42] Jack Simons. Migratory Labour, Migratory Microbes. Occupational Health in the South African Mining Industry: The Formative Years 1870–1956. Unpublished manuscript, 1960, p. 18.

[43] Dr Watkins-Pitchford *Evidence before Miners' Phthisis Working of Acts Parliamentary Select Committee 1916 SC 10-15 1915 Third Report AN 4923*, p. 685–686.

[44] *The Prevention of Silicosis on the Mines of the Witwatersrand.* Pretoria: Government Printer, 1937, p. 224.

Sanatorium. In contrast, few black miners were diagnosed with simple silicosis, and the far larger number who developed tuberculosis or silicosis with tuberculosis were repatriated.[45] Here too there was a dramatic change in disease form, with tuberculosis replacing silicosis as a major cause of death.[46]

In a letter to the Minister of Mines in July 1914, the Chair of the Miners' Phthisis Prevention Committee, R.N. Kotze, noted that tuberculosis was a very serious threat on the mines and that it was important to prevent the employment of infected men. Kotze also warned that tuberculosis was a complicating factor in pneumonia. 'A large number of natives die of tuberculosis following a pneumonic attack and amongst tropical natives especially tuberculosis runs a very acute course.'[47] Kotze identified two types of cases: there were the men who brought infection with them from rural areas and those who contracted tuberculosis while on the mines. Kotze believed that more thorough medicals were necessary to exclude infected recruits. The shift in the disease forms was accompanied by the emergence of contradictory elements in the medical discourse. South African physicians continued to use the term miners' phthisis to describe occupational lung disease, a term which recognised the role of dust in active tuberculosis. And yet that same body of specialists rarely made reference to dust as a causal factor in black miners' tuberculosis.

Kotze's concerns were taken up by the Miners' Phthisis Prevention Committee. In its annual report of 1916, the Committee noted that although the term 'silicosis' accurately defined the occupational factor in miners' disease, the more general term 'miners' phthisis' better described its composite character.[48] When tuberculosis invaded a silicotic lung, destructive and obliterative changes soon followed. The disease altered type, and

[45] McCulloch 'Hiding a Pandemic', pp. 835–848.

[46] Dr A. Watt 'Personal Experiences of Miners' Phthisis on the Rand 1903 to 1916'. In *Silicosis, Records of the International Conference held at Johannesburg 13th–27th August 1930*. London: ILO, 1930, p. 595.

[47] Appendix 10 Letter from R.N. Kotze, Chair of the Miners Phthisis Prevention Committee to the Minister of Mines, Johannesburg, 4th July 1914, *in General Report of the Miners Phthisis Prevention Committee Johannesburg 15th March 1916*. Pretoria, The Government Printing & Stationery Office 1916, p. 142.

[48] *General Report of the Miners Phthisis Prevention Committee Johannesburg 15th March 1916*. Pretoria: The Government Printing & Stationery Office, 1916, p. 10.

the symptoms of a simple silicosis were aggravated and to a large extent merged in an acute or chronic pulmonary tuberculosis. Tuberculosis was not merely a terminal phenomenon in miners' phthisis: it may intervene at any point. 'Fibrosis caused by silica dust is life threatening, in that when it has progressed to a certain stage tuberculous infection is almost inevitable.' In white miners it usually did so late in the disease process; with black miners it was frequently associated with a slight degree of fibrosis.[49]

Researchers in the US and the UK favoured a more discrete terminology to distinguish between silicosis and tuberculosis. In August 1925, Professor Kettle from the UK told a meeting of the Mine Medical Officers Association in Johannesburg that he was reluctant to use what he considered to be a 'somewhat obscure nomenclature'. Professor Kettle went on: 'To me, Miners' Phthisis would mean phthisis developed in a miner. In a good deal of the literature one finds no clear distinction between silicosis and tuberculosis; and Miners' Phthisis seems to be the general term employed to express either silicosis or tuberculosis combined with silicosis. So that there is a difficulty, as far as I'm concerned, as to what the condition is actually that we are investigating.'[50] Despite objections such as these, the term miners' phthisis continued to be used in the South African legislation until 1956 when it was replaced by pneumoconiosis, thereby bringing South African usage into line with international practice. The fact that miners continue to use the term phthisis attests to the ongoing significance of tuberculosis in the South African setting.[51] Given the fact that 'subclinical silicosis', or silica exposure which cannot be diagnosed in living subjects, still doubles their likelihood of contracting tuberculosis, the less precise term is, paradoxically, in some ways more scientifically appropriate.[52]

[49] *General Report 1916*, pp. 12; 15. Under the National Insurance (Industrial Injuries) Act of 1951 tuberculosis became a prescribed occupational disease for British health workers or research and laboratory staff coming into contact with patients or infective material.

[50] Professor Kettle presentation to the Mine Medical Officers' Association, Johannesburg in Proceedings of the Transvaal Mine Medical Officers' Association, August 1925, Vol. V, No.4, p. 4.

[51] Jaine Roberts. *The Hidden Epidemic Amongst Former Miners: Silicosis, Tuberculosis and the Occupational Diseases in Mines and Works Act in the Eastern Cape, South Africa*. Westville: Health Systems Trust, June 2009, p. 18.

[52] Eva Hnizdo and Jill Murray. 'Risk of pulmonary tuberculosis relative to silicosis and exposure to silica dust in South African gold miners'. *Occupational and Environmental Medicine*, vol. 55, no. 7, 1998, pp. 496–502; Ehrlich et al., 'Subradiological silicosis', 2018.

Critics of the Compensation System

The South African Miners' Phthisis Acts saw the country lead the world in monitoring and compensating hard rock miners. And yet, within six months of the passing of the first piece of legislation in 1912, its effectiveness was being questioned. The South African Secretary for Native Affairs was sure that only a small proportion of black miners with silicosis were receiving the compensation to which they were entitled. According to the official returns for the year ending 21 July 1912, only 287 cases were notified from a total workforce of over 200,000. The Secretary noted that the disproportion between the number of awards made to white and black miners was striking, especially since by now black miners did the actual drilling and therefore had the highest dust exposures. To address that problem, he wanted a government medical officer to oversee compensation cases. In addition to examining miners, the officer could advise on sanitation and be responsible for compiling statistics.[53]

In October 1913, the Director of the Native Labour Bureau wrote a brief review of the 1912 Miners' Phthisis legislation.[54] Under the Act medical officers were required to report all cases of miners' phthisis to the mine managers who in turn were to notify the Secretary for Native Affairs. The data showed that on some mines with a small workforce there was a comparatively large number of cases, while on some large mines, such as Crown Mines and East Rand Proprietary, there were practically none. The Minister of Native Affairs concluded that the system of referrals was not working and suggested that periodical examination of black miners be introduced.

The mining companies' insurers too had reservations about the medical procedures in place. Much to the annoyance of the Rand Mutual Assurance Company, which held policies for the major mining houses, every year a sizeable number of recruits would break down after brief

[53] Letter from Secretary for Native Affairs, Pretoria, upon Appointment of Medical Officer to the Secretary to the Interior, Pretoria, 20th October 1912, p. 3. SANA NTS 324, Dept. of Native Affairs, TB Commission & MP Act 1912.

[54] Letter from Director of Native Labour Bureau, Johannesburg, upon Position of Native Labourers under the Miners' Phthisis Act (No.19 of 1912), to the Secretary for Native Affairs, Pretoria, 16th October 1913. SANA NTS 324, Dept. of Native Affairs, TB Commission & MP Act 1912.

periods underground. For example, a migrant worker named Sikambele, who had been repatriated following an accident at the Modderfontein Deep in June 1911, was subsequently recruited by the WNLA and returned to Johannesburg in December 1912. He worked only one shift before being diagnosed with 'marked phthisis'. When he died on 16 March 1913, a claim for £30 was lodged by the Department of Native Affairs on behalf of his estate. Sikambele's was one of several cases in which claims were made by men who had worked for a month or less.[55] The secretary of Rand Mutual wrote in protest to the WNLA, pointing out that Sikambele had been under medical treatment at the WNLA's compound after his accident in 1911, and his illness should have been diagnosed at that time. The company wanted stricter pre-employment examinations. The Chamber's legal adviser noted that the cost of compensating black miners was rising fast, and he agreed there was need for more stringent entry medicals.[56]

For the mining houses, more thorough entry medicals presented a dilemma. Clearly, there were savings to be made in keeping men with pre-existing lung disease out of the mines, and in identifying early disease in serving miners who could then be quarantined and repatriated. On the other hand, enhanced case findings promised to increase the numbers referred for compensation, and therefore the cost of caring for miners at the WNLA hospital while their claims were being assessed. Mounting numbers of awards might not only cost money, but also raise serious problems with the HCTs about the mines' safety. To compound the problem, many mines faced periodic shortages of labour, and preferred to employ as many recruits as possible, whatever their health status, in order to avoid costly production cuts.[57]

On their part, mine medical officers tended to approach the issue of medical examinations as a technical problem compounded by lack of

[55] Letter from Managing Secretary, Rand Mutual Assurance Company, Johannesburg, to The Secretary, WNLA, 20th March 1913. WNLA 138 Phthisis: Allotment of Natives January 1913 to May 1915, TEBA Archives, University of Johannesburg.

[56] Memo from Mr G.E. Barry, Legal Advisor to the Chamber of Mines Compensation for Miners' Phthisis in Native Labourers under Act 19 of 1912, 18 December 1912. WNLA 138/2 Miners' Phthisis Amendment Act June 1912 to February 1916, TEBA Archives, University of Johannesburg.

[57] Packard, R. M. *White Plague, Black Labour: Tuberculosis and the Political Economy of Health and Disease in South Africa*. Berkeley: University of California Press, 1989, Chs 3 & 6.

staff and resources. Discussions at the MMOA highlighted concerns that cases of tuberculosis were not being picked up, the lack of a diagnostic standard and the intense pressure on mine doctors to keep a full complement of men underground. At a meeting in October 1921, for example, Dr Girdwood noted that most of the deaths at the WNLA compound were from tuberculosis. He asked his colleagues: if monthly weighing was reliable, why were there so many 'bad cases'?[58] The MMOA eventually sent a questionnaire to its members to gauge the scheme's usefulness. In 47 per cent of cases, the results showed, tuberculosis was discovered through weight loss, and in 15 per cent 'incidentally', usually when miners were admitted to hospital following an accident. In 38 per cent of cases tuberculosis was discovered following 'acute diseases', meaning that by the time of diagnosis those miners were dying.[59]

Even when a case had been picked up at weighing, the problem of diagnosis persisted. Under the Act of 1925, compensation for tuberculosis was based on the presence of tubercle bacillus in the sputum or marked physical incapacity. Because positive sputum was the exception rather than the rule in early stage infection, those prescriptions added to the problems of diagnosis. It was common for a miner to have tubercle bacilli in his sputum and physical signs suggestive of tuberculosis but return a negative X-ray plate. Conversely, a miner might display radiological evidence of disease without appreciable physical symptoms. Dr F. Retief spoke at length on the challenges of diagnosis at a MMOA meeting in February 1940. He noted that a single sputum sample in a suspected tuberculosis case was often inconclusive, and that repeated examinations should be made. Where physical and radiological tests were suggestive of infection, it might be necessary to do as many as six or seven examinations before the bacilli were found.[60] Medical officers who carried out hundreds of examinations each week were incapable of following such thorough procedures. In effect, given the available medical resources, the legislative definition of TB dramatically decreased its statistical visibility.

[58] Dr Girdwood in discussion *Proceedings of the Transvaal Mine Medical Officers' Association*, Vol. 1 No. 7 October 1921, p. 3.

[59] Dr Butt, in discussion *in Proceedings of the Transvaal Mine Medical Officers' Association*, Vol. V, No. 5 September-October 1925, p. 7.

[60] F. Retief. 'The Clinical Side of Tuberculosis'. *Proceedings of the Transvaal Mine Medical Officers' Association*, 19, no. 215 1940, p. 237.

Miners' Phthisis Commissions and Medical Examinations

Many of the individual criticisms of the compensation system were taken up during the numerous Miners' Phthisis Commissions. Although researchers such as Drs Watkins-Pitchford, Andrew Watt, A.J. Orenstein and L.G. Irvine, who appeared as expert witnesses, seldom referred at length to disease among black miners, the Commissions were often critical of the conduct of mine medicals.[61] The 1919 Commission into Miners' Phthisis, for example, concluded that the medicals of black miners were inadequate. It also found that black miners with tuberculosis, including men who in some cases had spent months in hospital, were being discharged without a medical certificate. 'We are informed' the Commission wrote, 'that in practice numerous cases have occurred where compensation could not be legally claimed'. Where a miner died at the hospital, it was common for the death certificate to state 'general tuberculosis' when in fact the miner had died from pulmonary tuberculosis. Such a diagnosis prevented surviving dependents from claiming compensation. The situation with living miners was no better, and the Commission found that often there was no exit medical as required by law.[62]

The final twenty pages of the Commission into Miners' Phthisis of 1929 are devoted to black miners.[63] The Commission heard applications from the Department of Native Affairs and the Director of Native Labour for improved benefits. Since there was no information on the circumstances of beneficiaries, the Commission refused to recommend any change to the legislation.[64] Two of the Commissioners, Mr W. Becker and Mr W. Boshoff, dissented. Mr W. Boshoff was particularly critical of the WNLA medicals, and he questioned whether periodic examinations

[61] See Appendix F in *Report of the Miners' Phthisis Commission 1902–1903*. Pretoria: Government Printer, 1903.

[62] *Report of the Commissions of Inquiry into the Working of the Miners' Phthisis Acts*. Cape Town: Government Printer, 1919, pp. 11–12.

[63] *Report of the Miners' Phthisis Commission of Enquiry 1929–30*. Union of South Africa. Pretoria: The Government Printer, 1930, p. 15.

[64] *Report of the Miners' Phthisis Commission of Enquiry 1929–30, Part Two*. Union of South Africa. Pretoria: The Government Printer, 1930, p. 83.

were in fact being carried out. 'Some of these natives died in the mines [from tuberculosis] without their [sic] having ever been examined during their working period.' Boshoff went on: 'I am convinced that the figures giving the deaths from tuberculosis among natives on the mines do not reflect the true position. Black miners were repatriated as soon as they were discovered to have tuberculosis and so the majority of deaths do not take place on the mines.' Boshoff recommended, in vain, that initial, periodic and final medicals be conducted by government medical officers.[65] The same criticisms were repeated by the Stratford (1943), Allan (1950) and Beyers Commissions (1952). Dr G.W.H. Schepers, who worked as an intern at the Silicosis Medical Bureau from 1944 to 1952, made similar comments about mine medicals.[66] Jaine Roberts, in her 2009 study of living miners from the Eastern Cape, found that almost 90 per cent of former miners had not been given an exit medical. The mines' failure to conduct those examinations over such a long period of time, she speculated, was suggestive of a coherent policy.[67]

In response to criticisms, the WNLA took every opportunity to publicise its success in detecting and removing infected men from the mines. In December 1941 correspondence with Mr K. Lambert Hall, the Chief Secretary in Zomba, for example, William Gemmill described in some detail the WNLA's management of health and compensation.[68] All recruits were medically examined before entering the mines and at regular intervals during their contracts. At discharge, black miners were weighed and those with significant weight loss were admitted to hospital for examination, which included an X-ray. Any suspicious cases were then referred to the Bureau for compensation. At the WNLA hospital, any repatriates suspected of tuberculosis were X-rayed. The system was such that it was all but impossible for a miner with tuberculosis or silicosis to be repatriated without receiving an award.

[65] 'Reservations by Mr W. Boshoff' *in Report of the Miners' Phthisis Commission of Enquiry, Part Two*, p. 101.

[66] McCulloch, 'Hiding a Pandemic'.

[67] Roberts, *The Hidden Epidemic Amongst Former Miners*, pp. 151; 154.

[68] Note *Native Labourers and Miners' Phthisis* from William Gemmill, WNLA General Manager (Tropical Areas) Salisbury to Mr K. Lambert Hall, Chief Secretary, Zumba, Nyasaland 20th December 1941 Emigrant Labour Governors Permits. M 2/3/19. Malawi National Archives.

Gemmill explained that within South Africa, a number of X-ray clinics had been established in the Native Territories at which former miners could be examined and their films sent to the Bureau in Johannesburg. It was the duty of District Surgeons, Magistrates and Native Commissioners to publicise the rights of former miners to compensation. Men who had worked on the mines and became ill were instructed to report to their District Surgeons. By implication, if such a system was not available in Nyasaland, it was up to the government to create one. The only problem was that the health services Gemmill described never existed.

The X-ray Puzzle

Although South Africa's gold mines led the world in using radiography to monitor a working population, the technology was not routinely used with black miners until the mid-1950s. That is puzzling, given the acknowledged failure of mine medicals to pick up tuberculosis in serving miners. The 1912 Commission into Miners' Phthisis and Pulmonary Tuberculosis, which featured an X-ray survey of miners, was probably the first study of its kind, and formed the centre-piece of a report which was a landmark in the science.[69] The Miners' Phthisis Prevention Committee found that while an X-ray provided the most reliable single piece of evidence, an accurate diagnosis required a radiograph, a work history and a clinical examination.[70] From 1916, state physicians laid the foundation of modern practice when the Bureau introduced routine X-rays for the entry, periodic and exit medicals of *white* miners. Soon after the establishment of the Tuberculosis Research Committee in 1926, an X-ray plant was installed at the WNLA compound in Johannesburg, but its use

[69] *See Report of a Commission into Miners' Phthisis and Pulmonary Tuberculosis*. Cape Town: Government Printer, 1912. For a history of the use of X-ray technology in Britain see Joseph Melling. 'Beyond a Shadow of a Doubt? Experts, Lay Knowledge, and the Role of Radiography in the Diagnosis of Silicosis in Britain, c.1919–1945'. *Bulletin of the History of Medicine*, Vol. 84, 2010, pp. 424–466.

[70] *General Report of the Miners Phthisis Prevention Committee Johannesburg 15th March 1916*. Pretoria: The Government Printing & Stationery Office, 1916, p. 13.

was limited to that very small number of *black* miners referred to the Bureau for compensation.[71]

At the beginning of 1930, WNLA trialled the routine use of X-rays at entry examinations, but the scheme soon ran into problems. The workload of mine doctors was already overwhelming, and there were simply too many plates to review. Following protests from the MMOA, the trial was ended prematurely.[72] In 1938, a departmental committee recommended that where there was a suspicion of disease, the final examination should include an X-ray. The technology was introduced on most mines and according to the Bureau: 'There can be no doubt that the installation X-ray plants on practically all the mines in recent years has been and will be in the future of the greatest value in the early detection of tuberculosis … among "native labourers"'.[73] However, because of cost, the procedure was reserved for those cases under review for compensation, a decision which rested with the mine medical officers. The GPC estimated that to conduct radiological examinations at every major point of recruitment would require twenty plants at a cost of £100,000, and an additional £40,000 per annum to operate them.[74] At a number of different forums, A.J. Orenstein not only reiterated the issue of cost and capacity but argued against the use of X-rays on medical grounds. According to him, to X-ray every black miner at regular intervals would require increasing the number of medical officers four- or five-fold. Moreover, he was sure that if X-rays were introduced for black miners, other methods of detection would be neglected, thereby leading to more early cases being overlooked.[75]

---

[71] A.P. Cartwright. *Doctors of the Mines: A History of the Work of Mine Medical Officers*. Cape Town: Purnell and Sons, 1971, p. 128.

[72] Letter from A. Percival Watkins, Transvaal Mine Medical Officers Association to General Manager, Transvaal Chamber of Mines, Johannesburg, 28th November 1930. WNLA 20L Diseases and Epidemics Tuberculosis February 1923 to December 1930. TEBA Archives.

[73] *Report Miner's Phthisis Medical Bureau for the Three Years Ending 31st July 1941*. Pretoria: Government Printer, 1944. UG No. 18, 1944, p. 28.

[74] Memo A.T. Milne, General Manager to G.P.C. Members, 27th September 1957, Memorandum to the Round Table Conference on Vulnerable Mines: Recruiting and Medical Examination of Mine Native Labourers, p. 5. TEBA Archives, WNLA 14/3, Mass Miniature Radiography General, Dec. 1956 to Jan. 1958.

[75] Report of the Medical Committee appointed to Enquire into the Deaths of Certain Mine Natives 8th February 1926 to 26th May 1926, pp. 5–6, 13. GES 1005 9 17 A TB in Mines Natives SANA.

In response to criticism from the Tuberculosis Research Committee about the conduct of entry examinations, in June 1926 William Gemmill instructed mine medical officers on the screening for tuberculosis.[76] Officers were to perform their own initial examinations and not rely on the medicals already conducted at the WNLA compound. They were also to screen long-service men every three months. Within two weeks of Gemmill's directive, Dr J.F. Young of the WNLA Hospital made a presentation to the Mine Medical Officers Association's monthly meeting. He announced that the WNLA was conducting an experiment involving 2000 black miners to determine whether X-rays should be used routinely at entry medicals.[77] The experiment was authorised by the GPC, of which A.J. Orenstein was a permanent member and which he often chaired. Dr Young told his audience: 'I do not want you to take too much notice of the X-ray findings … as we are by no means experts [at reading X-rays]. For example, if a recruit had a shadow on his lung he should not be rejected. A *special test* had already been carried out by an expert and the results were "extraordinary".' Orenstein explained that the *special test* with 100 recruits had shown that X-rays could not supplant a stethoscopic examination because, even in the hands of an expert, the results were totally unreliable as an indicator of lung disease.[78] It was a most curious discussion. According to the Proceedings, none of those present questioned Orenstein's characterisations of the technology as worthless, even though under the various Miners' Phthisis Acts from 1916 the Bureau used X-rays as its principal diagnostic tool with white and black miners in assessing compensation claims.[79]

---

[76] W. Gemmill, General Manager Gold Producers' Committee Circular No. 59/26, cited in *Proceedings of the Transvaal Mine Medical Officers' Association*, June 1926, Vol. VI, No. 2, p. 1.

[77] See Dr J.F. Young. 'Explanation of Methods Adopted at WNA. Hospital to Carry Out New X-ray Tests, With a Practical Demonstration of Radioscopy of Cases'. In *Proceedings of the Transvaal Mine Medical Officers' Association,* June 1926, Vol. VI No. 2, p. 3. See also Circular from L.S. Raymond, Secretary, Medical Committee on Tuberculosis, Gold Producers' Committee, Transvaal Chamber of Mines, to Members of the Medical Committee on Tuberculosis, 1st March 1927. SANA, 1928, GES2596 5/54A, League of Nations, TB Research.

[78] Dr Orenstein, discussion in Proceedings of the Transvaal Mine Medical Officers' Association June 1926, Vol. VI, No. 2, p. 4.

[79] For an overview of radiography on the mines see Isidore Donsky. A History of Silicosis on the Witwatersrand Gold Mines, 1910–1946. PhD, Rand Afrikaans University, 1993, pp. 27–30.

Despite the GPCs' opposition to X-rays on the grounds of costs, some mines turned the technology to their advantage. Dr Peall, the Senior Medical Officer of Randfontein Estates Mines, noted that from its installation in May 1930 to December 1936, the plant at Randfontein examined 15,237 recruits, or just over 10 per cent of all miners on entry. Amongst the rejects were 59 cases of tuberculosis and a further 142 of silicosis, who were then repatriated without compensation. If employed and then subsequently compensated, those men would have cost Randfontein £10,050. But perhaps the major economic benefit came from the additional 252 cases suggestive of tuberculosis and 138 cases of possible silicosis who were also rejected. Dr Peall was pleased that the X-ray plant, which had cost £2000, had paid for itself handsomely.[80] In effect, at Randfontein Estates Mines the technology was used to prevent the employment of men with early stage lung disease rather than to identify miners for compensation. The Chamber's Legal Adviser, G. Barry, later acknowledged that the exclusion, with or without the use of medical technology, of experienced miners likely to suffer from occupational lung disease was in effect an industry-wide policy: 'There is a tendency on the part of mine medical officers to reject black recruits with considerable underground history as such natives may soon be certified to be silicotics, and the responsibility for compensation rests with the employer who has last signed him for underground work'.[81]

In 1938 a Miners' Phthisis Bureau committee recommended that where there was evidence of tuberculosis, the final examination should include an X-ray. That suggestion was introduced with some success: 'There can be no doubt', the MPMB wrote, 'that the installation X-ray plants on practically all the mines in recent years has been and will be in

[80] Dr P.A. Peall. 'Physical Examination and Scientific Management of Mine 'Native Labourers'. In *Proceedings of the Transvaal Mine Medical Officers' Association*, November 1937, Vol. XVII, No. 191, p. 52.

[81] Memo from G. Barry, Legal Adviser to H. Wellbeloved, The Chamber, 26th March 1942. Appended to Memo to Member of the Gold Producers Committee 8th April 1942 subject: Miners' Phthisis Commission: Pooling of Native Compensation NRC 390 1&2 Miners Phthisis Compensation for Natives 1942–1949 TEBA Archives.

the future of the greatest value in the early detection of tuberculosis'.[82] The decision to take an X-ray was in the hands of mine medical officers rather than the Bureau. Curiously, despite the use of the technology, there was no rise in compensation rates. In that same year, the WNLA conducted a small X-ray study of 435 long-service miners. Of that group 11 were identified as having tuberculosis, 6 tuberculosis with silicosis and 6 silicosis, giving a rate of over 5 per cent for previously undetected cases.[83] In 1938 the mines employed more than 300,000 migrant workers, the majority of whom served multiple contracts, suggesting there may have been thousands of infected men working underground.

Conclusion

Compensating tuberculosis as an occupational disease cost the mines little in terms of payments to white miners. However, that concession by the Chamber to the MWU created a number of ongoing problems for the WNLA. In particular, the legislation redefined the status of tuberculosis for the HCTs, the Colonial Office and later for the ILO, all agencies on which the WNLA depended in recruiting from the north.

The first perceived health crisis in gold mines led to a swathe of legislation designed to reduce dust levels, improve mine medicine and provide compensation. It saw the creation of the SAIMR and the Bureau, but also the state handover the conduct of mine medicals to employers. Finally, the crisis led to the appointment of A.J. Orenstein as head of sanitation in the largest of the mining houses, Rand Mines Ltd, a position from which he came to dominate mine medicine for half a century.

In theory, the South African legislation was the most progressive in the world. The legislation was a product of the racialised labour system which enabled the powerful white MWU to gain concessions regarding medical

[82] *Report Miner's Phthisis Medical Bureau for the Three Years Ending 31st July 1941.* Pretoria: Government Printer, 1944. UG No. 18, 1944, p. 28.

[83] Annual Report, 1937 Witwatersrand Native Labour Association p. 9. BNA S305/9 Annual Reports, Payment of Accident Compensation to Mine Natives.

care and compensation for occupational disease that trade unions elsewhere were unable to achieve.[84] The Miners' Phthisis Acts were also unique in terms of the other labour sectors. Outside of South Africa's gold mines, there was little state regulation of factories, and virtually none of farms or the domestic sector, where large numbers of black women worked.[85]

The challenges and flaws in diagnosing occupational lung disease in a huge workforce were understandable. But those challenges were quite distinct from the mines' management of tuberculosis. The WNLA's refusal to use X-rays at exit examinations or to issue notifications to local health authorities were incongruous in terms of the legislation. They were also incongruous in terms of the existing medical knowledge. Having reluctantly agreed to compensate tuberculosis, the Chamber was relentless in opposing any suggestion that the mines were a source of infection.

The lack of reliable data makes it impossible to quantify the mines' impact in spreading infection. The picture is made more complex by the variety of factors at play, such as land hunger, droughts and the expropriation and relocation policies of the apartheid state about which Randall Packard has written with such precision.[86] We can be sure, however, that the mines influenced the spread of tuberculosis in three particular ways. Dust exposure greatly expanded the pool of men suffering from tuberculosis who, when diagnosed, were repatriated to rural areas where little or no biomedical care was available. The below-subsistence wages offered by the mines impoverished labour-sending communities, and this in turn created the ideal social setting for infection to take hold and to spread. Finally, because most infected miners did not receive compensation, the resources of their home communities were further depleted by having to care for men who could no longer work.

[84] Penrose, 'Medical Monitoring & Silicosis in Metal Miners'.

[85] *Report of the Commission of Enquiry on Occupational Health*. Pretoria: Government Printer, 1976, pp. 95–96.

[86] See Packard, *White Plague, Black Labour*.

References

Cartwright, A.P. *Doctors of the Mines: A History of the Work of Mine Medical Officers*. Cape Town: Purnell and Sons, 1971.

Chanock, M. *The Making of the South African Legal Culture, 1902–1936: Fear, Favour and Prejudice*. Cambridge: Cambridge University Press, 2001.

Dembe, Allard E. *Occupation and Disease: How Social Factors Affect the Conception of Work-Related Disorders*. New Haven: Yale University Press, 1996.

Donsky, Isidore. A History of Silicosis on the Witwatersrand Gold Mines, 1910–1946. PhD Rand Afrikaans University, 1993.

Dormer, B.A. and F.J. Wiles. 'Tuberculosis in the Bantu'. *The South African Medical Journal*, Vol. 20, No. 10, 25th May 1946, pp. 262–264.

Ehrlich, Rodney, Jill Murray and David Rees. 'Subradiological Silicosis'. *American Journal of Industrial Medicine*, Vol. 61, No. 11, November 2018, pp. 877–885.

Hnizdo, Eva and Jill Murray. 'Risk of Pulmonary Tuberculosis Relative to Silicosis and Exposure to Silica Dust in South African Gold Miners'. *Occupational and Environmental Medicine*, Vol. 55, No. 7, 1998, pp. 496–502.

McCulloch, Jock. 'Hiding a Pandemic: Dr G. W. H. Schepers and the Politics of Silicosis in South Africa'. *The Journal of Southern African Studies*, Vol. 35, No. 4, December 2009, pp. 835–848.

McCulloch, Jock. *South Africa's Gold Mines and the Politics of Silicosis*. Oxford: James Currey, 2012.

McIvor, Arthur. 'Germs at Work: Establishing Tuberculosis as an Occupational Disease in Britain, C.1900–1951'. *Social History of Medicine,* Vol. 24, No.4, 2012, pp. 812–829.

Melling, J. 'Beyond a Shadow of a Doubt? Experts, Lay Knowledge, and the Role of Radiography in the Diagnosis of Silicosis in Britain, c.1919–1945'. *Bulletin of the History of Medicine*, Vol. 84, No. 3, 2010, pp. 424–466.

Packard, R.M. *White Plague, Black Labour: Tuberculosis and the Political Economy of Health and Disease in South Africa*. Berkeley: University of California Press, 1989.

Penrose, B. 'Medical Monitoring & Silicosis in Metal Miners: 1910–1940'. *Labour History Review*, Vol. 69, No. 3, 2004, pp. 285–303.

Roberts, Jaine. *The Hidden Epidemic Amongst Former Miners: Silicosis, Tuberculosis and the Occupational Diseases in Mines and Works Act in the Eastern Cape, South Africa*. Westville: Health Systems Trust, June 2009.

Simons, Harold Jack. Migratory Labour, Migratory Microbes. Occupational Health in the South African Mining Industry: The Formative Years 1870–1956. Unpublished manuscript, 1960.

Swanson, Maynard W. 'The Sanitation Syndrome: Bubonic Plague and Urban Native Policy in the Cape Colony'. *Journal of African History*, Vol. 18, No. 3, 1977, pp. 387–410.

Weindling, P. (ed.) *The Social History of Occupational Health*. London: Croom Helm, 1985.

Wikeley, N.J. *Compensation for Industrial Disease*. Aldershot: Dartmouth Publishing, 1993.

5

Lifting the Ban on the Recruitment of Tropical Labour: 1933–1945

The global ban on the recruitment of Tropical labour, imposed in 1913, was arguably the single most important state intervention in the mines' history. It was certainly the only intervention aimed specifically at protecting migrant labour—and it was strongly opposed by the industry. To have the ban lifted, the Chamber of Mines needed to persuade the South African government and the British Colonial Office that conditions on the mines were safe and oscillating migration was beneficial to labour. The Chamber also had to convince the ILO, which, under the leadership of Dr Luigi Carozzi, took an active interest in migrant workers and monitored the data coming out of Johannesburg.[1]

The Chamber had a great deal at stake in regaining access to Tropical workers. In his annual address for 1936, the Chamber's President noted there had been an adequate labour supply during the year, with the numbers employed increasing. The labour force consisted of 56 per cent South

[1] See 'Introduction', International Labour Office (ILO). *Silicosis: Records of The International Conference Held at Johannesburg 13–27 August 1930.* Studies and Reports, Series F (Industrial Hygiene) No. 13, Geneva 1930, pp. 1–3. See also Thomas Cayet, Paul-Andre Rosental and Marie Thebaud-Sorger. 'How International Organisations Compete: Occupational Safety and Health at the ILO, a Diplomacy of Expertise'. *Journal of Modern European History*, Vol. 7, No. 2, 2009, pp. 191–92.

J. McCulloch, P. Miller, *Mining Gold and Manufacturing Ignorance*,
https://doi.org/10.1007/978-981-19-8327-6_5

Africans, 19 per cent from the Protectorates and 25 per cent from Portuguese East Africa. However, the mines' labour needs would rise over the coming five years, and this made Tropical recruiting essential. The President acknowledged that Tropical workers were unaccustomed to industrial life and therefore more prone to disease and accidents. He was confident, however, their employment was 'not fraught with any undue danger'.[2] William Gemmill was equally forthright about the mines' need for Tropicals. South African men were reluctant to work underground, and around a quarter of the recruits offering themselves to the NRC were rejected on medical grounds. That forced the mines to recruit abroad. The average cost of bringing men from Nyasaland to the Rand, including medicals, transport, rations and compulsory repatriation, was £2 15 s.[3] At that time, the gold mines employed 300,000 black miners and the Gold Producers Committee estimated that soon they would require an additional 100,000. Those recruits could only come from Northern Rhodesia and Nyasaland, which at that time were supplying less than 2000 men.[4]

Immediately after the ban was imposed in 1913, the Chamber commissioned the international experts Almroth Wright, William Gorgas and Lyle Cummins to find a solution. It was an expensive exercise. Sir Almroth Wright's mission to Johannesburg cost £14,905 and General Gorgas' investigations (considered at length in the next chapter) £6815.[5] That investment was largely wasted: Wright failed to develop an effective vaccine, and the expert recommendations of Gorgas and Cummins were largely ignored. Rather than take up Gorgas' suggestion to reorganise the mine medical system and abolish the use of migrant labour, the Chamber

---

[2] Address of the Chairman, Annual Meeting, Transvaal Chamber of Mines, 30th March 1936. BNA S344/7 Recruiting North of 22 degrees South Latitude, 1934, pp. 9; 11.

[3] William Gemmill and Mr. Wellbeloved, Chamber of Mines. Evidence presented before the Native Labour Committee, Johannesburg 13th December 1937, p. 5. SANA K 356 Native Labour Commission 1937.

[4] Memorandum, Gold Producers' Committee, Chamber of Mines, November 1936. SANA, 1936–1946, HKL325 80/168 Shortage of Native Mine Labour, High Commissioner.

[5] Dr Samuel Evans, 'Re Pneumonia Research Work 1910–15', Enclosed with Confidential Circular, No.145/28, from W. Gemmill, General Manager, GPC, Johannesburg, 28th August 1928. Subject: Research Work- Pneumonia, p. 3. TEBA Archives, TEBA 133/1, Medical Research: Sir Almroth Wright's investigations concerning Pneumonia among Mine Natives.

turned to lobbying, redefining the understanding of risk and further vaccine research.

The lobbying reached particular intensity in 1933. Throughout that year, the WNLA and William Gemmill used the daily press to promote an experimental pause in the ban. In August that year, *The Star* reported that as the result of medical improvements, the ban on recruiting from north of latitude 22° south would soon be lifted.[6] The mines would benefit from a supply of labour from the British territories, as sufficient men could not be obtained in the Union. The High Commissioner, Sir Herbert Stanley, spoke with the Southern Rhodesian Prime Minister, Godfrey Huggins, who was opposed to organised local recruitment for the Transvaal mines. The Northern Rhodesia administration initially also opposed the WNLA recruiting locally, but it soon changed its mind.[7]

The Resident Commissioner of Bechuanaland, Charles Rey, supported the WNLA. His only concern was that the number of men available in the Protectorate would far exceed the quota.[8] In 1933, he wrote to the High Commissioner: 'The recruitment of these additional thousand men will undoubtedly be of much assistance to the Territory. It will help the native to get a little money which they badly need, and will enable the Administration to get in a certain amount of additional Hut Tax, which they need no less badly.'[9] There were other benefits. To facilitate transport to the mines, the WNLA had started clearing the route from the Tati to Kazunguls, a distance of nearly 500 kilometres, at its own expense. Rey hoped that network would open up the Game Reserves for tourists. The High Commissioner agreed. Tropical recruiting gave men in the Protectorate a means of paying their tax.[10]

[6] 'Field for Mine Labour. Recruiting Work of Latitude 22° S. Restriction Likely to Be Removed.' *The Star* (Johannesburg), 21st August 1933.

[7] Letter from Sir Herbert Stanley, High Commissioner's Office, Cape Town, to Lt Col CF Rey, Mafeking, 6th December 1933, p. 2. BNA S344/4 Recruiting North of Parallel 22 degrees.

[8] Letter from C.F. Rey, Resident Commissioner, Mafeking, to Mr. W. Gemmill, Esq., 29th August 1933. BNA S344/3, Recruiting Corporation, Native, 1933.

[9] Letter from C. Rey, Resident Commissioner, Mafeking, 7th November 1933 to The High Commissioner, Cape Town. BNA S344/3, Recruiting Corporation, Native, 1933.

[10] Letter from Sir Herbert Stanley, High Commissioner's Office, Cape Town, to Lt Col C.F Rey, Mafeking, 6th December 1933, p. 2. BNA S344/4 Recruiting North of Parallel 22 degrees.

The experiment began in January 1934 with 1006 Bechuana and 1009 men from Northern Rhodesia. They were housed in compounds containing not more than 20 men and inoculated against pneumonia with Lister's not very effective vaccine.[11] The early recruiting programme was poorly organised and there was much discontent among men who presented themselves to the WNLA office at Francistown, as almost half failed the entry medical.[12] Men under the age of 20 or over 40 years were rejected outright.[13] There was also discontent among some local officials. The Resident Magistrate at Serowe, G. Nettleton, was critical of the WNLA's failure to publicise the recruitment drive. If notice had been given, he would have made a Hut Tax tour. Nettleton made it 'very clear [to defaulters] that they were not being coerced into going to the Mines or anything else but if they failed to do something towards paying their Hut Tax they would have to take the consequences'.[14]

A second problem was ensuring that Bechuana were employed. Following discussions in Salisbury, Gemmill informed Rey that the Southern Rhodesian government had no objections to their workers crossing the border to WNLA camps in Bechuanaland. The only difficulty Gemmill could see was to prevent the small experimental quota being swamped by Rhodesians. He had instructed the local WNLA recruiter to reserve the quota almost entirely for Bechuana. Gemmill was confident that if the experiment was successful, he would be able to reach an arrangement with the Southern Rhodesians as well.[15] Much to the

[11] Transvaal Chamber of Mines, Gold Producers' Committee, Results of Tropical Native Experiment January 1934–May 1935, 8th July 1935, p. 1. BNA S344/5 Recruitment North of 22 degrees S. Health Stats 1934.

[12] Letter from Resident Magistrate, G. Nettleton 10th March 1934, to the Government Secretary, Mafeking. BNA S344/3 Recruitment of Natives from North of 22°. Recruiting Corporation, Native, 1933.

[13] Investigation by Dr I. Schapera on the effects of Mining Employment on Native Economic Life—Memorandum submitted by the Witwatersrand Native Labour Association in Respect of Bechuanaland, North of 22 degrees Latitude, p. 4. Botswana National Archives, DCF 941943, Enquiry into Economic Aspects of Immigration.

[14] Letter from G. Nettleton, Resident Magistrate, Serowe, 3rd April 1934, to the Assistant Resident Commissioner, Mafeking. BNA S344/3 Recruitment North of Latitude 22. Recruiting Corporation, Native, 1933.

[15] Letter from W. Gemmill, Transvaal Chamber of Mines, Johannesburg, to Lt Col C.F. Rey, Resident Commissioner's Office, Mafeking, 3rd January 1934. BNA S344/4. Recruiting North of Parallel 22 degrees.

annoyance of Salisbury, however, the WNLA was soon recruiting Southern Rhodesian workers who had crossed the border.[16]

By October 1934, the experiment had been going on for ten months and Charles Rey wrote a memorandum in support of the WNLA. He noted that the gold mines were practically the only wage labour open to the Territory's men. While it might be ideal to retain the majority as a healthy peasantry, to do so would involve building up an export-oriented primary industry, and this in turn would require vast public expenditure on securing a reliable water supply and transport infrastructure. Rey was satisfied that the WNLA initiative fully conformed with the recommendations of the League of Nations Committee on Native Labour. In addition, the Association's presence made it possible for the government to safeguard the interests of migrant workers. Any attempt to prevent men from going south would be an interference with their freedom. 'Experience has shown that conditions of work on the Witwatersrand Mines are such that the majority of labourers return from the Mines infinitely better nourished and consequently of better physique than they were on arrival. A very small proportion inevitably contract Miners' Phthisis but as far as is humanly possible every means known to mining medical science are used to minimise this risk.'[17]

In January 1935, *The Rand Daily Mail* reported that the lifting of the ban was imminent. The initial batch of recruits had stood up admirably to mine work and the most stringent health tests had been completed.[18] The Chamber was looking to the future, when further expansion of mining would require increasing the labour supply. The government had sanctioned the recruitment of a further batch of workers from the north, conditional upon an assurance from the mines that their labour requirements could not be met from within the Union. Soon after, Gemmill completed a tour of Bechuanaland, Northern Rhodesia and Nyasaland.

[16] Letter from C.H. Rodwell, Government House, Salisbury, Rhodesia, to Sir Herbert Stanley GCMG, High Commissioner, Cape Town, 27th April 1934. BNA S344/4 Recruiting North of Parallel 22 degrees.

[17] Memorandum from Lt Col C.F. Rey, Resident Commissioner, Bechuanaland Protectorate, Mafeking, 22nd October 1934, pp. 2–3. BNA S116/2 Labour Native, 1935.

[18] 'More Natives from North: Leave to Recruit Batch of 3000'. *The Rand Daily Mail*, 29th January 1935.

In an interview with *The Cape Times*, he boasted that the experiment had been so successful and the death rate so low that the gold mines hoped to recruit several thousand men from Nyasaland. He also revealed that in due course the WNLA would require men from Southern Rhodesia.[19]

Initially, the Principal Medical Officer for Bechuanaland, Dr Hamilton Dyke, supported the WNLA as he felt it would be a great mistake to deprive men of the opportunity to supplement their meagre incomes. Mining was also beneficial to health. 'This is reflected in the well-nourished and healthy appearance of most of the men when they return to their homes after a period of work on the Rand Mines. They are generally better conditioned than the men who have not been to work on those Mines—provided, of course, that those returning with Tuberculosis are excluded.'[20] He wanted the administration to provide medical examinations of recruits before their departure. The Resident Commissioner endorsed Dyke's proposal.[21]

Initial Data on the Experiment with Tropical Recruiting

In July 1935, the Chamber produced the first data on the experiment with Tropical recruiting. The original batch of just over 2000 Nyasa and Bechuana translated into an average employment of 1759 men. The mortality rate from respiratory disease among the Bechuana 'Tropical' miners was 6.15 per 1000, compared to 3.24 per 1000 for Bechuana 'non-Tropicals' and 2.57 per 1000 for 'All Natives'. In addition, there were five deaths from accidents and a total of 88 men in the experiment were repatriated, 63 because of disease and 25 as the result of accidents. Although the mortality rates were two to three times higher than for non-Tropical labour, the Chamber was confident that further recruitment was 'not

[19] 'Nyasaland Natives for the Rand? Mr. Gemmill's Tour'. *The Cape Times*, 9th August 1935.

[20] Memorandum by Dr Dyke, Principal Medical Officer, Mafeking, 15th August 1934. Botswana National Archives, Health Deputation MPS 1934 S392/5, p. 14.

[21] Letter from Resident Commissioner's Office, Mafeking, 16th August 1934, to The High Commissioner, Pretoria. BNA, Health Deputation MPS 1934 S392/5.

fraught with any undue danger'. Over time 'natives from non-industrialised regions' would adjust to the rigours demanded of modern labour.[22]

Soon after the release of the initial data, Dr Orenstein, the Superintendent of Sanitation for Rand Mines, sent an explanatory note to the Bechuanaland government. The corrected mortality data show that up to May 1935, 12 of the Bechuana 'Tropical' miners died, giving a total mortality rate of 14.74. In addition, 37 men were repatriated on account of disease, 5 with pulmonary tuberculosis and a further 11 because of accidents. However, no deaths which occurred amongst natives after they had left the mines were recorded.[23] The returns were disturbing, especially when the mortality and repatriation data were combined. There was to be little improvement over time. As of January 1937, the experiment was three years old and the annual death rate among 'Tropical Natives' from disease was still 14.14 per 1000 per annum, compared to 6.87 for 'All Natives'.[24]

In a confidential memorandum to Gemmill, Orenstein acknowledged that the data suggested recruitment from the North should cease. However, certain factors needed to be borne in mind. The experiences on the Witwatersrand mines and the copper mines in the Belgian Congo showed that among Tropicals the mortality in the early stages of employment was very high and then gradually fell. Furthermore, comparisons with miners drawn from regions with greater resistance to infection were unfair. A better comparison was between the mortality rates of Tropicals under home conditions (which supposedly were very high) and those of men employed on the mines.[25] Orenstein's argument had no empirical

[22] Transvaal Chamber of Mines, Gold Producers' Committee, Results of Tropical Native Experiment January 1934–May 1935, 8th July 1935, pp. 1–2. BNA S344/5 Recruitment North of 22 degrees S. Health Statistics 1934.

[23] Letter from Dr A.J. Orenstein, Chief Medical Officer, Rand Mines, Johannesburg, to the Financial Secretary, Bechuanaland Protectorate, Mafeking, 12th October 1935. BNA S344/5. Recruitment North of 22 degrees S. Health Stats 1934. See also Transvaal Chamber of Mines, Gold Producers' Committee, Results of Tropical Native Experiment January 1934–May 1935, 8th July 1935, p. 1. BNA S344/5 Recruitment North of 22 degrees S. Health Stats 1934.

[24] Memorandum from W. Gemmill, General Manager, Chamber of Mines. Subject: Native Labour: Mortality of Tropical Natives, 3rd March 1937. TEBA Archives, WNLA. 17/1. Statistical data and prepared for special purposes. From December 1922 to March 1957.

[25] Confidential Memorandum from Dr Orenstein, Health Department, Rand Mines, Limited, to General Manager, Transvaal Chamber of Mines, 3rd March 1937, pp. 2–3. Subject: Tropical Natives. TEBA Archives, WNLA. 17/1. Statistical data prepared for special purposes. From December 1922 to March 1957.

foundation. As he had admitted while giving expert testimony before the Low Grade Ore Commission in 1919 and repeated in his confidential memo, no data on rural death rates existed.[26] In addition, mine recruits were a physical elite unrepresentative of their home communities.

Despite the weakness of his argument, Orenstein had important allies. E.N. Thornton, the Secretary for the South African Department of Public Health who monitored the experiment, noted that the mortality rate per annum among Tropicals was almost double that of all black miners. The Department was, however, satisfied that the mines were safe. 'There is not, in the opinion of the department, any reason to believe that objection can now properly be raised on the grounds of public health to the waiving completely of the ban on the recruitment of natives from north of latitude 22 degrees.'[27] By end of the 1930s, the Chamber had a regional headquarters in Salisbury and the key elements of its northern strategy were in place. British governments rarely challenged the claims that the benefits of mine labour far outweighed the costs, and generally supported the Chamber's demands for access to men from Bechuanaland, Northern Rhodesia and Nyasaland.[28]

The Salisbury Agreement

By the mid-1930s, the effects of the Great Depression were receding and Southern Rhodesian mines and white farms wanted access to labour from the north.[29] The Nyasaland government preferred to channel labour through the WNLA, since it offered capitation fees and better pay and conditions. It also believed the WNLA quotas were the only way to regulate the movement of men south. The likelihood of lifting the ban provided an impetus for an agreement between the three northern

[26] Dr Orenstein evidence before the Low Grade Ore Commission, 1919, pp. 399–400.

[27] Statement on Health Conditions by E.N. Thornton, Secretary for Public Health and Chief Health Officer for the Union, Pretoria, 3rd December 1936, p. 1. SANA, 1936–1946, HKL325 80/168 Shortage of Native Mine Labour, High Commissioner.

[28] Alan Jeeves 'Migrant Labour and South African Expansion, 1920–1950', pp. 73–92.

[29] See David Johnson. 'Settler Farmers and Coerced African Labour in Southern Rhodesia, 1936–46'. *Journal of African History,* Vol. 33, No. 19, 1992, pp. 111–128.

governments. It also gave the WNLA an opportunity to exploit tensions between Southern Rhodesia and its neighbours.

The Salisbury Agreement, signed in August 1936 between the Rhodesias and Nyasaland (now Zimbabwe, Zambia and Malawi), established two principles which must inform the setting of the labour quotas for South Africa's gold mines: the preservation of the social structure of native life and the prior claims of the territories to labour. Once those principles had been satisfied, the Rand mines could recruit workers.[30] Under the Agreement, the WNLA was permitted to recruit 8500 men per annum from Nyasaland. In return, the mines agreed not to engage Nyasa workers resident in the Union. Disputes were to be decided in London by the Secretary of State for the Colonies.

The Salisbury Agreement was designed to regulate the flow of labour to the advantage of the northern governments. However, it proved fragile because of the conflicting interests it was designed to reconcile. The Agreement did not, for example, stop clandestine emigration to South Africa. The contentious issues were addressed in a series of meetings between representatives from the Rhodesias and Nyasaland. The Standing Committee on Migrant Labour, which met in Salisbury in December 1937, agreed that the unregulated flow of labour had reached alarming proportions in part because of illicit encouragement by the Witwatersrand gold mines.[31] It wanted the South African government to help end clandestine immigration. That issue was not resolved and became the source of serious tension between Southern Rhodesia and Pretoria.

The need for uniform work conditions in the territories, and in particular the treatment of workers in Southern Rhodesia, was another contentious issue. At a 1939 meeting of the Standing Committee on Migrant Labour, for example, the Northern Rhodesia representative suggested that the Southern Rhodesia regulations be changed to limit the work day on its mines to eight hours in accord with ILO prescriptions. The conduct of medical examinations and the treatment of repatriates were also discussed, following complaints from Nyasaland about the repatriation

[30] 'Aide Memoire', Sept 1936. PRO Dominions Office (DO) 35/829/R17/18. See also Chapter 9.

[31] See Memorandum of Proceedings of a Meeting of the Standing Committee held at Salisbury on the 3rd and 4th December 1937. Standing Committee Migrant Labour Agreements. LB 1/4/1 Malawi National Archives.

of dying men. The Southern Rhodesia representative claimed that its Medical Director always notified the Nyasaland authorities of miners repatriated because of tuberculosis. In a policy reminiscent of the WNLA, the Committee agreed that 'subject to medical assent that they were fit to travel, incurables who expressed a wish "to go home to die" should be allowed to do so'.[32] At the next meeting of the Committee in October 1939, Nyasaland wanted the Salisbury Agreement reviewed to recognise the WNLA's quota. That proposal was opposed by Southern Rhodesia. The Salisbury government opposed WNLA recruiting in Nyasaland because it was blocking its own access to labour,[33] and would have restricted the flow of labour to its own mines and farms.[34] Despite such differences, the Agreement operated until the outbreak of Second World War, when the Nyasaland government restricted and then suddenly suspended WNLA recruitment because of a domestic labour shortage.[35]

The 1937 Abraham Report on Nyasaland

By October 1936, when the Tropical experiment was two years old, the Governor of Nyasaland commissioned a review of labour migration. J.C. Abraham, a Senior Provincial Commissioner, was asked to report on Nyasa employed in South Africa and Southern Rhodesia. Abraham's brief was to review wages, work conditions, housing, health and welfare, including that of agricultural labour in the Northern Transvaal. In order to compare conditions with those in the Union, he visited a number of

[32] Minutes of Meeting of Standing Committee, Salisbury, 22nd August 1939, pp. 8–9. PRO CO 525/185/1 Native Labour: Recruitment for the Union of S. Africa and S. Rhodesia.

[33] A review of the agreement is presented in a Memo from A.R. Thomas, the Colonial Office the History of Labour Negotiations 25/4/1940 PRO CO 525/185/1 Native Labour: Recruitment for the Union of S. Africa and S. Rhodesia.

[34] Minutes of Meeting of Standing Committee, Salisbury, 25th and 26th October 1939. PRO CO 525/185/1 Native Labour: Recruitment for the Union of S. Africa and S. Rhodesia.

[35] Gemmill, General Manager, (Tropical Areas) WNLA Memorandum Setting out the Association's Case for an Increase in its Nyasaland Quota. Salisbury, 7th July 1944. M2/24/30 National Archives Malawi.

mines and farms in Southern Rhodesia.[36] Abraham was also to report on the experimental employment of Tropicals by the WNLA.

Abraham estimated there were 20,000 Nyasa resident in the Union and another 70,000 in Southern Rhodesia. Each year, about 8000 Southern Rhodesians went to the Union where the wages were higher, and their places were taken by Nyasa who would work for less. Abraham was greatly impressed by WNLA's operations and in particular by its Johannesburg Depot, where 'the smooth efficiency of the organisation is as delightful to see as is the consideration with which all natives are treated'. He visited Crown Mines, where he found the conditions underground clean and comfortable, and noted that water was used to lay the dust, thereby preventing miners' phthisis. The first repatriates were returning to Nyasaland and Abraham observed that they 'looked exceptionally fit with the average weight gain of 14 lbs'.[37]

William Gemmill provided Abraham with data on the 550 Nyasa engaged by the WNLA over the previous two years. The official death rate was 11.76 per 1000 per year compared with 6.6 per 1000 for all black labour. Dr Orenstein reassured Abraham that a high death rate was to be expected during the first decade and then it would fall as recruits acclimatised to aggregation in compounds. Abraham was satisfied: 'No expense is spared, and whatever science can do is being done, to ensure the maximum degree of health for tropical natives employed by the Rand Mines Group, which employs all the Nyasaland natives under the experiment.' Abraham added that the mines were not solely responsible for the high mortality rates: he had heard comments in Johannesburg about the 'comparatively disappointing physique' of the Nyasaland recruits.[38]

Having accepted Orenstein's explanation, Abraham reported to the Governor: 'Nyasaland must expect a higher morbidity and mortality rate than that of natives from South African territories until they become similarly accustomed to aggregation and a degree of immunity from its incidental disease is achieved.' He recommended that recruiting should

[36] J.C. Abraham. *Report on Nyasaland Natives in the Union of South Africa and in Southern Rhodesia.* Zomba: Government Printer, 1937, p. 2.

[37] Abraham, *Report on Nyasaland Natives,* pp. 20; 22; 18.

[38] Abraham, *Report on Nyasaland Natives,* pp. 14–17.

not only continue but be expanded.[39] The distribution list for Abraham's Report included the Under Secretary of State, the Colonial Office, the Chief Secretary at Zomba, the Parliamentary Library, London and the ILO in Geneva. A review copy was also sent to *The Times* in London.[40]

There are a number of reasons why Abraham's report presented such a favourable picture of the WNLA. The Rand mines *did* provide better wages, living conditions and medical care than did South Africa's asbestos and coal mines, or the mines in Southern Rhodesia.[41] Abraham had no medical qualifications and no specialist knowledge of occupational diseases associated with hard rock mining. He did not speak with miners but relied instead upon Orenstein and Gemmill and the official data. Perhaps most importantly, he adopted a narrow view of disease, focussing on pneumonia rather than silicosis or tuberculosis, a serious problem about which Gorgas had already warned the Chamber in 1914. Abraham's report contains a single reference to miners' phthisis. Implicit in his account was a Darwinian model of primitive or unseasoned labour versus industrial labour, in which the weak were of necessity culled by infection. The same model underpinned the public presentations by Orenstein, Abraham's key informant.

In March 1937 the acting Governor of Nyasaland, K.L. Hall, wrote to the Secretary of State, W. Ormsby-Core, endorsing Abraham's findings. 'I can confirm', Hall wrote, 'the favourable impression gained by Mr. Abraham of conditions on labour on the Rand Associated Mines. The members of the Emigrant Labour Committee also were satisfied that labour was better looked after there than anywhere else in British Africa.' Hall had visited the Crown Mines in Johannesburg and applauded '[t]he almost ideal conditions under which Nyasaland natives work. The extent and variety of the food ration was remarkable, the sanitary arrangements would be the envy of any European club in Nyasaland and the Zomba European Hospital accommodation and facilities would hardly be acceptable for a native out-patient dispensary on the Rand.' Hall was also

[39] Abraham, *Report on Nyasaland Natives,* p. 17.

[40] Letter from K.L. Hall, acting Governor, Zomba, Nyasaland to W. Ormsby-Core, Secretary of State for the Colonies, 13th March 1937, p. 8. PRO CO 525/167/1 Native Labour.

[41] See, for example, evidence from 'Nyasaland Native Labour in Southern Rhodesia', by Capt. G.N. Burden, Nyasaland Labour Officer, Salisbury, 1938. PRO Nyasa 1936 CO 525/173/11.

impressed by the system of mine medicine. He found the South African Institute for Medical Research splendidly equipped and up to date, although he suggested that it would be an advantage to have accurate statistics on the number of Nyasa employed in the Union. The initial medical exam was vital if rejects at Johannesburg were to be kept to a minimum. 'I saw natives from the Protectorate who were supposed to have passed a local preliminary examination of such patently poor physique that they would never be finally accepted. I have no doubt that explains the comparatively high mortality among the Nyasa ... the Rand has the advantage over every other employer of labour whether elsewhere in the Union or in Southern Rhodesia of [offering] better wages and ideal conditions.'[42]

In the wake of Hall's visit, in September 1937 William Gemmill invited S.S. Murray from the Chief Secretary's Office at Zomba to tour the Randfontein Estates mines in Johannesburg, which employed 1500 Nyasa. Murray was particularly impressed by the quality of mine medicine and the low mortality rates.[43] He found that everything possible was done for the well-being, comfort and recreation of miners.[44] The hospital arrangements were equal to those in European hospitals, and X-rays were used freely in cases of accidents.

The 1938 Burden Report on Nyasa Workers in Southern Rhodesia

The WNLA was pleased with Abraham's report, which it used to justify an expansion of recruiting. In December 1937, an agreement was signed between the Nyasaland government and the WNLA for 8500 men. The WNLA paid the rail fares to Johannesburg, provided free medical care, a deferred pay scheme and compulsory repatriation after a maximum

[42] Letter from K.L. Hall, acting Governor, Zomba, Nyasaland to W. Ormsby-Core, Secretary of State for the Colonies, 13th March 1937, pp. 3–4; 8. PRO CO 525/167/1 Native Labour.

[43] Report by S.S. Murray to the Chief Secretary, Nyasaland, 8th October 1937, p. 1. PRO CO 525/167/1 Native Labour.

[44] Letter from S. Murray, Nyasaland government, to the Chief Secretary to the government, Zomba, 8th October 1937. PRO Nyasaland, CO 525/220/12, Migration of Labour, 1950.

period of 18 months. According to Gemmill, those conditions conformed to the ILOs Convention No. 50. Provision was also made for the collection of Hut Tax and the payment of a Pass Fee to the Nyasaland government. In addition, the WNLA promised to purchase all the Nyasaland rice available for issue on the mines to Tropical labour. Northern Rhodesia and Nyasaland would not only benefit financially, but: '[t]hrough the improvement in the health and physique of the adult male Native population that would follow'.[45]

Abraham gave Nyasaland and the WNLA what they wanted. For reasons which are unclear, in September 1938 Captain G.N. Burden, the Nyasaland Labour Officer stationed in Salisbury, was commissioned to provide a further report on the work and living conditions of Nyasa in Southern Rhodesia. In the event, his account provided additional support to the WNLA. Burden visited mines, farms, road camps, brickworks, locations and factories for a study which was conducted over six months. He also visited the Cam and Motor Mine in Southern Rhodesia, which J.C. Abraham had inspected some months earlier.[46] It was the biggest gold mine in the country, employing 2000 men, more than half of whom were Nyasa. The death rate for all recruits was around 16 per 1000 per annum while the rate among Nyasa was over 20 per 1000, but Abraham did not question the mine's safety.[47]

Burden reported that migrant workers usually arrived in small groups numbering less than ten. They received a free issue of food, provided by Southern Rhodesian officials at various points *en route*. On entry, their fingerprints and particulars were recorded, and a pass issued. In 1937 the mining industry employed just over 30,000 Nyasa.[48] There were a further 25,000 on farms and an estimated 20,000 employed in the towns. The city of Salisbury was a magnet, with many migrants finding work and

[45] The Rhodesian and Nyasaland Royal Commission 1938. 'Statement by the Gold Producers' Committee of the Transvaal Chamber of Mines', pp. 5–6; 8. PRO CO 525/173/2 Labour: Employment of Natives on the Witwatersrand Gold Mines Johannesburg Agreement, 1938.

[46] See 'Southern Rhodesia' in Abraham, *Report on Nyasaland Natives*, pp. 6; 50–69.

[47] Abraham, *Report on Nyasaland Natives*, p. 52.

[48] There were 25,764 Nyasa on the gold mines, 2725 on chrome mines and 1951 on the Shabanie asbestos mines. See Appendix D to Report, 'Nyasaland Native Labour in Southern Rhodesia', by Capt. G.N. Burden, Nyasaland Labour Officer, Salisbury, 1938. PRO Nyasa 1936 CO 525/173/11.

accommodation as servants. Others lived on the east side of the city in hovels. The conditions on white farms were oppressive, with work beginning at dawn or soon after and continuing until dusk. Farm labourers were paid as little as 10s a month and were subject to the whim of white farmers. Labour conditions on the mines varied greatly. Some of the bigger mines compared favourably with the Rand, but there were also large mines where the facilities were poor. The wages for experienced miners were a third of those offered on the Rand and the non-payment of wages was common. Burden found that the real value of wages depended on the conditions under which migrants lived and the rations they received. He also found that in their disputes with managers, foreign workers were 'defenceless'.[49]

Abraham had created the impression that there were some minor problems in the treatment of migrant labour which the Southern Rhodesian government was doing its best to address. Burden found that Nyasa were cruelly exploited, and that Southern Rhodesia was doing little to improve their situation. The Colonial Office endorsed his report: 'Here we see the contract system at its worst. This is the sort of thing which has set the ILO against contract labour. These contracts are made between private employers and defenceless employees. They have practically nothing in common with the desirable contracts made between the Witwatersrand Native Labour Association and (in fact, though not in form) the Governments of Nyasaland and Northern Rhodesia.' The Colonial Office pointed out that if the government of Nyasaland expected the Secretary of State to defend the contract policy in Parliament and at ILO Conferences, it must ensure that such abuses ceased. It also questioned why, despite repeated requests, it had received Burden's report seven months late. London believed the Report had been suppressed by Nyasaland.[50] On its part, the Southern Rhodesia government did not accept Burden's report as accurate, and lamely instructed field officers to

[49] Report, 'Nyasaland Native Labour in Southern Rhodesia', by Capt. G.N. Burden, Nyasaland Labour Officer, Salisbury, 1938, pp. 8–10; 15–16; 18; 20. PRO Nyasa 1936 CO 525/173/11.

[50] Letter from F.J. Pedler, Colonial Office, London 4th January 1939, in reply to H. Swanzy's communication regarding Captain Burden's 1938 Report. PRO Nyasa 1936 CO 525/173/11.

ensure conditions improved, where necessary.[51] In March 1940, the Secretary of State for the Colonies, Malcolm MacDonald, acknowledged receipt of Burden's report which he had read 'with great interest'. However, in view of Burden's criticisms of the Southern Rhodesia government, he vetoed its publication.[52]

Abraham and Burden's reports both favoured the WNLA. Abraham praised the work and living conditions on the Rand mines, while Burden's criticisms of Southern Rhodesian labour markets were to the benefit of South Africa's mines. Instead of protecting migrant labour from the north, Burden argued, London gave white settlers in Southern Rhodesia what they wanted: namely access to cheap and unprotected labour. Both reports were also characterised by a number of notable omissions. They contain virtually no reference to tuberculosis, silicosis or the mines' repatriation policies. There was also no reference as to what constituted an acceptable death rate.

Medical Warnings of an mpending Tuberculosis Crisis

Alongside favourable reports on the experiment with recruitment of Tropical labour was a series of reports from medical officers about a sudden rise in the number of tuberculosis cases due to infected men returning from the mines. Those reports presented clear warnings about the impact of contracted migrant labour on public health. During 1933, a total of 348 tuberculosis cases were diagnosed in Bechuanaland. Of those, over two-thirds were pulmonary, with the numbers being far lower in the northern districts. The Principal Medical Officer, Dr Hamilton Dyke, noted that until the end of 1933 men had not been recruited from the North for work on the gold mines. He was greatly concerned, as the local population was susceptible to infection and their powers of resistance

[51] Report on Mine Conditions, forwarded by C. Bullock, Secretary for Native Affairs, Southern Rhodesia, for Nyasaland government, 27th September 1938. PRO Nyasa 1936 CO 525/173/11.

[52] Letter from Malcolm MacDonald, Secretary of State for the Colonies, London, to K.L. Hall, Acting Governor, Zomba, Nyasaland, undated, March 1940. PRO Archives, Nyasaland, CO 525/185/5, Native Labour, Report by Capt. Burden on Nyasaland Natives.

were lowered by a poor diet. Having read Dyke's Report, the Resident Commissioner Charles Rey admitted that the incidence of tuberculosis was disturbing. 'The distribution of Tuberculosis suggests that a good deal of infection originally is introduced by men who have worked on the Rand Mines and would indicate the necessity for compulsory medical examination of all Mine recruits before they leave the Territory, and also that the standard of physical fitness required of such recruits be higher than it is at present.'[53]

In July 1935, the Resident Magistrate at Molepolole, S.L. Forster Towne, reported that tuberculosis was increasing rapidly in the district and no precautions were taken when men who had been infected on the mines returned home. There was no system of notification and the men themselves did not realise the danger they posed to their families. 'Would it not be possible' he asked, 'to raise the standard of medical examination when the men are recruited? Also, a notification from the Mines of the return of these men and their early treatment in some special place and not allowing them to return home without being treated.'[54] The District Commissioner's Office in Mafeking feared the Protectorate would become a bed of disease if preventive measures were not taken. As a matter of urgency, it wanted the government to fund isolation wards to halt the spread of infection.[55] In Francistown, the Medical Officer was concerned that the disease was increasing rapidly. Men who left to work in South Africa or Rhodesia became infected and returned home to their families. When questioned, young tuberculosis patients, including children, would often recount that a relative, living in the same hut, had recently returned from South Africa with a cough.[56]

[53] C.F. Rey, Resident Commissioner's Office. Extracts on Bechuanaland Protectorate Annual Medical and Sanitary Report for the year 1933, Mafeking, 18th June 1934, p. 1. BNA S438/2/1. Tuberculosis in the B.P, 1938–39.

[54] Report of Effects on the Tribe of Withdrawal of its Men for Labour in Other Territories, 15th July 1935, p. 1. Extract from S.L. Forster Towne, Resident Magistrate Molepolole, BNA S438/2/1. Tuberculosis in the B.P., 1938–39.

[55] Letter with Extract enclosure from District Commissioner's Office, Mafeking, 18th December 1935 to the Resident Commissioner. BNA S438/2/1. Tuberculosis in the B.P., 1938–39.

[56] Letter from Medical Officer, Francistown, to the Principal Medical Officer, Mafeking, 21st August 1936, p. 2. BNA S438/2/1 Tuberculosis in the B.P., 1938–39.

One of the most detailed reports on a looming health crisis was from the Resident Medical Officer at Serowe. The district had an area of more than 100,000 square kilometres, and a single medical officer served its population of over 100,000. Of those, 20,000 to 25,000 constituted the floating population of Serowe, the rest being scattered in small, often inaccessible villages, where the incidence of tuberculosis was impossible to gauge. The Resident Medical Officer offered two reasons for the steady increase in the number of pulmonary cases during his seven years in Serowe. The first was a prolonged drought which resulted in poor crops and a scarcity of milk. The other factor was the return of infected men from the mines. Some may have contracted tuberculosis while underground or have started their underground work with latent infection which flared up as a result of exposure to silica dust. Repatriated men were not re-examined on their return, so there was no data to corroborate his suspicions. To identify the cause would require full details of the number and percentage of men who were repatriated with lung disease. It was important that those men report to the District Medical Officer before returning to their homes so that some follow-up scheme could be initiated.[57]

The Bechuanaland Administration had no control over the WNLA data. However, in October 1935 the Department of Health did produce its own review of the impact of recruiting on public health. The Medical Officer at Gaberones was in no doubt that the incidence of tuberculosis was rising, and that in most cases the infection originated in the gold mines: 'As regards women and children, these contract the disease through contact with the men who have returned from the mines and developed the disease there.' While on the mines, men were well fed and housed, and the disease lay dormant. After fulfilling a contract, a miner returned home where, because of poverty, he began to lose weight and his resistance was lowered. As a result, the disease, dormant for perhaps years, flared up and the patient died a few months later. 'In most cases, once the disease becomes active, it is a rapidly spreading and very infectious one,

[57] Impressions on Tuberculosis and Venereal Disease in the Bamangwato Reserve, Resident Medical Officer, Serowe, 30th March 1936, pp. 1–2. BNA S438/2/1 Tuberculosis in the B.P., 1938–39.

with the result that those living in close contact with him are liable to be infected, and this is by no means an infrequent occurrence.'[58]

The length of mine contracts varied from 6 to 12 months, but migrants were allowed to work on indefinitely provided they remained fit. The longer a man worked underground, the more likely he was to become infected. It was unusual for a miner to develop even a dormant pulmonary lesion after he had served a single contract, but common after he had served two. 'In fact, one cannot help noticing that the vast majority of natives who develop active pulmonary tuberculosis are those who have completed more than two contracts.' The standard of fitness required by the WNLA and the NRC was very high, and it would be impossible to make it any higher. The Gaberones Medical Officer proposed that recruits should not spend more than 18 months underground during a working life and that period should be divided into two terms of nine months, with an interval of at least a year. He also wanted the mines to notify local authorities of all miners repatriated with tuberculosis. Such a policy was imperative as: '[p]ractically every native who goes to the mines and works there over a certain period of time develops pulmonary tuberculosis. He becomes infected on the mines.'[59]

In October 1936, 18 months before the ban on Tropical recruiting was lifted, Dr J.W. Sterling, Principal Medical Officer of Bechuanaland, wrote a memo on tuberculosis. The number of pulmonary cases, he noted, rose from 180 in 1931 to 290 in 1935. The figures were based on patients seen at hospitals and clinics. This, Sterling acknowledged, was a poor guide as most patients did not seek the aid of European doctors. 'There is no doubt that this disease was originally introduced from the Mines, but now cases are frequently met with in women, children and young adults who have never been to the Mines but who, in quite a number of cases, have had relations or friends who at some time worked on the Mines. This bears out the fact that the disease is now contracted in the Territory itself.' Like the Gaberones Medical Officer, Sterling believed

[58] Letter from Medical Officer, Gaberones, to The Principal Medical Officer, Mafeking, 9th October 1935, p. 1. BNA S438/2/1 Tuberculosis in the B.P., 1938–39.

[59] Letter from Medical Officer, Gaberones, to The Principal Medical Officer, Mafeking, 9th October 1935, pp. 1; 3–4. BNA S438/2/1 Tuberculosis in the B.P., 1938–39.

that *every* man who worked on the gold mines for a certain period developed tuberculosis. Sterling noted that the spread of infection was also governed by conditions in the Protectorate. Frequent droughts caused malnutrition and the Bechuana custom of living together in large villages influenced the disease rate. As tuberculosis was of comparatively recent origin, the population lacked immunity, and Sterling expected a rapid increase in the number of deaths.[60]

In his official correspondence, the Bechuanaland Resident Commissioner, F.C. Rey, usually disregarded Dr Sterling's advice and played down the spread of tuberculosis from the mines. In a letter to the High Commissioner in November 1936, Rey admitted that his administration might have done more to secure medical examination of men before they left for Johannesburg, and to provide medical attention on their return. But to do so would have imposed an additional burden on his already overworked administrative and medical staff. What did real harm, according to Rey, was the uncontrolled and unregulated exodus of men seeking work other than on the mines. Those migrant workers were neither as well-nourished nor so well paid. The Protectorate's enormous borders made it impossible to control the flow of labour.[61]

There were several possible reasons for Rey's behaviour. Whenever a medical officer raised the issue of tuberculosis or miners' phthisis, William Gemmill responded with data from the Bureau showing the rates of lung disease were low and the mines safe. The Protectorate's medical department had neither the capacity to examine repatriated miners nor the capacity to commission its own research. More important still, Rey viewed the reintroduction of recruiting as a purely economic matter which was 'of enormous importance' in enabling men to earn wages and the government to collect tax. There was a high rate of defaults and the hut tax received in 1933, the year prior to the experiment, was only a

[60] Dr J.W. Sterling, Principal Medical Officer, Mafeking, Memorandum on Tuberculosis, Venereal Diseases and Leprosy in the Bechuanaland Protectorate, 26th October 1936, pp. 2–4. BNA S392/6 Adjunct Health & Medical, 1936.

[61] Letter from C.F. Rey, Resident Commissioner, Mafeking, to Sir William Clark, 24th November 1936, pp. 2–4. BNA S392/6 Health & Medical, 1936.

quarter of the expected £40,000.[62] Whatever perspective one adopts, the fact remains that the Bechuanaland administration depended on labour migration.

Political Contests Over Tropical Recruiting

In addition to the mounting medical evidence of spreading disease, there was political opposition in London to WNLA's recruiting. The British Social Hygiene Council (Council) took an active interest in migrant labour. In March 1934, its deputation met in London with the Secretary of State for the Dominions, J.H. Thomas, and the Under Secretaries Malcolm MacDonald and Dr Drummond Shiels. It was an important meeting, as the experimental introduction of Tropical labour, inoculated with Lister's vaccine, was under way, and there was a great deal at stake for both the Chamber and the Protectorate.

Dr Shiels noted that tuberculosis in the Bechuanaland Protectorate appeared to be on the increase and that most of the new infections had occurred on the gold mines. The poor physique and poor diet of the Bechuana added to their susceptibility to disease. Sir Basil Blackett, President of the Council, replied that the population's well-being was at risk from the heavy recruiting and the lack of supervision. Of a recent batch of 500 men recruited for the mines, from which the recruiter had already discarded obvious weaklings, another 33 per cent were rejected on medical grounds. Four years after Britain had signed the ILO's Forced Labour Convention, there had been several cases which infringed it. The local administration lacked the necessary knowledge and experience to create a labour organisation. Blackett suggested that the cost of establishing such an organisation and expanding the medical service might be met by a capitation fee paid by the mines.[63] He also wanted a general commission into the social effects of migrancy.

[62] Letter from C.F. Rey, Resident Commissioner's Office, Mafeking, to Sir Herbert Stanley, High Commissioner, Cape Town, 8th December 1933, p. 2. BNA S344/4 Recruiting North of Parallel 22 degrees.

[63] Dominions Office, Notes of a deputation received by the Secretary of State for the Dominions from the Parliamentary Committee of the British Social Hygiene Council, the Parliamentary

The meeting reached no agreement and Blackett continued to lobby Thomas, suggesting the cost of appointing Commissioners to monitor the health of migrant workers be met by the Chamber of Mines. Thomas was unsympathetic.[64] The examination of recruits was so strict that only those in excellent health and of first-class physique were accepted. Modern methods of diagnosis on the mines led to the discovery of lung fibrosis at an early stage. The diet and housing conditions of recruits in their home settings were, he believed, more dangerous than was employment on the mines. Recruits were 'fattened up' before they were sent to work and most labourers returned from the mines better nourished and in better health than they were on arrival. While a very small proportion contracted miners' phthisis, every effort was made to minimise the risk.[65] The poor physique of recruits, Thomas concluded, was due to the inadequacy of water supplies, the general economic conditions in the Protectorate and a poor diet.

In early July 1936, the Secretary of State received a second deputation from the British Social Hygiene Council. The notes for the meeting include a number of extracts from the Bechuanaland Protectorate's annual medical reports, which showed a looming crisis. Major Orde Browne was unconvinced by the Council's arguments, but he did concede that the Protectorate needed an anti-venereal campaign and better supervision of migrant labour.[66] That, however, was not the end of the matter. Four months later there was a meeting in Mafeking between the Resident Commissioner, Charles Rey, and a British delegation which included Major Orde Browne and Dr Drummond Shiels. The deputation suggests how seriously the issue of health and migrant labour was taken in London. Much of the discussion was about the importance of combatting venereal disease and tuberculosis. Shiels argued that the

Medical Committee and the British Social Hygiene Council, on Monday 26th March 1934. Note taken by Treasury Reporter. BNA, Health Deputation MPS 1934 S392/5, pp. 7–10.

[64] Letter J.H. Thomas, Secretary of State for the Dominions to Sir Basil Blackett, President of the British Social Hygiene Council 22nd March 1935. BNA, Health Deputation MPS 1934 S392/5.

[65] Note 23rd July 1935, enclosed with Letter from J.H. Thomas, Secretary of State, Dominions Office, London, to Sir Basil Blackett, the President of the British Social Hygiene Council, Victoria Embankment, London, 23rd July 1935, p. 1. BNA S426/5. Labour: Native 1945.

[66] Notes of Meeting Between Secretary of State and a Deputation from the Parliamentary Committee of British Social Hygiene Council, London, 16th June 1936, relating to Health Matters in the Bechuanaland Protectorate, BNA S392/6 Adjunct. Health & Medical, 1936.

unsatisfactory position in the Protectorate was largely due to malnutrition and the remedy lay in improving the water supply. Major Orde Browne's perspective was rather different: he wanted enhanced supervision of migrant labour.

Charles Rey did not share Orde Browne's concerns. Unlike migrant workers from other parts of Africa, recruits did not have to trek long distances as they were drawn from places close to the railway. 'I do not, however, agree with the necessity for the creation of machinery for supervision of migrant labour, nor do I consider that this is a real problem so far as this Territory is concerned.' The movement of labour across the border for employment other than in the mines was far less satisfactory in terms of nourishment, health, wages and general control.[67] Rey's stance was at odds with his Principal Medical Officer and also with other members of his administration. One senior officer remarked that neither the Territory nor the dependants of men recruited to the mines received any benefit from their employment. All too often their wages were spent on the Reef, or men returned with useless items such as gramophones.[68]

Importing workers from the north was expensive but it offered the mines several advantages. The signing of specified quotas with colonial governments guaranteed a supply of labour and it did so at pay rates which depressed wages for South African recruits. For that reason, the ban drew the Chamber into direct conflict with the South African Department of Native Labour. In 1922, William Gemmill told the Mining Industry Commission that for purely political reasons, the government was refusing to allow a trial of 5000 Tropical recruits. The ban had been imposed because of the high mortality rate from pneumonia, but with Dr Lister's anti-pneumonic vaccine that danger had passed. 'We [the Chamber] consider that the Government cannot maintain a most absurd restriction.'[69]

[67] Notes by C.F. Rey, Resident Commissioner, Mafeking, 25th November 1936, on the Deputation to the Secretary of State, 16th June 1936, from the Parliamentary Committee of the British Social Hygiene Council, pp. 2–3. BNA S392/6. Health & Medical, 1936.

[68] Minutes, Assistant Resident Commissioner to C.F. Rey, Resident Commissioner, 14th February 1936, p. 1. Subject W. Gemmill/ Press Cutting. BNA S344/9 Recruitment of Tropical Natives, 1936–7.

[69] William Gemmill, Mining Industry Commission, 16th May 1922. (15th day), p. 1364. For the full evidence by W. Gemmill see pp. 1357–1409. SANA, K161, Minutes of Evidence.

Colonel S.A.M. Pritchard, the Director of Native Labour who also appeared before the Commission, emphatically disagreed. Tropical recruiting, he argued, was both unnecessary and expensive. Pritchard pointed out that in 1921 alone, the mines had spent £489,000 in capitation, levies and distribution fees. There were, however, ample recruits available within South Africa to meet the mines' needs. Pritchard's main objection to lifting the ban was that the Chamber would use imported labour to depress wages in the Union. Pritchard had also seen reports of high phthisis rates among men who had returned from the mines. The situation was particularly disturbing, as medical officers in the Territories were not sufficiently *au fait* with silicosis to produce reliable data. He was sure that the death rates would be a good deal higher in Tropical recruits. Pritchard did not believe that importing Tropical labour was in the interests of South Africa and he referred to Gemmill's proposed experiment as 'the thin end of the wedge'.[70] Pritchard suspected that by importing foreign labour, the mines were able to export and thereby hide the occupational disease which would otherwise have been visible.

Negotiations About Medical Examinations and Repatriations of Black Miners

From 1916, the Miners' Phthisis Acts made exit medicals compulsory, but in the case of black miners those medicals did not include a chest X-ray. Three years before the ban on Tropical recruiting was lifted, the issue of exit medicals again became the focus of top-level concern. In 1934, Dr Hamilton Dyke's proposal that a medical examination of black miners should be compulsory and include an X-ray prior to repatriation was endorsed both the Bechuanaland Resident Commissioner and Sir H.J. Stanley, the British High Commissioner in South Africa. Stanley wanted the mines to take responsibility for such examinations.[71] The

[70] Col. S.A.M. Pritchard, Minutes of Evidence 1922 Mining Industry Commission, 30th June 1922, pp. 3865; 3867. SANA, K161, Minutes of Evidence.

[71] Letter from Sir H.J. Stanley, High Commissioner's Office, Pretoria, to Colonel C.F. Rey, Resident Commissioner's Office, Mafeking, 3rd October 1934. BNA S426/5 Labour: Native 1945.

Secretary of State for the Dominions recommended the notification and follow-up of tuberculosis cases repatriated from the mines.[72] Charles Rey also wanted legislation to make the medical examination of recruits compulsory *before* they were dispatched *to* the gold mines, and he intended to discuss the matter 'in a preliminary way' with William Gemmill during his visit to Johannesburg. Rey also promised the High Commissioner that he would raise the possibility of the medical examination of mine labourers prior to their repatriation.[73] The matter dragged on for months as Rey had difficulty in obtaining information from the WNLA. The correspondence makes clear that none of these issues had been formally negotiated with the Chamber before the resumption of Tropical recruiting.

It appears that Rey was intimidated by Gemmill, who in turn was sensitive to what he perceived as lack of cooperation from the northern governments. Soon after recruiting recommenced, Gemmill complained to the Resident Commissioner's office in Mafeking that the WNLA's efforts in opening up Northern Bechuanaland were 'looked upon with suspicion' in some official circles. The WNLA had already spent several thousand pounds on the scheme which, when completed, would enable the WNLA to tap the whole of Bechuanaland and in addition obtain labour from adjacent areas. It would be one of the biggest development projects in the Protectorate and Gemmill cautioned: 'You will appreciate, however, that operations of this description are only possible if we have the cordial co-operation of the Administration.'[74]

In August 1935 the High Commissioner's Office asked Rey for an update on his year-long discussions with Gemmill on medical examinations.[75] Rey in turn wrote not to Gemmill but to his subordinate at the

---

[72] See note in J.W. Stirling, Principal Medical Officer, Mafeking, 24th February 1939, Extract, Annual Medical and Sanitary Report, 1938, BNA S438/2/1 Tuberculosis in the B.P., 1938–39.

[73] Letter from Colonel C.F. Rey, Resident Commissioner's Office, Mafeking, to Sir Herbert Stanley, 30th October 1934. BNA S426/5 Labour: Native 1945.

[74] Letter from William Gemmill, Transvaal Chamber of Mines, Johannesburg, to Major R. Reilly, Mafeking, 18th May 1935. BNA S344/9 Recruitment of Tropical Natives, 1936–7.

[75] Letter from W.H. Clark, High Commissioner's Office, Pretoria, to Mr. E. Reilly, 23rd August 1935. BNA S426/5 Labour: Native 1945.

Chamber, Mr. Wellbeloved, asking for clarification.[76] Wellbeloved assured Rey that, as provided under the Miners' Phthisis Acts, every migrant worker was medically examined prior to repatriation. He gave no details as to the form those medicals took and whether, as with white miners, they included an X-ray.[77] Charles Rey had little understanding of either the miners' phthisis legislation or the function of the Bureau, and he tended to believe what he was told by the WNLA and the NRC. During the final months of 1935, Ray had a tortuous correspondence with the Chamber about exit medicals. Rey believed that such examinations would best be done in Johannesburg by experts at the Bureau.[78] He was not aware that the Bureau examined very few black miners.

William Gemmill was unimpressed by Ray's suggestions. Such examinations would be expensive and cumbersome, and in the vast majority of cases wholly unnecessary. Such a proposal had been fully examined by a Parliamentary Select Committee in 1925, which found it impracticable. Gemmill pointed out that the Bechuanaland Administration could always conduct its own medicals of suspected tuberculosis cases. If any such patients were identified, they could be referred to the Bureau in Johannesburg for further examination. Gemmill then outlined the procedures he claimed were used routinely in South Africa. If a miner in the Native Territories was suspected of suffering from tuberculosis or silicosis, he was examined by the District Surgeon, who reported to the Bureau. If fit to travel, the man could be sent to Johannesburg for examination. The costs were paid by the Bureau, that is by the state rather than by the mines. While that procedure applied to South African recruits, in two cases it had been used with black workers resident in Basutoland. Gemmill concluded: 'I should be obliged if you would let me know, in the light of this information, what procedure you would like adopted in respect of

[76] Letter from C.F. Rey, Resident Commissioner, Mafeking, to Mr. Wellbeloved, the WNLA, 2nd October 1935. Botswana National Archives, S426/5 Labour: Native 1945.

[77] Letter from the Native Recruiting Corporation, Ltd., Johannesburg, to Colonel C.F. Rey, Resident Commissioner, Mafeking, 4th October 1935. BNA S426/5. Labour: Native 1945.

[78] Letter from Colonel C.F. Rey, Resident Commissioner, Mafeking, to W. Gemmill, General Manager, Transvaal Chamber of Mines, 18th December 1935. BNA S426/5. Labour: Native 1945.

Bechuanaland Natives.'[79] Gemmill was no doubt aware that the Protectorate had a single X-ray machine and no funds to return men to Johannesburg for examination. He was also aware that the procedures he described for monitoring black miners in South Africa rarely occurred.

Despite Gemmill's assurances, protests about repatriations continued. In May 1936 the Government Secretary of Bechuanaland, H.H. Price, wrote to Gemmill about the repatriation of several dying men. The District Commissioner at Kanye reported that three men repatriated with lung disease had died within days of their return. They were in Price's view unfit to travel, but no notification had been received from the WNLA. A fourth man was repatriated with pulmonary tuberculosis, and Price asked whether he had received compensation.[80] Price reminded Gemmill that where migrant workers suffered from silicosis or tuberculosis, their particulars should be sent to the nearest District Commissioner and a copy forwarded to the Principal Medical Officer at Mafeking. Price also requested that notification of men repatriated for other disabilities be forwarded so that treatment could be made available. Gemmill's response has disappeared, but we do know that nothing changed.

At the beginning of the Experiment with Tropical recruiting, the WNLA had agreed to provide the Principal Medical Officer with full particulars of all Bechuana miners rejected on arrival in Johannesburg and then repatriated. In addition, all repatriated tuberculosis cases were to be sent to a Medical Officer for examination and treatment. In practice that did not happen. Out of the 94 cases notified in 1937 and 1938, the WNLA and the NRC forwarded documentation for only 12.[81] When confronted about the lack of notifications, the WNLA claimed that for some years certificates had been sent routinely to District Officers in Bechuanaland.[82]

[79] Letter from William Gemmill, General Manager, Transvaal Chamber of Mines, to the Government Secretary, Bechuanaland Protectorate, Mafeking, 21st December 1935, pp. 1–3. BNA S426/5. Labour: Native 1945.

[80] Letter from H.H. Price, for Government Secretary, to William Gemmill, the General Manager, Transvaal Chamber of Mines, Johannesburg, 8th May 1936, p. 1. BNA S426/5 Labour Native 1945.

[81] J.W. Stirling, Principal Medical Officer, Mafeking, 24th February 1939, Extract, Annual Medical and Sanitary Report, 1938, BNA S438/2/1 Tuberculosis in the B.P., 1938–39.

[82] Letter from District Superintendent, Witwatersrand Native Labour Association, Limited, to the Government Secretary, Bechuanaland Protectorate, Mafeking, 11th December 1935. BNA S426/5 Labour: Native 1945.

That claim was false. The numbers compensated by the Bureau were small, which makes the WNLA's refusal to issue notifications curious. Was it an oversight or did the WNLA and the NRC have a policy of not providing such information? We do know that during the 1930s and 1940s, the lack of notifications was a major concern for local medical officers.

Conclusion

The ban on tropical recruiting was imposed in 1913 as a response to what both the South African government and the British Colonial Office saw as an unacceptably high death rate among miners from tropical regions. The ban would be lifted if, as the mining houses promised, the mortality rate could be brought down to a satisfactory level. Precisely what that meant became the focus of a complex process of definitional manoeuvring. In his report from October 1935, for example, the Medical Officer at Gaberones presented a clear warning that increased recruiting by the gold mines was a threat to public health. As a number of other health professionals had done, he identified the cumulative exposure to silica dust as a major risk. While prolonged work on the mines saw men develop some immunity to bacterial pneumonia, it increased their dust exposures and therefore their risk of contracting tuberculosis. Orenstein disagreed. Those deaths, he argued consistently in a variety of forums, were due to the racial inferiority of black miners and not because of working in a phthisis-producing industry. The remedy was acquired immunity which could only be gradually gained by prolonged exposure to industrial labour.[83] Orenstein's arguments gained wide acceptance among the white public but were not sufficient to have the ban lifted. For that purpose, scientific data was perceived to be needed.

In the two decades after the ban was imposed, much changed. The mining companies had substantially reduced (but certainly not eliminated) deaths from infectious pneumonia by building changing houses, improving conditions in the compounds and providing miners with more adequate rations. They did much less to alter those structural

[83] For a more detailed discussion of Orenstein's work, see Chapter 11.

features, such as dust exposures, which could not easily be engineered out of the mines. They also strenuously denied the threat that tuberculosis and repatriations posed to labour-sending communities. In the process, the impact of tuberculosis as an occupational disease was rendered largely invisible. The statistics on mortality rates reflected this dynamic. Not only were the figures inconsistent from year to year and document to document, but they also systematically ignored much of the evidence. The data used to measure risk was based solely on deaths from disease and accidents on the mines and excluded deaths at the WNLA compound, where many miners succumbed to illness and injury before they could be repatriated. It also excluded those who died in transit, or after returning home. Were those deaths included, the WNLA's own data showed a mortality rate from disease between three and five times higher than for other black miners.[84] The Union government's decision in 1937 to permanently lift the ban followed what it viewed as a successful experiment with Nyasa and Barotse. That decision was underpinned by the failure of London and the colonial administrations to distinguish between the risks of pneumonia, which killed miners at work, and tuberculosis and silicosis, from which most died after leaving the mines.

After the ban was lifted, the WNLA took immediate steps to recruit in Nyasaland, Barotseland (now the Western Province of Zambia) and Northern Bechuanaland, not only tapping the local populations but offering employment to workers in transit from Angola and Namibia.[85] In addition to its headquarters at Francistown, the WNLA established rest and feeding camps at various points along the recruiting trail. To the end of 1944, the WNLA had invested more than £35,000 on depots and camps and more than £55,000 on road construction and maintenance.[86]

[84] Memorandum submitted by the WNLA in Respect of Bechuanaland, North of 22 degrees Latitude, Investigation by Dr I. Schapera on the effects of Mining Employment on Native Economic Life, p. 2. BNA DCF 941943, Enquiry into Economic Aspects of Immigration.

[85] Letter from Management, Witwatersrand Native Labour Association Limited, Johannesburg, to the Government Secretary, Bechuanaland Protectorate, Mafeking, 8th September 1939. BNA S344/10/1, Recruitment of Labour North of Parallel 22 degrees South, 1936.

[86] T.W. Jones, WNLA District Manager, Francistown, 21st July 1945, 'Notes on the Northern Bechuanaland Organisation of the Witwatersrand Native Labour Association Limited Tropical Areas Administration', p. 1. BNA S344/10/1, Recruitment of Labour North of Parallel 22 degrees South, 1936–1945.

The transport system established by the Association had a wider economic significance. Each year its trucks carried not only recruits and repatriates but around 8000 fare-paying passengers, and conveyed goods to the benefit of government, traders and local customers.

References

Cayet, Paul-Andre Rosental and Marie Thebaud-Sorger. 'How International Organisations Compete: Occupational Safety and Health at the ILO, a Diplomacy of Expertise'. *Journal of Modern European History*, Vol. 7, No. 2, 2009, pp. 191–92.

International Labour Office (ILO). *Silicosis: Records of The International Conference Held at Johannesburg 13–27 August 1930*. Studies and Reports, Series F (Industrial Hygiene) No. 13, Geneva 1930.

Jeeves, Alan H. *Migrant Labour in South Africa's Mining Economy: The Struggle for the Gold Mines' Labour Supply 1890–1920*. Kingston and Montreal: McGill-Queen's University Press, 1985.

Johnson, D. 'Settler Farmers and Coerced African Labour in Southern Rhodesia, 1936–46'. *Journal of African History*, Vol. 33, No. 19, 1992, pp. 111–128.

6

The Research Community, Risk and Evidence: 1912–1932

South Africa's first silicosis crisis helped to make Johannesburg a world centre for research into occupational lung disease. From as early as 1902, science was a powerful tool in defining risk and shaping the legislative response. The complexion and dynamics of the research community came to exert a profound effect on the nature of scientific studies it produced, and the evidence it tendered to the various commissions of enquiry and parliamentary select committees.[1]

The closely knit research community, centred on the South African Institute of Medical Research (SAIMR) and the Miners' Phthisis Medical Bureau (Bureau), was created by the gold mining industry and the state between 1912 and 1916. Its members included L.G. Irvine, A. Sutherland Strachan, F.W. Simson, Wilfred Watkins-Pitchford, Anthony Mavrogordato, Spencer Lister, A.J. Orenstein and Andrew Watt. For the next half-century, the research agenda was set by the Chamber. The intimate relationship between the Chamber and the research community is epitomised by SAIMR. The Institute was founded in 1912 by the Witwatersrand Native Labour Association and the government in

[1] See also Jock McCulloch. *South Africa's Gold Mines and the Politics of Silicosis*. Oxford: James Currey, 2012.

J. McCulloch, P. Miller, *Mining Gold and Manufacturing Ignorance*,
https://doi.org/10.1007/978-981-19-8327-6_6

response to the high mortality rates from pneumonia and silicosis, which threatened recruitment from the north. In 1919, the SAIMR had a staff of seven; by 1926 that had grown to 75. William Gemmill was the WNLA's long-term representative on the SAIMR's Board.[2]

The SAIMR had two divisions. The Research Division's primary focus was on silicosis and pneumonia, while the Routine Division carried out the diagnosis and treatment of miners and conducted medico-legal investigations. The Routine Division's funding came from services to the mines and government departments. In 1919, the Division carried out over 60,000 diagnostic investigations.[3] The Institute's first Director was Wilfred Watkins-Pitchford, who in 1916 also became the first Director of the Bureau. Both positions were prestigious, and his combined salaries made him the highest paid health officer in South Africa.[4] Despite his status, Watkins-Pitchford had limited authority: it was the Chamber which set the Institute's research agenda. In addition, Watkins-Pitchford's role at the Bureau was part-time, and his duties were purely administrative. He did not, for example, examine patients. The appointment of one person to head both the SAIMR and the Bureau raised concerns about the independence of the two organisations. The Bureau had no qualified pathologist and relied entirely for such reports on the SAIMR, thereby further blurring the boundaries between the two. As one Johannesburg specialist pointed out, it was important that in settling disputed compensation claims, an independent authority should tender evidence to the Bureau.[5]

The Miners' Phthisis Medical Bureau, founded in 1916, was the first state organisation of its kind, and it conducted all entry, periodic and exit medicals of white miners. It also adjudicated compensation awards for

[2] Annual Report, 1937 Witwatersrand Native Labour Association, p. 4. BNA S305/9 Annual Reports, Payment of Accident Compensation to Mine Natives.

[3] Marais Malan. *In Quest of Health: The South African Institute of Medical Research, 1912–1973.* Johannesburg: Lowry Publishers, 1988, p. 35. For a history of the SAIMR see J.M. Malan. 'History of the South African Institute for Medical Research'. *The South African Medical Journal*, Vol. 76, No. 7, October 1989, pp. 374–382.

[4] Letter from the Office of the Secretary for Public Health to The Secretary, Public Service Commission, Pretoria, 17th April 1924. SANA, 1926, GES 3069, 237/53, Watkins-Pitchford.

[5] Statement submitted by Dr Pratt-Johnson, to the Miners' Phthisis Acts [Young] Commission 1929, p. 4. SANA, K162 Submissions.

both white and black mine workers. It was from this limited data set that it compiled the official disease rates. The Bureau's primary role was to promote health and safety, but it had no capacity to conduct research. The Bureau interns were also poorly paid. In 1945, a full-time intern's commencing salary was £1000 per annum, with £50 increments to a maximum of £1350.[6] By comparison, a District Surgeon in the Transkei had a gross income of between £1500 and £2000.[7]

We have an account of the Bureau's daily operations by Professor Arthur J. Hall, who was a delegate to the 1930 Silicosis Conference in Johannesburg. Hall found the Bureau was under-staffed and that the seven full-time clinicians devoted their mornings to examining the more than 30,000 *white* miners who attended annually. After each man's work and medical history was taken, he was X-rayed and given a full clinical examination. The examinations were completed by midday, and after lunch the clinicians met to consider the cases. Because of the volume of work, the process was very rapid. Initially, Hall questioned why there was no similar supervision of *black* miners, but later accepted the explanation by the Bureau Director: since black miners worked on contracts for only a few months, they were not at risk from silicosis. Hall was also assured that blacks were subject to regular medical reviews during their service.[8]

As General Manager of Tropical Areas for the Chamber, William Gemmill routinely negotiated labour quotas with colonial governments. During one such assignment, Gemmill explained the Bureau's role in the following way: 'Every Mine Medical officer is by law an officer of the Miners Phthisis Medical Bureau, and is subject to the instructions of the Bureau in all Miners Phthisis matters affecting natives employed by the Mines.'[9] In practice, the workloads of Bureau interns meant there was no such supervision. According to Dr Gerrit Schepers, who was a Bureau

[6] Advertisement, 'Vacancies for Members of The Miners' Phthisis Medical Bureau, Johannesburg'. *The South African Medical Journal*, Vol. 19, No. 16, 25th August 1945, p. 300.

[7] Dr A.R.R. Mears, 'The Part-Time District Surgeon in the Transkeian Territories'. *The South African Medical Journal*, Vol. 19, No. 21, 10th November 1945, p. 403.

[8] Arthur J. Hall MD. 'Some Impressions of the International Conference on Silicosis held at Johannesburg, August 1930'. *The Lancet*, 20th September 1930, pp. 655–656.

[9] William Gemmill, General Manager Tropical Areas, Note for Mr K. Lambert Hall, Secretary, Nyasaland, Northern and Southern Rhodesia Inter-Territorial Conference, Salisbury, 20th December 1940. Malawi National Archives. Emigrant Labour Governors Survey M2/3/19.

intern from 1946 until 1952, at no time did the Bureau supervise or evaluate mine medicals.[10] Oluf Martiny, who served as a medical officer at the WNLA compound from 1954 until the early 1980s, made similar comments.[11]

In political terms, the SAIMR and the Bureau were designed to achieve two outcomes. They were to placate the imperial authorities about the safety of the mines for migrant labour, and they were to placate the MWU about the safety of the mines for its white members. How the Institute's first Director defined the Bureau's role is significant. In a memo from 1917, Watkins-Pitchford wrote: 'The Bureau assumes that its province in relation to the Mine Medical Officers is solely to secure the removal of cases of Pulmonary Tuberculosis from underground work systematically and without unnecessary delay.'[12]

Mine medicine was a bifurcated system in which the Bureau and the mine medical service operated separately.[13] In September 1917, the Minister for Mines asked Watkins-Pitchford several questions about the duties and capacities of mine medical officers. Watkins-Pitchford admitted that he had no first-hand knowledge of mine medicine. To answer the Minister's questions, he would have to investigate the work of medical officers, whether they were full or part-time, the number of miners each supervised, the nature and efficiency of the mine hospitals, and the sickness and death rates at each mine.[14] In summary, after six years as Director of the SAIMR and a year as Director of the Bureau, which was in theory responsible for the supervision of mine medical officers, Watkins-Pitchford was not able to provide the most basic information about the workings of the Miners' Phthisis Act in regard to medical supervision of 90 per cent of the workforce.

[10] See Jock McCulloch. 'Hiding a Pandemic: Dr G.W.H. Schepers and the Politics of Silicosis in South Africa'. *The Journal of Southern African Studies*, Vol. 35, No. 4, 2009, pp. 835–848.

[11] Interview with Dr Oluf Martiny, 27th April 2011.

[12] Dr W. Watkins-Pitchford MD, Chairman the Medical Bureau, Memorandum on The Medical Supervision of Native Mine Labourer, 21st October 1917, p. 2. SANA, Mines & Industries, MNW 394, MM 2599/17, Medical Supervision of Mine Natives, 1917.

[13] McCulloch, *South Africa's Gold Mines*, pp. 33–54.

[14] Dr W. Watkins-Pitchford MD, Chairman the Medical Bureau, Memorandum on The Medical Supervision of Native Mine Labourer, 21st October 1917, p. 1. SANA, Mines & Industries, MNW 394, MM 2599/17, Medical Supervision of Mine Natives, 1917.

Despite its heavy workloads and apparent lack of coordination, the Bureau enjoyed a glowing international reputation. In March 1937, *The Lancet* commented on what it termed 'the Bureau's outstanding role' in awarding compensation and in controlling and preventing disease. 'The value of the records and experience of an established organisation such as this Bureau in assessing the effect of new technical measures is inestimable, and Dr Irvine [the Director] and his colleagues can be assured that their work is appreciated not only in the Transvaal but throughout the world wherever silicosis is a problem.'[15]

The Rand system of medical surveillance and research was stable over time, and from its foundation until the early 1970s the SAIMR had only four Directors. They were Dr Watkins-Pitchford (1912–1926), Dr Spencer Lister (1926–1939), Dr E.H. Cluver (1940–1959) and Dr J.H.S. Gear (1960–1973).[16] That stability extended to the workloads of Bureau interns, which were much the same in 1950 as they had been during the First World War. Because of the lack of staff, the clinical examinations of white miners at the Bureau were always cursory. During November 1949, for example, each of the eight interns was required, on average, to carry out 30 clinical examinations in a morning.[17]

The other unchanging element was the limited circle of specialists who were employed by the state or the mining houses, and who dominated the science. Mine medicine in South Africa was more of a club than an open society, and a handful of men controlled the discipline. In December 1955, the Department of Mines approached the British specialist, Professor E.J. King, with the offer of appointment as Director of the newly created Pneumoconiosis Research Unit. Professor King declined, and a special sub-committee was appointed to make recommendations. Dr Orenstein was elected Chair of that Committee, which in turn

[15] 'Silicosis in the Rand Gold Mines', *The Lancet*, 27th March 1937, pp. 764–765.

[16] See J.F. Murray. 'History of the South African Institute for Medical Research'. *The South African Medical Journal*, Vol. 37, No. 16, 20th April 1963, pp. 389–395; Malan, 'History of the South African Institute for Medical Research', pp. 374–82.

[17] *Report of the Commission of Enquiry Regarding the Occurrence of Certain Diseases, other than Silicosis and Tuberculosis, Attributable to the Nature of Employment in and about Mines*. Pretoria: Government Printer, 1951, p. 15. U.G. 22/1951.

nominated him as Director of the PRU. Orenstein was duly appointed to the position.[18]

A range of physicians saw at first hand the impact of silicosis and tuberculosis on black goldminers. These included company doctors who conducted pre-employment and periodic medicals at the WNLA compound or at individual mines, and the interns at the Bureau who carried out compensation reviews. District medical officers and missionary doctors in rural South Africa and in labour-sending states such as Basutoland (Lesotho) and Swaziland (eSwatini) saw the effects of mining on migrant workers. In addition, there were scientists at the SAIMR in Johannesburg who conducted research. Finally, there were medical specialists in the Department of Health who treated miners for tuberculosis and other diseases. Among that range of physicians, mine medical officers were best placed to identify occupational disease. However, their ability to intervene was compromised by two factors. First, their crushing workloads left them little time or energy for tasks not immediately related to their job. In addition to treating miners injured in accidents, they were responsible for daily sick parades, the examination of recruits, compound inspections and monitoring of work conditions underground.[19] The other factor concerned divided loyalties to patients and to their employers, and the perceived consequences of criticising the companies.

The Mine Medical Officers' Association

At a meeting in 1921, the Chamber created the Transvaal Mine Medical Officers' Association (MMOA) to represent the medical practitioners who worked on the mines. At the inaugural general meeting, a draft constitution was adopted. The first elected officials included A.J. Orenstein as Vice-President and Dr A.I. Girdwood of the WNLA as Secretary. From

[18] Report of Meeting of the Pneumoconiosis Research Advisory Committee held on the 6th December 1955 in the Boardroom of the South African Institute for Medical Research, p. 2. TEBA Archive, WNLA 14/3, Mass Miniature Radiography, General File, March 1954 to January 1956.

[19] High Court of South Africa (Gauteng Local Division, Johannesburg), Case No. 08108/2013, Bongani Nkala et al and Harmony Gold Mining Company Limited, Anglo American South Africa Limited (AASA) et al, 28th May 2014, pp. 57–58.

its inception, the Association was supported by the Chamber, with the WNLA and the NRC funding the publication of the monthly proceedings.[20]

Apart from dealing with mine accidents and serious infections, the principal duties of a mine medical officer were to conduct entry, periodic and exit examinations. The Miners' Phthisis Act of 1925 made the employment of full-time medical officers compulsory, thereby to a degree professionalising the service. Medical officers were to examine every black recruit and certify him free from tuberculosis and fit for employment. They would also carry out periodic and exit examinations.[21] In his commemorative history of the MMOA, A.P. Cartwright notes that the health of the black mineworker became almost an obsession with the mining companies. A balanced diet and 'the healthy lives' they led while employed on the mines usually resulted in a weight gain and an improvement in their general health. 'It is of importance to many of the African communities south of the Equator that so many of their men return to their villages after a sojourn in South Africa in far better physical condition than most of them were when they left home.'[22]

Despite the Chamber's claims about the excellence of medical care, the mines' health service had its critics. The medical corps consisted of a small number of senior officers supported by a group of young graduates who used the mines as a stepping stone to a better career. In the period between 1929 and 1935, for example, 13 out of 34 medical officers resigned from the service.[23] The workloads were no doubt a factor: a full-time medical officer on a medium sized mine would be responsible for 6000 men.[24] The workloads of the five full-time physicians at the WNLA

[20] President's Report 1921–1922 in *Proceedings of the Transvaal Mine Medical Officers' Association*, Vol. 1 No. 12, April 1922, p. 4.

[21] *Tuberculosis in South African Natives with Special Reference to the Disease Amongst the Mine Labourers on the Witwatersrand*. South African Institute for Medical Research, Johannesburg, March 1932, p. 80.

[22] A.P. Cartwright. *Doctors of the Mines: A History of the Work of Mine Medical Officers*. Cape Town: Purnell and Sons, 1971, pp. 1–3.

[23] Dr F. Daubenton. 'Training and Specialisation of Mine Medical Officers'. In *Proceedings of the Transvaal Mine Medical Officers' Association*, Vol. XV No. 164, May 1935, p. 76.

[24] On occasions, it could be as high as 12,000. See Dr Williams in discussion, *Proceedings of the Transvaal Mine Medical Officers' Association*, Vol. XV, No. 164, May 1935, p. 81.

hospital were also onerous. In addition to responsibility for more than 250 hospital patients, each doctor examined between 300 and 1200 black miners a day.[25] The corps' status was such that they were referred to by their Johannesburg colleagues by the derogatory term 'Kaffir Doctors'.[26]

The Miners' Phthisis Commissions held between 1919 and 1952 routinely commented on the inadequacy of the medical examinations and the obstacles black miners faced regarding compensation. The 1919 Commission, for example, was highly critical of WNLA's medical officers for making it difficult if not impossible for the relatives of a deceased miner to receive an award. Services for living miners were no better. The Commission found that medical examinations were poor. Often there was no exit medical as required under the Act, not even for seriously ill men who had spent months in hospital. Without a certificate, a miner could not lodge a claim for compensation.[27] The findings of the 1919 Commission were endorsed by the Young (1930), Stratford (1943), Allan (1950) and Beyers (1952) Commissions. They are also consistent with recent research by Jaine Roberts on former miners from the Eastern Cape. Among her cohort of 205 men, 85 per cent had not received an exit examination as required by law.[28]

Despite their limitations, the medical care and state regulation of South African mines were unique. From 1912 the industry was subject to a swathe of legislation covering most aspects of work, compounds, rations and medical care. By contrast, there was little regulation of factories, and virtually none of farms or the domestic sector where large numbers of black women were employed. The work conditions were far worse on the coal and asbestos mines, and they were certainly hazardous in smaller industries, especially in factories with less than 500 employees. There, pre-employment and regular health examinations were rare, and

[25] 'Reservations by Mr W. Boshoff' in *Report of the Miners' Phthisis Commission of Enquiry, Part Two*, p. 100.

[26] Dr Miller, President's Address, in *Proceedings of the Transvaal Mine Medical Officers' Association April 1939*, Vol. XVIII, No. 206, p. 147.

[27] *Report of the Commission of Inquiry into the Working of the Miners' Phthisis Acts*. Cape Town: Government Printer, 1919, pp. 11–12.

[28] Jaine Roberts. *The Hidden Epidemic Amongst Former Miners: Silicosis, Tuberculosis and the Occupational Diseases in Mines and Works Act in the Eastern Cape, South Africa*. Health Systems Trust, Westville, June 2009, p. 81.

post-employment examinations non-existent. The availability of large numbers of migrant workers made it cheaper for employers to replace injured men and women rather than make workplaces safer. That pattern continued well into the apartheid era. As recently as 1976, the Erasmus Commission found that over 70 per cent of South Africa's eight million workers were inadequately covered by occupational health and safety legislation.[29] In 1983, just 81 inspectors were responsible for worker safety in more than 35,000 factories.[30]

Dealing with Risk

High mortality rates and the prospect of a ban on tropical recruiting led to the Miners' Phthisis Commissions of 1902, 1912 and 1914. Those enquiries were bolstered by the work of international experts contracted by the Chamber. In evaluating the industry's response to the ban, it is important to identify how the evolving contemporary science viewed the risks associated with gold mining.

In March 1913, under pressure from London, the government in Pretoria barred the Chamber from further recruiting in the north. In order to regain access to Tropicals, the Chamber needed to reduce miners' death rates and to convince the British and South African governments that the mines were relatively safe. To that end, it commissioned Sir Almroth Wright and Spencer Lister in a futile attempt to find a vaccine for pneumonia.[31] The Chamber also turned to Major W.C. Gorgas, Surgeon-General in the US Army, who had achieved fame for his work on the Panama Canal. Gorgas arrived in Johannesburg in December 1913. His brief was to investigate the causes of the high mortality rates from pneumonia and to find a remedy.

---

[29] *Report of the Commission of Enquiry on Occupational Health*. Pretoria: Government Printer, 1976, pp. 95–96.

[30] See Anthony Zwi, Sharon Fonn and Malcolm Steinberg. 'Occupational Health and Safety in South Africa: The Perspectives of Capital, State and Unions'. *Social Science and Medicine*, Vol. 27, No. 7, 1988, pp. 691–702. For a review of occupational health and safety from the mid-1980s see Frederick Sitas. Occupational Health Services in South Africa: A Pilot Survey of Manufacturing Industries, Dissertation for the Degree of Master of Science in Medicine, Johannesburg, 1986.

[31] See Malan, *In Quest of Health*, pp. 95–112.

W.C. Gorgas was a scrupulous researcher and his report on the ecology of the mines was the first of its kind. In fact, there was no comparable research until the eve of majority rule. Gorgas found that the fatalities from pneumonia varied between individual mines. They also varied according to the origin of labour. Most cases occurred during the first months of service. Fatalities were highest among miners from Nyasaland, and lowest in recruits from the Cape, suggesting that immunity was a factor. Gorgas was highly critical of the mine rations which were inadequate for men performing hard labour. 'I have never seen so large a proportion of the ration supplied by one article as is here supplied by mealie meal', Gorgas wrote.[32] The compounds were crowded and full of litter and miners wore wet and soiled clothes. The bucket system of disposing of waste added to the unsanitary conditions.

Gorgas modelled his recommendations on his Panama experience, where the scattering of workers from crowded barracks into single huts dramatically reduced the death rate from infectious disease. 'And for the sanitation of pneumonia I would urge a similar measure on the Rand. Place your negro labourers in individual buildings, and bring in and place with them their families.' Married men would form a permanent, skilled and efficient workforce. Gorgas noted that while the industry was spending a million pounds a year on recruitment, the construction of family locations would save the greater part of that expense. By housing labour in huts and providing an adequate diet, Gorgas was confident that in a year or two, immunity to infection would be greatly increased.[33]

The Gorgas report is also a landmark in knowledge about tuberculosis and its synergy with silicosis. Gorgas argues forcefully that pneumonia and tuberculosis on the mines were linked by the conditions which gave rise to them, namely malnutrition, overcrowding and poor hygiene. Transmission was usually by means of the expectorated sputa of diseased men which, once dried and airborne, could enter the lungs of co-workers. In addition, miners' phthisis lowered a workman's resistance to

[32] *Recommendation as to Sanitation Concerning Employees of the Mines on the Rand made to the Transvaal Chamber of Mines*. W.C. Gorgas, Surgeon-General, United States Army; Chief Sanitary Officer, Isthmian Canal Commission 1914, p. 349.

[33] *Recommendation as to Sanitation*, pp. 345–47; 355.

tuberculosis. There was a high rate of infection: during 1912, over 1100 tuberculosis cases were repatriated by the WNLA. The annual mortality rate of 5.65 per 1000, while lower than for pneumonia, was significant. In comparison, the mortality rate from tuberculosis among men, women and children in London in 1911 was 1.03 per 1000; in New York City it was 1.67 per 1000. Given these facts, Gorgas predicted that 'for the future, present conditions continuing, tuberculosis will cause you more trouble among natives than does pneumonia at present'. As with pneumonia, the most important preventative measure was the replacement of the compound system.[34]

Dr Darling, who assisted major Gorgas, carried out a number of post-mortems in Johannesburg. In each of one set of 11 consecutive post-mortems of miners diagnosed as succumbing to pneumonia, the underlying cause of death was found to be tuberculosis.[35] Importantly, Gorgas noted, most men with miners' phthisis died from incidental tuberculosis implanted upon a silicotic lung, rather than from silicosis itself. To Gorgas, the hygiene of silicosis was obvious: that of laying the dust, so that the particles of silica will not be floated in the air where they could be breathed. Gorgas also wanted regular medical examinations and the exclusion of infected men from the mines.[36]

In his report, Gorgas recommended two structural reforms. The first was an end to the migrant labour system under which single men lived in compounds. The system was, he believed, inefficient in its use of labour and came at a high human cost. Gorgas suggested miners become a permanent workforce and live with their families in village settlements adjacent to the mines. The second proposal was to centralise the mine hospitals into one or more large facilities to achieve economies of scale and service the industry as a whole. Neither recommendation was adopted, and pneumonia remained a serious problem. According to the

[34] *Recommendation as to Sanitation*, pp. 343–44.

[35] Cited by Dr Alexander Orenstein in evidence before the Miners' Phthisis Commission, Minutes of Proceedings 27th May 1930, p. 564. SANA K105 Miners Phthisis Commission Correspondence Vol. 3.

[36] *Recommendation as to Sanitation*, pp. 344–345.

official data, between 1933 and 1938, it accounted for between 29.65 and 37.87 per cent of deaths from disease among miners.[37]

There were several reasons for the Chamber's rejection of the Gorgas' recommendations. A stabilised work force was anathema to employers as it would have required wages capable of supporting a family.[38] It also went against the entrenched government policy, fiercely supported by the white electorate, of preventing black urban settlement. Centralising the hospital system had the obvious advantage of reducing duplication and costs. As noted earlier, the industry achieved similar aims in successfully centralising and streamlining the recruiting system with the setting up of the WNLA in 1896, and the NRC in 1912. However, centralisation of the hospital system would have enabled the state to exercise greater control over mine medicine and to collect more accurate data on mortality and morbidity rates. It might also have resulted in men with tuberculosis or silicosis dying in Johannesburg rather than in distant rural areas.

Soon after the release of the Gorgas Report, the newly appointed Miners' Phthisis Prevention Committee (MPPC) wrote to the Minister of Mines about what it viewed as a looming crisis. It was particularly concerned at the rate at which tuberculosis converted an otherwise early case of silicosis into a life-threatening disease which was killing large numbers of miners. The Committee's grim warning came after the passing of the first Miners' Phthisis Acts and coincided with the release of the Tuberculosis Commission Report in 1914. According to the MPPC, every precaution should be taken to ensure that infected men were not recruited, and that conditions be improved to reduce the spread of disease. The Committee recommended that all recruits rejected on the Rand because of tuberculosis should have their passes endorsed to that effect. They should be repatriated wherever possible, or otherwise assessed by a government medical officer with a view to their treatment. Compound bunks should be designed to allow for thorough cleansing and a steam steriliser should be available for each compound. The room from which a

[37] See *Proceedings of the Transvaal Mine Medical Officers' Association*, Vol. XIX No. 214, January 1940, pp. 229–230. The frequent misclassification of TB as pneumonia found by Dr Darling at autopsies might have inflated this figure.

[38] See Charles H. Feinstein. *An Economic History of South Africa: Conquest, Discrimination and Development*. Cambridge: Cambridge University Press, 2005, pp. 67–70.

patient had been removed should be immediately disinfected.[39] Once a month every room should be emptied and hosed out, the bunks boiled and all clothing and bedding sterilised. All room fixtures should be sprayed with a disinfectant and lime-washed. Underground tanks for drinking water should be covered and periodically emptied and cleansed. Ladder ways and incline shafts, travelling ways and stations should be systematically disinfected at least once a month. The question of supplying change houses should be given careful consideration. That final recommendation was the only one of the MPPC's suggestions that was acted upon.

The 1912 Tuberculosis Commission

The Tuberculosis Commission promised by Prime Minister Botha in August 1911 was appointed in February 1912. Its brief was to investigate the causes and prevalence of tuberculosis among all racial groups, and to report on the steps which the government should take to prevent the spread of disease. The Commission was also to identify the extent and causes of mortality in black goldminers and their susceptibility to pneumonia.[40] It was the first commission into tuberculosis, and it took evidence from more than 600 witnesses. Dr A. John Gregory, the Medical Officer of Health of the Cape Colony, was appointed Chairman. Dr Gregory was one of the first public health officers in South Africa to campaign for a national response to tuberculosis. Unlike those who usually chaired or served on such commissions, he had no allegiance to the mining industry.[41] Gregory was assisted by four commissioners, two of them closely linked to the mining industry. Dr G.A. Turner was the WNLA's Chief Medical Officer. Dr Charles Porter, the Medical Officer of Health for the Johannesburg Municipality, oversaw two inspectors on the mines

[39] Report from The Miners' Phthisis Prevention Committee, Johannesburg, to The Minister of Mines, Pretoria, 4th July 1914, pp. 1; 5. SANA, MNW 457, Miners' Phthisis on the Witwatersrand, Report of Committee, 1915.

[40] *Report: Tuberculosis Commission, 1914*. U.G. 34-'14, Cape Town: Government Printer, 1914. Union of South Africa, p. 1.

[41] Malan, *In Quest of Health*, pp. 119–120.

and claimed to have had more experience than anybody else in regard to mine sanitation. In eight years, he had never made a prosecution. The commissioners struggled to reach consensus, both during the Commission's sittings and in compiling its final report. Gregory twice resigned from the Commission on the grounds that Turner and Porter had in effect been asked to sit in judgement on their own work, only to be persuaded by the Minister to remain.[42] In the end, Gregory submitted a minority Report, and the two commissioners he clashed with responded with a supplementary statement.

The Commission's deliberations coincided with those of Gorgas. The industry had a great deal at stake in Gregory's report. The ban on Tropical recruiting was putting upward pressure on wages by reducing the labour supply. Because more miners were recruited within South Africa, it also threatened to make the actual disease burden more visible. In its final report from 1914, the Gregory Commission considered at some length historic evidence about the incidence of tuberculosis in southern Africa. It concluded that, given its ancient origins, the disease was almost certainly brought to the region by Arab traders. Over the last few centuries, it was on occasions described by European travellers, but never became widespread. This changed in the last 25 years. Many observers reported that tuberculosis was almost unknown among black South Africans in the past but recently had quickly become a 'most serious menace' in the large industrial centres.[43] The black industrial population consisted mainly of fit young men; the weak and ailing remained in the villages. The mines were one of the major sources of infection. All African labourers were adversely affected by the change from village life to unhygienic and crowded mine compounds.[44] The inhalation of 'silicous dust' was particularly harmful in predisposing miners to tuberculosis and pneumonia.[45] Of all the different groups working on the mines, Tropical recruits were particularly vulnerable. Subjected to an 'enormously increased risk'

[42] 'Minority Report by Dr Gregory on the Future Control of the Health of Mine Workers and Mine Sanitation on the Witwatersrand' in *Report: Tuberculosis Commission, 1914*, p. 260.

[43] *Report: Tuberculosis Commission, 1914*, pp. 21–37.

[44] *Report: Tuberculosis Commission, 1914*, pp. 123; 106–107.

[45] *Report: Tuberculosis Commission, 1914*, pp. 210; 273.

of infection, their mortality was almost five times that of miners from Natal and Zululand.[46]

When a miner became ill, he wanted to return home. As a result, the actual rates of sickness and mortality were hidden. The official annual mining death rates were confined to those men who died at the mines. A more accurate figure required adding those deaths which took place on the journey to the Rand, the deaths of men who waited to be allocated to individual mines or repatriated at the WNLA compound, and those of repatriated men who died before they reached home or soon after. As Gregory explained it with regard to statistics for the years 1910 to 1912, 'it is evident that while the death-rate has been going down, the repatriation rate has been steadily going up faster than the mortality has been diminishing, so that the total rate of wastage from death and disease increased in 1912 by 13 per cent over what it was in 1910. This well illustrates the fallacy involved if the health of mine natives on the Rand is considered merely on the recorded mortality and without reference also to repatriations.'[47] In contrast, tuberculosis rates in the white population were very low. Apart from mining, very few Europeans were employed in phthisis-producing occupations.[48]

Tuberculosis in blacks was notable for its rapid course, and for the frequency with which it was fatal. Of 524 deaths from the disease occurring on 22 gold mines taken at random, almost 20 per cent died within a week of admission to hospital. There was also evidence of widespread under-reporting. Dr G.A. Turner from the WNLA carried out post-mortems of 88 miners at the WNLA hospital where the recorded cause of death was silicosis; he found that 59 (or two-thirds) of the deceased also had tuberculosis.[49] None of the Commissioners, including Dr Turner, disputed the link between silica dust exposure and tuberculosis.

The Commission cited four reasons why black miners were so vulnerable to tuberculosis. The first was the absence of immunity acquired in early childhood, the second was living in unhygienic and overcrowded

[46] *Report: Tuberculosis Commission, 1914*, p. 198, Table 52.

[47] *Report: Tuberculosis Commission, 1914*, p. 204.

[48] *Report: Tuberculosis Commission, 1914*, pp. 197; 87–89.

[49] *Report: Tuberculosis Commission, 1914*, pp. 109–110.

compounds, and continuous and excessive labour. The most potent factor was the inhalation of silica dust, which depressed the powers of resistance to the bacillus. The final element was the pattern of recruitment, which possibly interacted with a 'tribal or racial proclivity to the disease'.[50] Local recruits worked on six-month contracts while Tropicals and East Coasters were typically recruited for 12 months, with most East Coasters returning again and again. The Commission found that repeated contracts were a factor in the production of silicosis, and therefore a 'powerful predisposing cause of pulmonary tuberculosis'. Tropical recruits, who often suffered from a series of other diseases, were unsuited for mine work, and the Commission strongly recommended that the ban should remain, and 'no further recruiting of Tropicals … be permitted unless and until the Government is fully satisfied that there is no reasonable likelihood of the recurrence of such mortality'.[51]

Gregory's findings confirmed those of William Gorgas, who made similar recommendations. Any scheme for combating tuberculosis must be based on the discovery and isolation of cases, the care of patients, and the safeguarding of co-workers and families from infection. That in turn required the breaking up of large compound dormitories into rooms housing not more than six miners, and sanatoria.[52] In his minority report, Dr Gregory recommended that a government Mines Board of Health take over the current uncoordinated, fragmented, overlapping and ineffective management of mine medicine and sanitary inspection. Its responsibilities would include conducting all medical examinations, reviewing ventilation, dust levels, sanitation both in the compounds and underground in the mines, compound conditions and rations. The Board would be funded by an annual levy on the industry and would have the power to appoint and remove mine medical officers and to initiate

[50] The commissioners and witnesses strongly disagreed about the racial susceptibility thesis, with some passages in the report supporting it and others pointing instead to straightforward environmental explanations. Chapter 10, pp. 118–121 of the Report summarises the commissioners' divergent views; Addendum A (pp. 285–290) and an appended paper by G.D. Maynard (pp. 323–343) discus the eugenic case at greater length.

[51] *Report: Tuberculosis Commission, 1914*, pp. 112; 114; 199–200.

[52] *Report: Tuberculosis Commission, 1914*, pp. 243; 275.

research.[53] Gregory's proposal to shift control of mine medicine from industry to the state was the most radical recommendation made by a Commission. It was strongly opposed by two of his fellow commissioners and never repeated. None of Dr Gregory's recommendations were acted upon and the practice of returning sick miners to their homes continued. The Commission findings were ignored by subsequent enquiries: A.J. Orenstein was scathing in his references to Gregory, whose report he termed 'of no value whatsoever'.[54] Today the Commission's Report is the only one of its kind not readily available in South African libraries. There are, for example, no copies at the Wits Medical Library or at the NIOH.[55]

The Tuberculosis Research Committee

Evidence that tuberculosis was being spread from the mines continued to mount. In February 1920, the Minister of Mines asked Dr L.G. Irvine, the Acting Chairman of the Bureau, to review the Bureau's operations. Irvine reported that during the previous five months, just under 1400 black miners had radiographic examinations at the Bureau. The number of cases of tuberculosis complicated by silicosis (297) was far greater than those with pure silicosis (78). 'The data given do not, of course, represent the total incidence of Tuberculosis amongst Natives working upon the Scheduled Mines. They obviously do not include any cases which have died from that disease in the Mine Hospitals, which during 1918 amounted to not less than 170 per 100,000 per annum.' The data was disturbing and, according to Irvine, reducing the tuberculosis rate was one of the major challenges facing the industry.[56]

[53] 'Minority Report by Dr Gregory' in *Report: Tuberculosis Commission, 1914*, pp. 262–263.

[54] Minutes of Meeting of Informal Conference re Tuberculosis in Natives, held at the New Law Courts, Johannesburg, on Tuesday, 31st October 1922, p. 2. SANA, MNW 634, MM 2906/22, Public Health Dept., Peter Allan, 1922.

[55] The copy used here comes from the Yale University Library in the United States.

[56] Report of the Miners' Phthisis Medical Bureau on the Working of the Miners' Phthisis Act No. 40, 1919, by Dr L.G. Irvine, Acting Chairman, Miners' Phthisis Medical Bureau (March 1920), pp. 27–28. SANA, MNW 528 2353/20.

Dr Irvine's concerns were shared by the Secretary for Public Health, Dr J.A. Mitchell, who in August 1922 directed a tuberculosis specialist, Dr Peter Allan, to begin the first study of tuberculosis and repatriated miners. Allan was asked to examine the extent of the disease, the level of resistance, what happened to infected men returning from the mines and the effect of repatriations on the general population. Mitchell was certain that the mines were spreading infection and he wanted a report within three months.[57] Allan visited several villages and towns in the Eastern Cape, including Libode, Umtata and Kingwilliamstown. In estimating the infection rate, he interviewed magistrates, district surgeons and mission stations. He also conducted a number of medicals. The NRC helped him to trace former miners. The records of the Umtata Hospital showed a high prevalence of tuberculosis, with around 10 per cent of all admitted patients suffering from the disease. According to Dr Drewe from the Holy Cross Mission, Flagstaff, infection was very common. In his interim report Allan commented: 'Much more time is required to investigate to what extent the repatriated native infects other natives.'[58]

Having received Allan's interim report, Dr Mitchell convened an informal conference to discuss tuberculosis. Held at the end of October 1922, it included the Director of Native Labour from the WNLA, Dr Girdwood; Dr A. J. Orenstein representing the MMOA; Dr Peter Allan; and the Director of the Bureau and the SAIMR, Dr Watkins-Pitchford. Dr Mitchell was in the chair. Mitchell told the delegates that when he first called the conference, it had been suggested that he had done so in response to Dr Orenstein's recent remarks accusing the government of doing nothing about tuberculosis. According to Mitchell, the government realised the importance of tuberculosis among miners and would cooperate with the industry to improve the situation. Mr Whitehead, representing the Native Affairs Department, promised to fully cooperate while Mr Villiers, General Manager of the NRC, said that his organisation had already instructed its agents to keep an eye on men returning from the mines.

[57] Letter from D.J.A. Mitchell, Secretary for Public Health to Dr Peter Allan, 4th November 1922. SANA GES 1001 6/17 Tuberculosis Survey 1922.

[58] Memo from Dr Peter Allan to Dr J.A. Mitchell, Secretary for Public Health 6th October 1922. SANA. GES 1001 6/17 Tuberculosis Survey 1922.

Dr Orenstein repeated his accusation that the government had done nothing to prevent the spread of disease. While Dr Allan's study was valuable, it was still unknown how the disease was acquired.[59] The MMOA had given the matter considerable attention but nothing could be done unless the necessary research was funded. Even if it cost £50,000 it would be money well spent as the very existence of the mines depended upon the labour supply. The meeting agreed it was important to understand where infection occurred, how many men developed tuberculosis on the mines and how many brought the disease with them. A month later, Dr Peter Allan attended a meeting of the Mine Medical Officers' Association in Johannesburg. With Dr Orenstein in the chair, Allan asked the meeting for estimates of how many men were repatriated with 'open' tuberculosis and how many had positive sputum? Dr Girdwood of the WNLA replied that there was no data but from the cases he examined the proportion of 'open' cases was high.[60]

Allan soon found that there were no statistics on death and disease for the Transkei, so that information had to be obtained from medical practitioners. The records of the Umtata Hospital showed that tuberculosis was prevalent. Tuberculosis or *sifubu* had been present for at least 40 years and it was more common than formerly, with many children suffering from chest complaints. The NRC provided Allan with a list of men who had been compensated and then repatriated. Of the 112 miners Allan traced, half had died within a year and another 15 per cent within two years. Of the 47 cases who were still alive, only 28 had recovered sufficiently to work. It appeared that in the Transkei tuberculosis was common apart from infection brought from the mines. At the Holy Cross Mission, for example, only 10 per cent out of a consecutive sequence of 42 admissions may have been infected on the mines. Allan concluded that without extensive examinations and diagnostic tests, it was difficult to determine the extent to which infection in the general population

[59] Minutes of Meeting of Informal Conference re Tuberculosis in Natives, held at the New Law Courts, Johannesburg, on Tuesday, 31st October 1922, p. 2. SANA, MNW 634, MM 2906/22, Public Health Dept., Peter Allan, 1922.

[60] Notes from Dr Peter Allan to Dr Mitchell, The Secretary for Public Health, Undated, November 1922, p. 1. SANA, GES 1002 6/17, Dr Allan's TB Survey 1922.

came from returning miners.[61] It was certain, however, that the danger of spreading infection was greatest with patients in whom the disease was advanced. The removal of such cases to a hospital was essential and wherever possible, Tuberculosis Dispensaries should be established.[62]

Dr Allan submitted his final report to Mitchell in 1924. The main themes were the paucity of morbidity and mortality data, and patients' lack of access to biomedical care. The report is notable, however, for its omissions. Allan does not mention exposure to silica dust, the synergy between silicosis and tuberculosis, the spread of infection in the mine compounds, the NRC's repatriation policies or the declining living standards and malnutrition in the Reserves. Every one of those factors had featured in Gregory's 1914 report.

Allan's survey led to the creation of the Tuberculosis Research Committee (TRC) in 1925. The Chamber agreed to fund an international expert to guide the Committee's research. Dr Lyle Cummins, famous for his theory of virgin soil infection, which explained high mortality rates in Africans by reference to their racial deficits, was appointed.[63] Dr Mitchell was concerned that Cummins would sandwich the work between visits to game parks and he preferred they 'get somebody else'.[64] The TRC members included Dr L.G. Irvine as the Department of Mines representative; Mr H.M. Taberer, Native Labour Adviser to the Chamber; Dr Mavrogordato from the SAIMR; Dr Watkins-Pitchford, Director of the Bureau; Orenstein and Mitchell. The TRC was conflict ridden and there was a battle over jurisdiction between the Departments of Public Health and of Mines. J.A. Mitchell was highly critical of the gold mines. Writing in December 1922, he told his counterpart that the main cause of the spread of tuberculosis was the mines. He wanted the

[61] *Report of Tuberculosis Survey of the Union of South Africa by Peter Allan, Medical Inspector, Department of Public Health.* Cape Town: Cape Times Ltd, Government Printer, 1924, p. 16.

[62] *Report of Tuberculosis Survey*, p. 33.

[63] Minutes of Meeting of the Tuberculosis Research Committee, Gold Producers' Committee, Transvaal Chamber of Mines, 22nd March 1927. SANA, 1928, GES2596 5/54A, League of Nations, TB Research.

[64] Letter from Dr J.A. Mitchell, Secretary for Public Health, to Dr L.G. Hayden, Member of the Research Committee and Department of Public Health, Pretoria, Subject: Tuberculosis Research Committee, 17th May 1927. SANA, 1928, GES2596 5/54A, League of Nations, TB Research.

Department of Mines to pay the cost of investigating and treating the disease in miners and their families.[65]

On his part, Orenstein insisted that he knew a great deal about the work and living conditions of migrant workers, and he was certain that the mines bore no responsibility for the spread of tuberculosis; a view which put him at odds with the Departments of Public Health and Native Affairs. Lyle Cummins was sure that silica dust played a major role in the tuberculosis rate and he wanted that question investigated. The only issue on which the Committee agreed was the urgent need for research. However, it took almost ten years before a report was completed.

Lyle Cummins, who attended the July meeting of the TRC in 1927, noted that among white miners, about 97 per cent of pulmonary tuberculosis cases were chronic or 'secondary', and only about 3 per cent acute or 'primary'. In the case of black miners those figures were reversed. The Committee agreed that insufficient systematic work had been done to offer an explanation. Cummins then raised two adjacent questions: 'Assuming that tuberculosis is an environmental disease, is mining on the Witwatersrand a phthisis-producing industry because conditions favour the lighting up of old foci, or because they render possible infections from without that would not occur apart from such conditions?' To find an answer, Cummins suggested a study comparing the prevalence of tuberculosis in villages that were directly associated with the scheduled mines and villages which were not. There should also be a follow up of open-tuberculosis 'repats' and their families. Such an enquiry should explain whether miners were spreading tuberculosis and whether a high prevalence was peculiar to the scheduled mines.[66]

Cummins believed it essential to compare the phthisis incidence and mortality and the ratio of tuberculous 'repats' to 'deaths' between Africans working in the gold mines and those working under the less strenuous and less dusty conditions in the coal and diamond mines. Such comparisons should take account of the length of time worked before phthisis

[65] Memo from J.A. Mitchell, Secretary for Public Health to the Secretary for Mines and Industries 30 December 1922. GES 1005 9 17 A TB in Mines Natives SANA.

[66] Minutes of the TB Research Committee, Gold Producers' Committee, Transvaal Chamber of Mines, signed by Sir Spencer Lister, Chairman, undated July 1927, pp. 1; 3. SANA, 1928, GES2596 5/54A, League of Nations, TB Research.

developed. It was necessary to obtain a series of post-mortems of men killed in mine accidents to estimate the extent to which focal infections were present in healthy recruits. In rural areas, the disease seemed to take a more benign course. If the clinical types were found to be more severe on the Rand, that would suggest that, in addition to native susceptibility, some factor of aggravation existed on the mines. That factor might be exposure to silica dust or in repeated exogenous re-infection, either in the mines or in the compounds. Those factors (which are still highly relevant today), required careful investigation.

Mavrogordato agreed with Cummins that gold mining was a phthisis-producing industry. He wanted the tuberculosis incidence and mortality on the Rand compared with that of phthisis-producing industries in Great Britain, rather than with that of the general African population. In such industries in Great Britain, the average age of mortality was in late middle-age and old age, the tuberculous process being, as it were, superposed on a silicotic fibrosis. In black miners, in contrast, death or disease from tuberculosis was common in young adults and occurred before silicosis had become marked. Cummins wondered if the cause was silica dust inhaled over a relatively short period: 'The question of the part played by silica dust in the tuberculosis of natives is of fundamental importance and every effort must be made to solve it.'[67]

Cummins noted a contrast between the tuberculosis in the mine hospitals and that seen in the Transkei. In Johannesburg, the disease took an acute and generalised form characteristic of recent exogenous infection: in the Transkei tuberculosis was of the 'modified' type found in communities where infection had been present for some time.[68] During his visit to the Transkei, Cummins discussed that problem with district surgeons and medical missionaries, including Dr Macvicar. All were sure that the disease was widespread and that chronic cases were becoming more common. 'It is undeniable, however, that many cases of "open" tuberculosis

---

[67] Memorandum from Professor Lyle Cummins, Subject: Questions re Tuberculosis in South African Races, February 1928, pp. 6–7. SANA, 1928, GES2596 5/54A, League of Nations, TB Research.

[68] A Report to the Tuberculosis Committee by S. Lyle Cummins, 'The Problem of Tuberculosis in Native Mine Workers on the Witwatersrand', undated, 1929, p. 4. SANA, 1928, GES2596 5/54A, League of Nations, TB Research.

are repatriated [from the mines] every year and such cases are necessarily a risk to others.' Cummins presented two research priorities. The first was to identify and remove carriers; the second was to conduct experiments in the prevention of tuberculosis on the mines. In the case of those repatriated, data on the period of survival might be collected by the NRC.[69]

Lyle Cummins' presentations to the Committee must have surprised Orenstein and Mitchell. The Chamber wanted Cummins as its international expert because of his virgin soil thesis, which attributed Africans' high mortality rates to their lesser civilisation. Having viewed the evidence, Cummins abandoned that thesis in favour of dust as the causative factor. Cummins identified several sophisticated avenues for research, none of which were subsequently taken up by the Committee. Indeed, his views about dust exposure were completely erased from the Committee's final report.

Orenstein was the only member of the Committee who denied outright that there was a major tuberculosis problem on the mines. He argued that over the previous 11 years, the average annual tuberculosis mortality rate was approximately 1.2 per 1000, resulting in 240 deaths per annum. He admitted that, in addition, a certain number ultimately died at home.[70] Orenstein was arguably the best-informed member of the Committee about tuberculosis in black miners. He was certainly the most experienced and, unlike the other members of the Committee, had access to two recently completed surveys of long-service miners. While the Committee was sitting, Dr Girdwood reported to the GPC, which Orenstein chaired, that the radioscopic examination of approximately 1800 long-service miners at the WNLA Compound had revealed around 8 per cent were suffering from tuberculosis. In an expanded study of 2023 long-service miners, half of whom were surface workers, Dr Girdwood found definite tuberculosis in 66 men and probable tuberculosis in another 59. In all, 18 per cent of the subjects had evidence of

[69] A Report to the Tuberculosis Committee by S. Lyle Cummins, pp. 16; 18–19.

[70] Letter from Dr A.J. Orenstein, Superintendent of Sanitation, Rand Mines Limited, Johannesburg, to The General Manager, Transvaal Chamber of Mines, Johannesburg, Subject: Medical Committee on Tuberculosis, 2nd December 1926. SANA, 1928, GES2596 5/54A, League of Nations, TB Research.

chest abnormalities.[71] If those results were representative, there were thousands of carriers working underground. The situation was grave, and the Health Committee of the League of Nations offered to assist the Chamber with a further investigation.[72]

The ILO co-sponsored the 1930 Silicosis Conference in Johannesburg in the hope of promoting workplace reform. In contrast, the Chamber viewed the Conference as a stage on which to convince the South African and British governments to lift the ban on Tropical labour. The Chamber invested heavily in persuading an international audience of experts that the mines were safe. Three years later, the ban was provisionally lifted.[73] In 1932, the final outcome of the TRC deliberations was published under the title *Tuberculosis in South African Natives*. Much of the report was written by Orenstein. It found that blacks were heavily tubercularised and that they would soon develop immunity like whites.[74] Dr B.A. Dormer, one of the leading specialists on tuberculosis, was unimpressed by the report and blamed Cummins. 'This theory of racial susceptibility in primitive people was, to use modern parlance, sold to the Committee by Professor Lyle Cummins, who was its expert adviser.'[75] Like the Chamber, Dormer had expected Cummins to reprise his virgin soil thesis and therefore blamed him for what he saw as sloppy science.[76]

---

[71] Circular from L.S. Raymond, Secretary, Medical Committee on Tuberculosis, Gold Producers' Committee, Transvaal Chamber of Mines, to Members of the Medical Committee on Tuberculosis, 8th March 1927. SANA, 1928, GES2596 5/54A, League of Nations, TB Research.

[72] Minutes of Meeting held 14th December 1926, Medical Committee on Tuberculosis, Gold Producers' Committee, Transvaal Chamber of Mines, pp. 1; 3. SANA, 1928, GES2596 5/54A, League of Nations, TB Research.

[73] See McCulloch, 'Air Hunger'. See also Jock McCulloch and Paul-André Rosental, with Joe Melling. 'Johannesburg and Beyond: Silicosis as a Transnational and Imperial Disease, 1900–1940'. In Rosental, Paul-André, ed. *Silicosis: A World History*. Baltimore: Johns Hopkins University Press, 2017, pp. 64–104.

[74] Tuberculosis Research Committee. *Tuberculosis in South African Natives with Special Reference to the Disease Amongst the Mine Labourers of the Witwatersrand*. Johannesburg: South African Institute for Medical Research, 1932. See also McCulloch, *South Africa's Gold Mines*, pp. 85–89.

[75] B.A. Dormer. 'Tuberculosis in South Africa'. *British Journal of Tuberculosis and Diseases of the Chest*, Vol. 50, No. 1, 1956, p. 54.

[76] Packard too blamed Cummins in his 'The Invention of the "Tropical Worker": Medical Research and the Quest for Central African Labor on the South African Gold Mines, 1903–36'. *The Journal of African History*, Vol. 34, No. 2, 1993, pp. 271–292.

Conclusion

The Commissions and allied investigations carried out from 1902 until the SAIMR's report of 1932 all reached the same conclusions. Dust exposure was a major hazard, as was tuberculosis for migrant labour. The causes of lung disease were environmental rather than racial and therefore could only be reduced by major investment in work and living conditions, and not by the 'seasoning of labour'. The ban should remain until the risks associated with silica dust had been removed. Gorgas and Cummins, who were recruited by the Chamber because of their status as world authorities, both identified silica dust exposure as a major hazard for migrant labour. Their recommendations were consistent with the then-ruling medical orthodoxy in Britain, the United States and Australia, as well as with findings of the South African Commissions of 1902, 1912 and 1914. And yet, until the advent of majority rule, these findings were obscured by the Chamber's orthodoxy regarding mine safety.

References

Cartwright, A.P. *Doctors of the Mines: A History of the Work of Mine Medical Officers*. Cape Town: Purnell and Sons, 1971.

Dormer, B.A. 'Tuberculosis in South Africa'. *British Journal of Tuberculosis and Diseases of the Chest*, Vol. 50, No. 1, 1956, pp. 52–60.

Feinstein, Charles H. *An Economic History of South Africa: Conquest, Discrimination and Development*. Cambridge: Cambridge University Press, 2005.

Malan, J.M. 'History of the South African Institute for Medical Research'. *South African Medical Journal*, Vol. 76, No. 7, October 1989, pp. 374–382.

Malan, Marais. *In Quest of Health: The South African Institute of Medical Research, 1912–1973*. Johannesburg: Lowry Publishers, 1988.

McCulloch, Jock and Paul-André Rosental, with Joe Melling. 'Johannesburg and Beyond: Silicosis as a Transnational and Imperial Disease, 1900–1940'. In Rosental, Paul-André, ed. *Silicosis: A World History*. Baltimore: Johns Hopkins University Press, 2017, pp. 64–104.

McCulloch, Jock. 'Air Hunger: The 1930 Johannesburg Conference and the Politics of Silicosis'. *History Workshop Journal*, Vol. 72, No. 1, 2011, pp. 118–37.

McCulloch, Jock. 'Hiding a Pandemic: Dr G.W.H. Schepers and the Politics of Silicosis in South Africa'. *The Journal of Southern African Studies*, Vol. 35, No. 4, December 2009, pp. 835–848.

McCulloch, Jock. *South Africa's Gold Mines and the Politics of Silicosis*. Oxford: James Currey, 2012.

Mears, A.R.R. 'The Part-Time District Surgeon in the Transkeian Territories'. *The South African Medical Journal*, Vol. 19, No. 21, 10th November 1945, pp. 402–405.

Murray, J.F. 'History of the South African Institute for Medical Research'. *The South African Medical Journal*, Vol. 37, No. 16, 20th April 1963, pp. 389–395.

Packard, R.M. 'The Invention of the "Tropical Worker": Medical Research and the Quest for Central African Labor on the South African Gold Mines, 1903–36'. *The Journal of African History*, Vol. 34, No. 2, 1993, pp. 271–292.

Roberts, Jaine. *The Hidden Epidemic Amongst Former Miners: Silicosis, Tuberculosis and the Occupational Diseases in Mines and Works Act in the Eastern Cape, South Africa*. Westville: Health Systems Trust, June 2009.

Sitas, F. Occupational Health Services in South Africa: A Pilot Survey of Manufacturing Industries, Dissertation for the Degree of Master of Science in Medicine, Johannesburg, 1986.

Zwi, Anthony, Sharon Fonn and Malcolm Steinberg. 'Occupational Health and Safety in South Africa: The Perspectives of Capital, State and Unions'. *Social Science and Medicine*, Vol. 27, No. 7, 1988, pp. 691–702.

Part III

The Impact of Lifting the Ban

7

Tuberculosis, Malnutrition and Mining in South Africa: 1903–1960

Tuberculosis was an occupational disease on the gold mines and therefore a cost of production for the mining houses. It was also a notifiable disease and subject to state monitoring. The spread of infection from the mines threatened the supply of migrant labour; over time, it came to represent one of the industry's key points of fragility. Among those who attempted to keep the issue in the public domain, the Mine Workers' Union (MWU), which claimed its white members and their families were at risk, was the most effective.

In the period 1903–1905, the official annual tuberculosis mortality rate for the Cape colony was more than double than that of London, with an alarming rate among Coloureds and blacks.[1] However, the data was unreliable. In the Cape, many cases in Europeans and Coloureds were never notified, while very few deaths from tuberculosis in blacks were reported. At the South African Medical Congress meeting in Pietermaritzburg in 1905, a Tuberculosis Standing Committee, including Dr. W. Watkins-Pitchford, was appointed. The Committee found that tuberculosis was most common among Coloureds in Town Locations

[1] D.P. Marais. 'The Prevention of Consumption'. *The South African Medical Record*, Vol. 10, No. 3, 10th February 1912, p. 40.

J. McCulloch, P. Miller, *Mining Gold and Manufacturing Ignorance*,
https://doi.org/10.1007/978-981-19-8327-6_7

and in mine compounds. While in the previous three years the mortality rate among whites had fallen by almost a half, the rate among blacks and Coloureds had risen sharply.[2] The Committee recommended immediate improvements to mine sanitation and Town Locations. Tuberculosis became a notifiable disease in the Cape in 1907, and throughout the Union under the Public Health Act of 1919. The Act, which was designed to limit infection among whites, introduced programmes for early diagnosis, treatment and notification, as well as improved housing, sanitation and education. It also established dispensaries, sanatoria and farm colonies.[3] There was no parallel programme for the majority of South Africans.

Following the Union, the growth of the hospital systems in South Africa's four provinces was uneven. In the rural areas, there was an acute lack of services. The several Native Territories covered an area roughly the size of Scotland with a population of about 1 million.[4] The Territories were divided into Magisterial Districts, each with around 40,000 people. Districts were divided into locations under a chief or headman. Most of the population were peasant farmers, with most men working on the mines to support their families. There was also a small salaried class of teachers, clerks and ministers of religion. In the Transkei, there were in all 13 Provincial and Mission hospitals, offering a total of just over 600 beds. Half of the Districts had no hospital. Most of the doctors were in the large towns. The rural areas were served by part-time District Surgeons, but many areas were inaccessible.

In 1919, the Union Department of Public Health was established, and Dr. J. Alexander Mitchell was appointed the first Chief Medical Officer. Mitchell was determined to improve medical care in the rural areas but was frustrated by a lack of funding and resources.[5] Before 1924, there

---

[2] 'Report of the Standing Committee on Tuberculosis of the South African Medical Congress'. *The South African Medical Record*, 25th February 1908, p. 54.

[3] J. Alexander Mitchell. 'The Problem of Tuberculosis in South Africa'. *The South African Medical Record*, Vol. 19, No. 12, 25th June 1921, p. 227.

[4] The Native Territories were areas set aside in several different parts of the country for black South Africans by the Natives' Land Act (No. 27 of 1913). This effectively allocated less than one-tenth of arable land to Africans and left over four-fifths of the more fertile land for whites, who made up less than a fifth of the population.

[5] Over time the Department did expand and by the mid-1950s it had a staff of over 9000, supported by 400 part-time District Surgeons and an annual budget of nearly £10,000,000. See 'Historical Survey of Medical Research in South Africa'. *The South African Medical Journal*, Vol. 31, No. 27, 6th July 1957, pp. 677–679.

were few beds for isolation and treatment; it was not until 1923 that the Nelsproot Farm Sanatorium was opened in the Karoo. Dr. Peter Allan, a Scottish-born specialist, was its first medical director. Treatment for whites and a small number of Coloured and black patients was provided, with an emphasis on education, hygiene and farm work. A second sanatorium was opened at Durban in 1938, bringing the number of beds available nationally to 834.[6]

In South Africa in 1936 there was on average one hospital bed for every 375 Europeans, and one bed for every 1783 non-Europeans.[7] In 1941, the population of the Transkei was over 1 million and there was one bed for every 300 whites and one for every 1900 blacks. In Australia the standard was 10 beds per 1000, and in English cities it varied between 6 and 9 per 1000. There were no domiciliary nursing services for black patients. Despite their contribution to public health, the missions struggled for funding, with only the Victoria Hospital at Lovedale in the Cape receiving a Provincial allowance.[8] Treatment outside the hospitals was expensive, with patients usually charged 5 s for an examination.[9]

Writing in 1937, Dr. Harry Gear of the Union Department of Public Health remarked that medical researchers were handicapped by the lack of data on the incidence and prevalence of the major diseases.[10] Public health officers and administrators were helpless because of the lack of reliable statistics, especially in relation to the non-European population. The one exception were the gold mines which had conducted, usually with other official and non-official co-operation, some of the most important epidemiological studies carried out in South Africa.

In 1943 a senior figure in the Department of Public Health, Dr. G.W. Gale, published a review of tuberculosis prevention in the Union. Gale, who as

---

[6] Dr. Peter Allan. 'The Tuberculosis Problem in the Union of South Africa'. *The South African Medical Journal*, Vol. 13, No. 21, 11th November 1939, p. 740.

[7] H.S. Gear. 'The Position of General Hospitals in South African Health Organisation'. *The South African Medical Journal*, Vol. 12, No. 17, 10th September 1938, p. 645.

[8] R.L. Paterson. 'Medical Missions in the Transkei'. *The South African Medical Journal*, Vol. 17, No. 10, 22nd May 1943, p. 154.

[9] Arnold Tonkin. 'A State Medical Service for Natives'. *The South African Medical Journal*, Vol. 15, No. 15, 9th August 1941, pp. 283; 286.

[10] H.S. Gear. 'A Plea for Improved South African Medical and Vital Statistics'. *The South African Medical Journal*, Vol. 11, No. 5, 13th March 1937, pp. 149–154.

a young physician had served at a mission hospital in the Eastern Cape, noted that the medical services among the poor of all races were inadequate, with lack of capacity being the major barrier to prevention. The tuberculosis death rate among whites in South Africa was lower than in most European countries and was falling.[11] However, the limited data suggested that the death rates among non-Europeans were disturbingly high. The purpose of isolation was to prevent the spread of the disease, but isolation required a sufficiency of beds, usually computed as a minimum of one bed for each death recorded annually. Under that formula the Union had just 13 per cent of the beds required for the 15,000 non-European deaths.

As did his colleague Harry Gear, Gale singled out the gold mines as a special case. He noted that gold miners were the least likely group to contract tuberculosis as they were a physical elite, and unlike other classes of migrant labour were well fed. For those reasons, Gale believed that the tuberculosis rates in gold miners should be the lowest for any class of migrant labour. In contrast, non-miners who moved to the towns were not pre-selected, they fed themselves and mostly poorly, they worked long hours, often took a long journey to work and lived in unhygienic conditions. Gale concluded: 'By all the canons of hygiene and of reason we may anticipate a torrent of tuberculosis among native urban populations during the next twenty years'.[12]

The Tuberculosis Crisis in the Transkei Native Reserve in the Eastern Cape

From Union in 1910 there were repeated warnings from a range of medical officers about a looming tuberculosis crisis in the Eastern Cape. In 1913 Dr. D. Melville, who had worked in the Native Territories for almost two decades, published a paper on tuberculosis. Of the first 3000

[11] Peter Allan. 'The Treatment of Pulmonary Tuberculosis from the Public Health Aspect'. *The South African Medical Journal,* Vol. 13, No. 9, 13th May 1939, p. 318.

[12] G.W. Gale. 'The Prevention of Tuberculosis in the Union, with Special Reference to Urbanised Natives'. *The South African Medical Journal,* Vol. 17, No. 20, 23rd October 1943, pp. 321–322.

patients Melville treated in 1897, only 1 per cent had tuberculosis. Of the first 3000 he treated in 1912, it had risen to 40 per cent, with the pulmonary form being dominant. Melville noted that there was a high death rate, with many men returning from the mines 'riddled with disease'.[13] Dr. Melville's published report was cited by the 1912 Commission into Miners' Phthisis and Pulmonary Tuberculosis (Medical) Commission.

Over the next twenty years, the Eastern Cape was transformed into a labour Reserve, with most families dependent upon mine wages. A report from the Transkei Division of the South African Medical Association from February 1943 offers a critique of living conditions and public health. In the Transkei the average family consisted of five members. Many had no cattle or arable land.[14] Over-stocking and soil erosion were so serious that in some Districts no milk was produced. Rural families rarely ate meat and malnutrition was a major factor in the high tuberculosis rate. In comparison, the Association noted that mine labourers were 'gloriously well-fed'. At any one time, seven out of ten men were absent on wage labour and even boys were sent to the towns. As a result, most children did not attend school. Glen Grey was typical of the Reserves. Of the 18,000 resident families, only 8000 had arable land. A family would earn on average just 4 s a month from animals, while the average holding would produce five bags of grain per family per year, meeting half of their needs. According to one medical officer, the Native Reserves were not agricultural areas: they were dispersed residential areas where a family made no more from the land than an urban family did from its backyard garden.[15]

Nine months later the former Secretary for Native Affairs, J.F. Herbst, reviewed the state of health in the Transkei. Herbst noted that the decline in public health was so extreme as to be visible in the people's physiques. Fifty years ago, locals were healthy and strong. Now, the only people fed well were men working on the mines. In the period from 1933 to 1937, out of the 400,000 men recruited for the mines, 69,353 had been rejected.

[13] D. Melville. 'Tuberculosis in the Native Territories'. *South African Medical Record,* Vol. 11, No. 18, 27th Sept 1913, pp. 394–395.

[14] 'Medical Problems of the Transkei'. *The South African Medical Journal,* Vol. 17, No. 10, 22nd May 1943, p. 159.

[15] 'Medical Problems of the Transkei'. *The South African Medical Journal,* Vol. 17, No. 10, 22nd May 1943, pp. 161–163.

In 1937 alone, there were 2752 tuberculosis notifications in the territory, but Herbst believed they represented a minute proportion of those infected. To counter the spread of infection, proper nutrition and medical care were vital, and there was an urgent need for medical officers, rural clinics and nursing sisters.[16]

Writing in May 1944, Arnold Tonkin, a physician at Umtata, described tuberculosis as the region's greatest health problem. Tonkin was frustrated that over the previous twenty years various surveys, including the work by Peter Allan from 1922, had identified a serious problem but there had been no public health response. The Umtata Hospital had only one specialist ward serving a population of 1,250,000.[17] The lack of facilities meant that most patients were sent home to die, and they often infected their families. Some local doctors blamed the mines and Tonkin had seen patients dying a month after being repatriated. However, he believed that malnutrition rather than the mines was the major cause of mortality.

Poverty, Malnutrition and the Spread of Tuberculosis

During the early years of mining, the rations of black miners were deficient in protein and fresh vegetables, and this resulted in a large number of deaths from scurvy and pneumonia. Although the Chamber acknowledged that improvements were essential if the high mortality rates were to be reduced, it initially did little. By 1906, the Native Affairs Department gazetted a standard diet for the mines which included meat, vegetables and mealie meal. Further improvements gazetted during the 1920s saw a further decline in deaths.[18] Malnutrition was also a major factor in the high rejection rates of mine recruits. By the late 1920s, its impact on infant mortality and tuberculosis among adults in the labour-sending

---

[16] J.F. Herbst. 'The Transkeian Territories'. *The South African Medical Journal,* Vol. 18, No. 8, *22nd* April 1944, p. 127.

[17] Arnold Tonkin. 'The Control of Pulmonary Tuberculosis in the Transkei'. *The South African Medical Journal,* Vol. 18, No. 8, 22nd April 1944, p. 128.

[18] Randall M. Packard. *White Plague, Black Labour: Tuberculosis and the Political Economy of Health and Disease in South Africa.* Berkeley: University of California Press, 1989, pp. 471; 166.

Reserves was obvious. The underlying causes were low family incomes and land hunger in the Reserves.

During the 1930s, both the Colonial Office and the League of Nations took an active interest in public health and nutrition. A Nigeria study by McCulloch in 1929 was followed by the Carnegie Commission on poverty in South Africa, which found that around half of *white* families could not afford adequate food. During that same period F. William Fox, a biochemist with the SAIMR, published several papers on malnutrition. The topics ranged from studies of specific regions such as the Transkei to reflections on food markets and government farming policies. In his initial studies, Fox found that there was little data on nutrition, especially among the black population. While conditions were worst in the Transkei and the Ciskei due to overpopulation, overgrazing and droughts, the most common problem was an inadequate supply of protein and in particular a lack of milk.[19]

Malnutrition was the most common reason for the rejection of mine recruits. In 1937, Fox was commissioned by the Chamber to carry out a study on public health in the Transkei and the Ciskei. He presented his report to the Native Labour Committee enquiry into malnutrition in December the same year. The rural diet, he noted, was simple and inexpensive, and when mealies, milk, meat and home-made beer were available it worked well. But any small change to such a diet could be disastrous. Fox noted that in two generations, the native population had changed from a semi-pastoral to a semi-agricultural one. The land was being degraded and people were consuming less meat and milk. In times of drought maize was expensive, machine-processed mealie meal had little nutritional value and scurvy was widespread. Fox estimated that in the Eastern Cape, a quarter of infants died at one year or under, a third under two years and half before fifteen. The situation was even worse in some urban areas where fresh milk, vegetables and fruit were expensive, and a large part of household incomes went on rent and transport to and from work.

[19] F. William Fox. 'Diet and Health in South Africa: Malnutrition'. *The South African Medical Journal*, Vol. 10, No. 1, 11th January 1936, pp. 25–36.

Long-term malnutrition meant that individuals wore out more quickly because their resistance to disease was compromised. The rising tuberculosis rate was the result. Fox told the Committee: 'Malnutrition is one of the most expensive diseases the country has to face and yet it is one of the easiest and cheapest to cure'.[20] Fox pointed out that the mines were so successful in feeding serving miners for 4d and 5d a day that most men put on weight while doing arduous work. Fox was also highly critical of the food markets, which he argued served the interests of producers and middle men. Food exports were subsidised at the expense of the home market. Butter produced in South Africa was subject to a levy, making it more expensive for local consumers. However, exported butter was given a bounty, making it cheaper for overseas consumers. Similarly, locally produced sugar sold at three or four times the price of exported sugar. In addition, there was a heavy duty on imports. Fox concluded that most malnutrition could be eradicated with some adjustment to prices, better systems of distribution and an education campaign.[21]

Douglas Back, who accompanied Fox in his fieldwork, was highly critical of the Native Affairs Department for promoting wool growing when there was an acute shortage of food. There was much soil erosion in the Ciskei due to over-stocking, and only the wealthy had sheep. Fresh milk was becoming increasingly scarce in winter, and in some places it was practically unobtainable. Black attacked the Chamber's claims that any increase in food production in the Reserves would reduce the labour supply to the mines. 'The constantly increasing demand for labour can only be met if the source of supply, which is the family unit in the reserves and locations, is stabilised by the production of an adequate food supply to maintain the family.'[22]

Three months later Dr. E.H. Cluver, at that time Deputy Chief Health Officer, presented similar evidence to the Native Labour Committee in Pretoria. Cluver explained that in the Transkei, malnutrition showed up

[20] Evidence by Dr. F.W. Fox and Mr. Douglas Back presented before the Native Labour Committee, Johannesburg 13th December 1937, p. 4. SANA. K 356 Native Labour Commission 1937.

[21] Fox, 'Diet and Health in South Africa: Malnutrition', pp. 35–36.

[22] Evidence by Dr. F.W. Fox and Mr. Douglas Back presented before the Native Labour Committee, Johannesburg 13th December 1937, pp. 6–7. SANA. K 356 Native Labour Commission 1937.

statistically in the tuberculosis rate. The native population lived increasingly on starch, and the number of men failing the mine entry medicals was increasing. Many of those who *did* pass had to be fattened before they were fit to work. Cluver told the Committee: 'If the supply of labour in the Union was properly nourished, there would be no necessity to recruit from outside the Union'. There were not enough cattle in the country to produce the meat and milk which the Health Department considered necessary to feed the population, and yet South Africa was exporting those products.[23] Cluver estimated that at least three-quarters of the population could not afford to feed themselves properly. He agreed with Fox and Back that the best way of combatting tuberculosis was by supplements of milk, and if milk was not available then meat once a week, along with fruit and vegetables.

One of the major barriers to Cluver's proposals were white commercial farmers, who were protected from foreign competition by export and loan subsidies. They received rebates on railway rates, were given assistance with seed and fertiliser and in bad years received grants for drought relief. In addition, they benefited from schemes which maintained domestic prices above the levels set by world markets. Feinstein estimated that between 1939 and 1940, at least £7,500,000 was transferred to white farmers, with much of that amount coming from consumers who paid inflated prices for maize, meat and milk.[24] Fox was the one scientist particularly critical of the government's marketing policies: 'While the Union producer is obliged to export a surplus [of maize] at a subsidised price, there is a large unsatisfied demand in the Native reserves which is a principal cause of much malnutrition'.[25] In July 1943, the findings of the Nutrition Council were released. The Council found that most South Africans had an inadequate diet because they lacked the income to buy

[23] Evidence by Dr. E.H. Cluver, Deputy Chief Health Officer before the Native Labour Committee by, at Pretoria 3rd February 1938, pp. 1–2. SANA K 356 Native Labour Commission 1937.

[24] Charles H. Feinstein. *An Economic History of South Africa: Conquest, Discrimination and Development.* Cambridge: Cambridge University Press, 2005, p. 142.

[25] F.W. Fox. 'Some Nutritional Problem Amongst the Bantu in South Africa'. *The South African Medical Journal,* Vol. 13, No. 3, 11th February 1939, p. 95.

protective foods. It agreed with Fox that the problems could be remedied by wage rises and some form of food subsidies.[26]

During his research in the Ciskei and Transkei, Fox found that the people were dependent on wages from the mines, while the absence of so many men reduced food production. Unsurprisingly, the physiques of men and women were deteriorating. The milk yield was extremely low and this was having a disastrous effect upon the children. The infant mortality rate in the first year was between 15 and 25 per cent. Dietary status was changing rapidly: an increase in the population was coupled with the falling productive capacity of both the land and the communities it supported. The margin of safety was very small. In the Ciskei conditions were particularly dire, with an alarming decrease in the amount of milk available. In the Towns and Locations low wages and high rents left little money for food. As a result, scurvy and pellagra were common, as was tuberculosis.[27]

The Lansdown Commission and the Costs of Oscillating Migration

The studies by Fox, Back and Cluver are representative of a literature in which tuberculosis is emblematic of poverty. In his survey of the Transkei and Ciskei in 1924, Dr. Peter Allan similarly commented on the part played by malnutrition in the spread of tuberculosis. The evidence from public enquiries covering the period 1924–1945 confirms the depictions of chronic malnutrition and declining public health. The Mine Native Wages (Lansdown) Commission of 1943 is a case in point. In assessing the merits of a pay rise for black miners, the Commission examined the migrant labour system and its impact on families and communities. Lansdown held public meetings in the Ciskei, the Transkei and Natal, where it heard evidence from District Medical Officers, miners and community leaders. The Commission also received a damming and

[26] J.F. Brock. 'Malnutrition in South Africa'. *The South African Medical Journal*, Vol. 17, No. 14, 24th July 1943, p. 221.

[27] Fox, 'Some Nutritional Problem Amongst the Bantu in South Africa', pp. 92–93.

comprehensive submission from the African Mine Workers' Union.[28] The Chamber was represented by Mr. W. Gemmill, and Mr. A.G.T. Chaplin appeared on behalf of the High Commission Territories. What was at stake was the social and economic impact of migrant labour.

William Gemmill warned the Commission that the gold industry was so vulnerable that a wage rise would place many mines in jeopardy. Gold was a waning asset, and the mine's inevitable decline would involve the extinction of many secondary industries, and with them the loss of many white jobs. For that reason, everything possible should be done to extend the life of the mines.[29] According to Gemmill, the existing wage was adequate 'having regard to the provision of [free] rations, housing and medical care'. In setting wages, the Chamber was entitled to make allowance for 'the full subsistence' migrant workers had from their landholdings in the Reserves. In fourteen months on the mines, a man earned sufficient to maintain himself and his family during his absence and to keep him 'in idleness for a further period of twelve months'. The system, Gemmill explained, had a wider national benefit as it underpinned segregation. Oscillating migration kept black families out of urban areas and therefore protected white communities from being 'swamped'. 'The policy [of migrant labour and fixed wages] is a coherent whole, and is the antithesis of the policy of assimilation and the encouragement of a black proletariat in the towns, divorced from its tribal heritage.'[30]

Lansdown agreed with the Chamber that initially men went to the mines to satisfy finite needs which could not be met from their holdings in the Reserves. However, that soon changed, and by the First World War family incomes in the Eastern Cape were so low that many men were working almost continuously on the mines. Since Allan's report, poverty, overstocking, soil erosion and land hunger had worsened, and the scarcity of

---

[28] 'Statement submitted to the *Witwatersrand Gold Mines Native Wages Commission*, by the African Mine Workers Union' in V.L. Allen. *The History of the Black Mineworkers in South Africa: Vol. 1, The Techniques of Resistance 1871–1948*. Keighley: The Moor Press, 1992, Appendix 2, pp. 428–470.

[29] *Report of the Witwatersrand Mine Native Wages (Lansdown) Commission on the Remuneration and Conditions of Employment of Natives on the Witwatersrand Gold Mines*. Pretoria: Government Printer 1943, pp. 1–8.

[30] *Report of the Witwatersrand Mine Native Wages Commission*, pp. 7–9

essential foods such as meat, milk and fats had become acute.[31] By the early 1930s, the Reserves were blighted by over-stocking and soil erosion, and the typical family diet was lacking in meat, milk and fats. Nearly half the population owned no cattle, and as a result of over-stocking the production of milk had fallen. There was evidence of high tuberculosis rates in the Transkei and Ciskei. The Commission cited at length the testimony of Dr. Mary McGregor, the physician in charge of the Umtata Health Unit in the Eastern Cape. She pointed out that nearly half of the babies born in the Umtata District died before the age of two, with most deaths due to malnutrition.[32] On the basis of such evidence, the Commission rejected the picture of rural bliss described by Gemmill and found instead a very different reality of grinding poverty, high infant mortality and untreated tuberculosis. Rather than providing funds for discretionary purchases, the Commission concluded that the system was driven by poverty. The Chamber's own data showed that four out of five men arriving at the WNLA compound had been on the mines before. The rising periods of service were due to the deteriorating conditions in the Reserves. Lansdown predicted accurately that in the years to come, an ever-increasing number of families would depend entirely on migrant work.[33]

The case for a wage rise presented to Lansdown was compelling. There had been no increase in black miners' pay since 1914. In contrast to whites, blacks received neither increments for higher skills nor enhanced pay for overtime. The cost of living had risen, while incomes from farming in the Reserves had fallen.[34] In 1940, 37,826 European mine employees received an average wage of £1.11.10 per shift, as well as benefits equal to £4.18.2 per month and paid leave. In contrast, the 344,897 non-Europeans received an average wage of just over 2 s 3d per shift, as well as a miserly cost of 'keep'.[35] The Commission rejected the Chamber's submission that wages were more than adequate. On the contrary, it found that if the average income of a Transkei migrant labourer from his

[31] *Report of the Witwatersrand Mine Native Wages Commission*, pp. 11–14.

[32] *Report of the Witwatersrand Mine Native Wages Commission*, pp. 14–15.

[33] *Report of the Witwatersrand Mine Native Wages Commission*, pp. 15–16.

[34] *Report of the Witwatersrand Mine Native Wages Commission*, pp. 6–7.

[35] Cited in Allen, '*The History of the Black Mineworkers in South Africa*', Vol. 1, p. 433.

land and stock was added to his wages, the total was at least 20 per cent below the minimum required to sustain a family. At best, migrant work funded the debt in which families lived. Despite the evidence it marshalled, the Commission's recommendation for a substantial wage rise was rejected by both the employers and the government. At the same time, while Lansdown dismissed much of Gemmill's testimony, it did not question the gold mines' importance to the national economy. Nor did it question the need for migratory labour, finding that any other policy would cause 'a catastrophic dislocation of the industry and consequent prejudice to the whole economic structure of the Union'. The Commission also found that periods in the Reserves lowered the incidence of miners' phthisis.[36]

Malnutrition and Race

With the introduction of apartheid, the regime of low mine wages and poverty was further entrenched. In 1952 the District Surgeon at Tabankulu in the Transkei, Dr. J.H. Jackson, published a small study based on a random group of patients who had attended his surgery over a month. Within that group, 38 per cent of males and 62 per cent of females suffered from pellagra. Among the men, mine recruits had by far the lowest incidence. Jackson noted that the average family ate meat every three to six months and that milk was often unavailable.[37] Twenty years later, E.H. Cluver identified the high incidence of scurvy and pellagra as indicative of widespread malnutrition among miners and their home communities. He noted that most families subsisted on a diet gravely deficient in protein and protective vitamins: 'The evidence of this deficiency is there for all to see: The distressingly high incidence in particular of tuberculosis in our homelands. Tuberculosis is a true deficiency disease.' Many young men who appeared fit when recruited suffered from an underlying malnutrition, and their poor health only became apparent

[36] *Report of the Witwatersrand Mine Native Wages Commission*, pp. 11; 19–20; 24.

[37] J.H. Jackson. 'Malnutrition in the Native in the Transkei'. *The South African Medical Journal*, Vol. 26, No. 5, 21st June 1952, pp. 501–502.

once they entered the mines. 'Tuberculosis was of course essentially a food-deficiency disease; the prevalent tubercle bacilli provided the trigger.'[38]

The famines of 1912, 1927 and 1946 in the Eastern Cape were not lethal in the sense that few people actually died of starvation. But they were damaging to public health as growing numbers of men, women and children in the Reserves became malnourished. A diet consisting mostly of carbohydrates meant that individuals were unable to adequately grow, lactate, work, bear a child or recover from illness.[39] Diana Wylie calls the pervasive experience of malnutrition 'starving on a full stomach'. The cause was the lack of land alongside mine wages which were insufficient to sustain a family. During the twentieth century the percentage of household budgets British families spent on food steadily declined, while among black South Africans those expenditures remained constant. Between 1948 and 1983, they stagnated at around 34 per cent, reflecting high food costs and low incomes.[40]

While a number of reports, such as the ones cited above, attributed malnutrition to political economy and the balance between population and land, a powerful public discourse located it instead in the cultural deficits of African families. By the 1940s, the components of what Diana Wylie terms the *malnutrition syndrome* entered official discourse to explain why Africans so readily fell victim to tuberculosis. African families, it was argued, were poorly nourished because mothers and wives were incapable of preparing or using food properly. That problem was compounded by the careless over-stocking of land which damaged the soil and reduced incomes. Some South African medical specialists promoted that orthodoxy, and it was used by governments to characterise black poverty as a cultural trait rather than the result of Pretoria's political and economic policies.[41]

---

[38] Dr. E.H. Cluver. 'Development of Health Services on the Gold Mines'. First A.J. Orenstein Memorial Lecture 18th September 1974, pp. 3; 21. Adler Medical Museum, University of the Witwatersrand.

[39] Diana Wylie. *Starving on a Full Stomach: Hunger and the Triumph of Cultural Racism in Modern South Africa.* Charlottesville and London: University Press of Virginia, 2001, pp. 9–10.

[40] Wylie, *Starving on a Full Stomach*, pp. 89; 36.

[41] Wylie, *Starving on a Full Stomach*, pp. 198–9; 239.

The Diagnosis and Treatment of Tuberculosis After the Second World War

In the immediate post-war period, the official mortality rates for tuberculosis showed dramatic variations between different racial groups. The data on blacks was confined to the urban areas. In 1947, the estimated annual mortality rate per 100,000 in the industrial cities of Cape Town, Port Elizabeth, East London, Durban and Johannesburg was 579.7 per 100,000.[42] The official estimated annual death rate in the Reserves of 100 per 100,000 was acknowledged as totally unreliable. In 1950, the annual rate for Europeans was 23 per 100,000, while among Coloureds it was 387.[43] Access to health care fell along the same racial lines.

In 1950, the Department of Health issued a schedule for treating tuberculosis. The new generation chemotherapy had come on stream. Simple primary infection should be given a standard regime for one year and progressive primary cases for two years. Chronic cases which had been diagnosed late should be segregated in sheltered settlements and treated, in some instances for the patient's lifetime.[44] All cases should be seen by a consultant, with intervals of eight weeks being optimum. Housewives, and in particular those with young children, should receive special consideration, as should teachers and nurses. Sputum testing and X-ray examinations were provided free by the Health Department. Under the Public Health Act, local authorities were required to conduct case-finding programmes using mass X-ray surveys, keep a register and follow up reported cases. They were also to provide an adequate number of isolation beds for acute cases and facilities for out-patient care. The known risks factors for infection included malnutrition and a lack of basic health knowledge. 'When it is remembered that "environment constitutes the real headquarters of tuberculosis"', a 1955 circular stated, 'it is clear that

[42] Letter from Smit, Secretary for Health, to Institute of Pathology, Pretoria, 5th June 1952. 'Tuberculosis Death Rates', SANA, GES 10092 401/17 TB General Correspondence 1937–1952.

[43] Letter from Smit, Secretary for Health, to Professor J. Barnetson Institute of Pathology, Pretoria, 5th June 1952. 'Tuberculosis Death Rates'. SANA, GES 10092 401/17 TB General Correspondence 1937–1952.

[44] Treatment of Tuberculosis, Departmental Circular No 3, undated 1950, p. 8 SANA GES 1090 401/17 TB Conference 1950s.

measures directed against these factors are in fact the most important in the fight against tuberculosis.'[45] The care of the family while the breadwinner was incapacitated was viewed as an essential aspect of treatment, and social assistance was available through invalidity and maintenance grants to European patients. Virtually none of that was done for black miners and their families.

The Secretary for Health, Dr. Le Roux, complained that in the Reserves many hospital beds were filled by patients who could not be discharged because there were no facilities to continue their treatment. Local authorities would not accept financial responsibility, while District Surgeons lacked the capacity to provide care, placing such patients in danger of relapse.[46] The Provincial, Municipal and Mission Hospitals were widely dispersed and in many cases a patient had to make a journey of 160 kilometres for treatment. In contrast to the public health sector, the mines had X-ray equipment, hospitals with specialist staff to provide treatment and they collected data.

Who Should Pay?

All the parties engaged in formal debates about occupational health agreed that the treatment of tuberculosis should be based on case finding and that in this respect the mines did far better than the state. From that point there was little agreement between the Chamber and the health authorities about preventing the spread of infection. One of the most contentious issues was who should pay the costs of detention and treatment of black miners identified with lung disease at an initial medical examination. In the debates, the Chamber consistently argued that tuberculosis was spread from the countryside into the mines; others, including the South African Secretary for Health, blamed the mines for the spread

[45] Circular No. 8 of 1955 401/17 from Department of Health to all Local Authorities and Magistrates in the Union, 1st September 1955. Subject: Tuberculosis Control, pp. 2–3. SANA, GES 1093 401–17, TB Conference, 1955.

[46] Letter from Dr. Le Roux, Secretary for Health, to The Chief of the Division of Tuberculosis Services, Durban, 21st March 1950, 'Post Institutional Treatment of Tuberculosis Patients'. SANA, GES 10092 401/17 TB General Correspondence. 1937–1952.

of the disease. Following their distinct logic, the NRC fought to ensure that the costs of providing patient care, education and notifications to local health authorities were shifted onto the public purse. Much less successfully, several representatives of state authorities argued that the mines had both the duty and resources to deal with the issues.

The case of a migrant worker named Headman Ntlandl who became ill while working on the Luipaardsvlei Mine in 1952 illustrates the dynamic at play. The miner was found to be suffering from tuberculosis and was awarded compensation of £98. His repatriation was ordered, and he elected to return home to Cradock in the Eastern Cape.[47] Ntlandl was acutely ill and on arrival in Cradock he was sent to the Municipal Isolation Block. In October he was discharged, and the Municipality demanded £80 from the NRC for his accommodation. The NRC settled the account on the understanding that under Act No. 57 of 1935 it would in turn receive a refund of seven-eighths of the costs from the Department of Health. According to the NRC, its liability for Ntlandl ceased when it paid his compensation. That story was repeated numerous times.

Although it believed the NRC should pay, the cost of hospitalisation and treatment of Republic recruits with tuberculosis was reluctantly covered by the Department of Health.[48] From July 1958, the costs of detaining South African miners with tuberculosis became the responsibility of the Bantu Administration. As from 1 March 1960, those costs were borne by the Department of Health.[49] The underlying logic behind that allocation of cost was that tuberculosis was brought to the mines from the rural areas rather than being acquired as the result of exposure to silica dust and infected co-workers.

---

[47] Letter from R. Gordon Turner, District Superintendent, to Native Recruiting Corporation Ltd., Queenstown, 13th November 1952. SANA, GES 1106, 409/178, 1930–1953, TB in Mine Natives & Native Labourers.

[48] Minute signed L.J. Lemmer, Secretary for Bantu Administration and Development to The Chief Bantu Affairs Commissioner, Johannesburg, 3rd October 1953. Subject: Pneumoconiosis Compensation Act 1962: Costs of Detention of Bantu Persons. SANA, TES 1002, 5/352/2/1, Mines: The Pneumoconiosis Compensation Act 1962 Expenditure.

[49] Memorandum entitled Pneumoconiosis Compensation Act of 1962: Costs of Detention of Bantu Labourers, dated 10th January 1964, p. 2. SANA, TES 1002, 5/352/2/1, Mines: The Pneumoconiosis Compensation Act 1962 Expenditure.

In 1953, the Department of Health began a tuberculosis control campaign. It was essential, it argued, that the Chamber agrees to keep patients in mine hospitals until they were no longer infectious. In April of the following year, the Secretary for Health wrote to the Secretary for Native Affairs about repatriations. It had long been the practice of the WNLA to repatriate miners with pulmonary tuberculosis. That practice arose because such cases required prolonged treatment, often lasting years, which was seldom effective.[50] The introduction of modern chemotherapy had changed the treatment outcomes dramatically, especially for early cases. Given the mines' regular medical and radiological examinations, in most cases tuberculosis was likely to be discovered in its earliest stages. If such cases were given adequate treatment by employers, there was every chance that patients would return home non-infectious. The Secretary pointed out that since all black mine workers were medically and radiologically examined before being declared fit for mine work, it was obvious that among active miners, in the great majority of cases, the disease was contracted during their employment or was attributable to the nature of mine work. If all these cases were treated until the disease was arrested, it would have a beneficial effect upon the Native areas, in that focal points of infection would no longer be seeded throughout the Territories.

The issues came to a head during a high-level meeting held at the end of January 1955. The meeting, convened to discuss tuberculosis, was held at the office of the Chief Native Commissioner, Mr. Smuts. Those present included senior staff from the Native Affairs Department; the Chief Native Commissioner; Dr. van Rensburg from the Department of Health; the Superintendent of the TB Hospital in Durban, Dr. Dormer; and the Superintendent of the TB Hospital in Umtata, Dr. Wyles. The Under-Secretary of the Department of Mines, Mr. Nel, and the Chairman of the Silicosis Medical Bureau, Dr. du Plessis, were also present. The Chamber was represented by Mr. James Gemmill and Dr. Orenstein. It was an important meeting which shows the various factions at play. The two

[50] Letter from the Secretary for Health, Pretoria, to The Secretary for Native Affairs, Pretoria, 12th April 1954. Subject: Tuberculosis Sufferers Repatriated by the Witwatersrand Native Labour Association. SANA, 1955, NTS 962 21/141 Native TB Patients Repatriation.

major issues were the limited public health resources and the mines' refusal to pay the cost of treatment.

Dr. van Rensburg from the Department of Health told the meeting that because it was impossible to treat all cases in hospitals, a home-treatment programme had begun. His Department was being informed of repatriations and was in turn advising local authorities. Steps were then taken to treat men at home. During the past sixteen months, an average of 68 tuberculosis sufferers were repatriated per month. Of those, around a third had positive sputum.[51] However, the system had begun to break down. Local authorities were not prepared to accept responsibility as the initial diagnosis had been made outside their area of jurisdiction. Consequently, they were unable to recover any portion of the treatment cost from the government. The Department of Health was anxious to know to what extent the Chamber was prepared to provide treatment for men before they were repatriated.

The Superintendent of the Tuberculosis Hospital at Umtata, Dr. Wyles, pointed out that just over 2 per cent of the Transkei population was suffering from tuberculosis and it was impossible to provide hospital accommodation for the estimated 25,000 cases. Domiciliary treatment was impracticable. Mr. Tindale from the Chamber conceded that the mines repatriated infective men to the Reserves but added that the incidence of tuberculosis on the mines was only one-half of that of the civilian population. If the industry was to embark on an expensive scheme, something should also be done for sufferers outside the mines.

The Chief Native Commissioner, Mr. Smuts, asked whether the Chamber would allow tuberculosis cases to be kept and treated in mine hospitals for longer periods. Dr. Orenstein replied that 1200 cases would have to be dealt with annually as *Foreign Natives* would have to be treated on the same basis as *Union Natives*. That would require approximately 350 dedicated beds. There were also other problems to address. Once they had been compensated, sick miners wanted to return to their homes. Orenstein pointed out that powers under Section 8(2) of the Contagious

[51] Minutes of Meeting held at the Office of the Chief Native Commissioner, Johannesburg, on 21st January 1955, with Representatives of the Departments of Native Affairs, Health, Mines and the Chamber of Mines in Connection with the Treatment of Mine Labourers Suffering from Tuberculosis, p. 1. SANA, 1955, NTS 962 21/141 Native TB Patients Repatriation.

Diseases Act would have to be invoked to force infected men to remain in hospital. However, Orenstein did concede that it was wrong to repatriate positive-sputum patients to the Reserves.

Dr. du Plessis, Chair of the Silicosis Bureau, said that the WNLA was installing new apparatus to X-ray all recruits, and this meant it could take steps to treat patients. Dr. Orenstein replied that while it was not the Chamber's responsibility to certify tuberculosis sufferers, it was willing to take medical histories and X-rays. Mr. Tindale, of the Chamber, stated that all new recruits were examined before they were employed and that only a very small percentage of infective cases were missed. In addition, all black miners were examined on discharge.[52] Mr. Tindale wanted to know whether, as the government subsidised local authorities, a similar scheme could be worked out to assist the mines, for instance, by providing drugs. Mr. van Rensburg, from the Department of Health, pointed out that responsibility for treatment rested with local authorities and not the Union Government. Employers were responsible for the treatment of their labourers and the seven-eighths of the government subsidy could only be paid to local authorities and not to industries. Dr. Dormer, Superintendent of the Tuberculosis Hospital in Durban, said that the local authority was responsible for the treatment of infectious cases. His Department requested that the mines treat patients until they were no longer infectious. After that the local authority would accept responsibility.

The meeting took place almost five years after effective treatment, which should have completely changed the management of tuberculosis on the mines, became available. This did not happen. Rather, what changed was the increased visibility of the mines' determination to externalise costs. The only major outcome from the meeting was that the Health Department agreed to consider what financial or other assistance could be given to the Chamber, should it decide to treat patients. In August 1958, the Gold Producers' Committee agreed to treat Union miners diagnosed with tuberculosis in mine hospitals. Owing to limited

[52] Minutes of Meeting held at the Office of the Chief Native Commissioner, Johannesburg, on 21st January 1955, with Representatives of the Departments of Native Affairs, Health, Mines and the Chamber of Mines in Connection with the Treatment of Mine Labourers Suffering from Tuberculosis, pp. 1–2. SANA, 1955, NTS 962 21/141 Native TB Patients Repatriation.

hospital accommodation, however, the period of treatment was restricted to a maximum of six weeks. A tariff of £1 per patient per day was to be charged to the state during the period of treatment.[53]

After consultation between the Department of Health and the Chamber of Mines, a new scheme for treatment began in March 1960. While, in principle, the Department of Health retained the right to decide when a miner should be discharged, in practice mine medical officers made that decision. While every effort was to be made to provide treatment for the requisite six-week period, the Department accepted that in some cases a lack of accommodation or some other special circumstances might prevent that from happening. A tariff of £1 per man per day was charged to the Department by the mining companies. Where a miner had been employed for six months or more prior to diagnosis, the Department of health accepted full liability for the cost of hospitalisation. Where the period was less than six months, the local authority was liable. In such cases, the Department paid the full cost of treatment in mine hospitals and thereafter recovered one-eighth of that amount from the local authority. The Department's liability ended on the day the miner was discharged from the hospital.[54]

The South African population was diverse, but in each region there were similar problems concerning case finding and treatment. A typical review of treatment regimens was written in June 1957 by the Medical Officer at Glencoe, a coal mining town in Western Natal. Tuberculosis was a serious problem in the District. There was little follow-up of cases and the Local Health Inspector could not cope with the workload. Radiological examinations were done at the Dundee Hospital, but in general District Surgeons had no access to such technologies. African patients were willing to enter hospital provided they were informed why prolonged treatment was necessary. Dr. J.K McKechnie seldom received notification that a patient was discharged, and there was no follow-up

[53] Letter from Assistant General Manager, Gold Producers' Committee, Johannesburg, to The Acting Secretary for Health, Department of Health, Pretoria, 9th August 1958. SANA, GES 412/17A, Health Department, Miners' Phthisis Regulations, Pneumoconiosis Act.

[54] Letter from P.J. Botha for Secretary for Health, to Chief Regional Health Officer, Cape Town; Durban; Pietersburg; East London; Johannesburg; and Bloemfontein, 20th January 1960. SANA, GES 412/17A, Health Department, Miners' Phthisis Regulations, Pneumoconiosis Act.

regarding treatment. 'One is often shocked to see patients discharged from hospital with open cavities.' There were no facilities for rehabilitation, and it was bad practice for a patient to go straight from a hospital bed to manual labour, which many did. Out-patient treatment was poorly organised. Attendances were irregular and there was no guarantee that medication was taken. In addition, Dr. McKechnie had experienced delays in obtaining drugs. The other major problem was the low fee paid to District Surgeons, which was a disincentive to treating such patients. At the outset McKechnie had received £7 10s per month, but that had been reduced to £2 10s, while the out-patient attendances had increased. In order to improve case finding, Dr. McKechnie proposed that local doctors be paid the notification fee of £2 6 s per patient as prescribed by the Public Health Act, without having to submit an itemised account.[55] It would also be helpful, he suggested, if the Public Health Department provided tuberculin testing kits for children.

The District Surgeon in a rural part of Natal described similar problems. Tuberculosis control depended upon initial hospitalisation and the continuation of treatment at out-patient centres or clinics. When the scheme was first introduced at Weenen in rural Natal, four beds were reserved at the Haviland Memorial Hospital. But as soon as a patient felt better, he would leave hospital. Invariably the disease would flare up, and he would die at home. The first problem, therefore, was to convince patients that treatment worked. Haviland had sixty to seventy tuberculosis patients at any time. The District Surgeon felt that unless the initial hospital treatment was thorough and of sufficient duration, the whole programme would be a failure.[56] An undernourished patient in poor condition, who had to walk many miles in all sorts of weather to a clinic, did not benefit from out-patient treatment. At best, his death was merely delayed. There was a very high percentage of pulmonary tuberculosis cases in the Weenen population. Of the estimated 30 new cases each month, very few were prepared to enter hospital. Nor were they

[55] Commentary on Tuberculosis Control from Viewpoint of District Surgeon and Medical Officer of Health for Glencoe, by Dr. J.K. McKechnie, D/S and MOH, Glencoe, undated, but later than June 1957. SANA, GES 1091 401/17, TB Conference, 1957.

[56] Letter from District Surgeon, Weenen, Natal, to Magistrate, Weenen, 20th September 1957. Subject: 'Tuberculosis Control'. SANA, GES 1091 401/17, TB Conference, 1957.

encouraged, as there was no accommodation. The District Surgeon at Weenen felt that the control programme was probably only of value to a very small proportion of cases. The majority remained undetected and were spreading tuberculosis. The lack of hospital beds was a serious barrier to the scheme's success.

The position was much the same at Vereeniging in the Gauteng province, where it was extremely difficult to exercise control over cases who were treated on an out-patient basis.[57] The advantage of keeping the patient at home was that he could be supplied with food, but often the food was shared by the rest of the family. Where the patient was a breadwinner and was in a hospital, the family often starved. The Medical Officer of Health at Kimberley added another common theme. The incidence of tuberculosis, he noted, was a true measure of the social and economic conditions. The tuberculosis notification and death rates in the local European population were amongst the lowest in the world, while the rates in non-European were amongst the highest. The number of deaths at Kimberly was falling but the morbidity rates were not. Mass radiography was not being used by the Board of Health.[58]

The Consolidation of an Imperfect System

The flurry of Miners Phthisis legislation in the first half of the twentieth century and the creation of the Bureau in 1916 gave South Africa and its gold mines a reputation as world leaders in tackling occupational disease. Yet many lingering problems on the mines remained—as did the mining houses' intransigence in face of criticisms. The problems included overcrowded compounds, the failure to diagnose lung disease and the management of tuberculosis. In the early 1920s, for example, the Van Rhyn Deep Compound employed an average of 3697 workers per month, even though it had compound accommodation for only 3238 men. In 1921,

[57] Letter from Dr. R. Bernstein, Medical Officer of Health, Public Health Department, Vereeniging, to Chief Regional Health Officer, Union Department of Health, Johannesburg, 3rd October 1957. Subject: 'Tuberculosis Control'. SANA, GES 1091 401/17, TB Conference, 1957.

[58] Dr. H.M. Wolfsohn, Medical Officer of Health, Kimberley, 'Tuberculosis in Kimberley', October 1957. SANA, GES 1091 401/17, TB Conference, 1957.

the Witwatersrand G.M. compound had accommodation for 3250 men yet employed 3800. While the majority were squeezed into the compound room, another 150 slept in a store room and a further 40 in the engine room. Such overcrowding was common.[59] When Dr. E.H. Cluver from the Public Health Department carried out a survey of the Witwatersrand mines in 1928, he found the unsanitary conditions on some mines 'deplorably short sighted' and 'unpardonable' in view of the previous investigations into the compounds and the feeding of miners. Cluver concluded that a central health organisation for the whole industry, whose establishment was originally raised by General Gorgas, would be more efficient in terms of both miners' health and the cost to the industry.[60] It would also help overcome problems arising out of the poorly defined and often overlapping jurisdictions. All the medical officers Cluver spoke with favoured a centralised system.

Cluver acknowledged that 'it is often quite impossible to detect early silicosis by the ordinary clinical examination carried out on boys about to return to their homes'. He was sure that as a result, many migrant workers left the mines with undiagnosed silicosis and without receiving the compensation to which they were entitled. 'The frequency with which natives coming to Johannesburg are found to be suffering from phthisis after having been employed on the mines indicates that a considerable number of natives must leave the mines at the completion of their contract period with unrecognised but compensable silicosis.' As a solution, Cluver suggested that the initial entry examination in the Native Territories be supervised by an officer from the Bureau. 'At this examination recruits found to be suffering from phthisis should, if it appears that they had previously worked on the mines, undergo further examination with a view to compensation.'[61]

[59] Julie J. Baker, 'The Silent Crisis': Black Labour, Disease, and the Economics and Politics of Health on the South African Gold Mines, 1902–1930. PhD Thesis, Queen's University, Ontario, 1989, pp. 162–163.

[60] Minute from Dr. E.H. Cluver, Assistant Health Officer, Department of Public Health, to the Secretary for Public Health, Pretoria, 4th December 1928. Subject: Health Conditions on the Witwatersrand Gold Mines. Desirability of Central Health Control for Whole Industry. SANA, GES 934708/13 Sanitary Conditions on Gold Mines, Correspondence GPC 1926.

[61] Letter from Dr. E.H. Cluver, Assistant Health Officer, Union Health Department, Johannesburg, to the Secretary for Public Health, Pretoria, 18th May 1933. Subject: Phthisis Return, Pass Office, Johannesburg. SANA, GES 1106, 1930–1953, TB in Mine Natives & Native Labourers.

The Secretary for Public Health, J.A. Mitchell, supported Cluver's findings and raised the issue with William Gemmill. Gemmill in turn referred the matter to the GPC. The Committee conceded that while the Department had the right to draw attention to any problem regarding hygiene, each Group of mines was at liberty to adopt whatever administrative measures it preferred. The current annual mortality rate from disease, the Committee emphasised, was around 12 per 1000, a figure which compared favourably with any industrial enterprise employing Coloured peoples anywhere in the world. In the Committee's view, Mitchell's criticisms were grossly insulting to the industry,[62] and Dr. Cluver's criticisms were not based on any evidence.[63] Orenstein also strongly opposed a centralised system. Presumably, this was because the costs saving flowing from reduced duplication would have been outweighed by depriving the Chamber and GPC of control over issues such as compensation rates and data collection.

Like all aspects of mine medicine, the use of technologies was racialised. W. Watkins-Pitchford published his final paper on the mines and dust in August 1929. He noted that the efforts to reduce the incidence of miners' phthisis in South Africa and the UK were largely confined to reducing the levels of silica exposure. He regretted the lack of effort made to identify and segregate those with infective tuberculosis within working populations.[64] The key factor in controlling the disease was the exclusion of tubercular workers by efficient initial and periodic medicals. Watkins-Pitchford advocated the use of X-rays in place of the stethoscope as the key diagnostic tool, and recommended post-mortem examinations as an important source of data. A number of his colleagues were equally critical of mine medicine. In 1942, the Stratford Commission reviewed the problem of tuberculosis. Dr. E.H. Cluver, who served on the Commission,

---

[62] Letter from O. Christerferson, President of the GPC, to the Minister of Public Health, 7th February 1928, SANA, GES 934708/13 Sanitary Conditions on Gold Mines, Correspondence GPC 1926.

[63] Letter from William Gemmill, General Manager, GPC, to J.A. Mitchell, Secretary for Public Health, Department of Public Health, Pretoria, 12th April 1929, p. 7. Subject: Health Conditions on the Witwatersrand Gold Mines—General Report on Compounds and so on. SANA, GES 934708/13 Sanitary Conditions on Gold Mines, Correspondence GPC 1926.

[64] W. Watkins-Pitchford. 'The Relationship of Dust and Pulmonary Tuberculosis'. *The Journal of the Medical Association of South* Africa, Vol. 3, No. 16, 24th August 1929, p. 453.

pointed out that there has been no improvement in the incidence of tuberculosis among employed black miners over the previous ten years, while the rate among European miners had fallen sharply. Cluver suggested that the widespread malnutrition in the Native Territories rendered many recruits susceptible to disease. He also attributed that result to the inadequacy of the initial examinations, which failed to exclude miners with active TB from employment.[65]

Criticism of mine medicine was driven by senior state medical officers such as J.A. Mitchell, E.H. Cluver and Peter Allan. It was also often addressed at the numerous Miners Phthisis Commissions. At these forums the Chamber, often represented by A.J. Orenstein, used the official data from the Bureau to argue, successfully, that the mines were in fact safe. Two other issues were not so easily resolved: the mines' policy of repatriating sick and dying miners, and its failure to provide local health authorities with notifications. As detailed in the next chapter, this problem was most significant in the British High Commission Territories.

In South Africa itself, by the early 1980s more than 40,000 new tuberculosis cases were notified annually, but that was understood to represent only a proportion of the actual numbers. One of the major problems was that many patients only sought treatment very late in the disease process due to a fear of losing a job and other economic pressures. For the same reasons, those in treatment often failed to complete the programme, and this led to drug resistance. As one commentator noted, a control regime should include a national education programme concentrating on areas of high prevalence. Every sufferer should know the fundamental facts about the disease, its treatment and how to avoid infecting others.[66] After decades of repeated warnings, South Africa still fell short of these aims. It was only with the settlement of the miners' class action in 2019 that a workable compensation and treatment regime, largely funded by the mining industry, was put into place.

[65] Minutes of Evidence, Miners' Phthisis (Stratford) Commission, Tuesday, 24th February 1942, p. 390. SANA, K105, Miners' Phthisis 1941–48, Vol. 2 Correspondence: Transcripts, Stratford Commission.

[66] S.R. Benatar. 'Tuberculosis in the 1980s, with particular reference to South Africa'. *South African Medical Journal,* Vol. 61, 4th September 1982, pp. 363–64.

References

Allan, P. 'The Treatment of Pulmonary Tuberculosis from the Public Health Aspect'. *The South African Medical Journal,* Vol. 13, No. 9, 13th May 1939, pp. 317–319.

Allan, P. 'The Tuberculosis Problem in the Union of South Africa'. *The South African Medical Journal,* Vol. 13, No. 21, 11th November 1939, pp. 740–742.

Allen, V.L. *The History of the Black Mineworkers in South Africa: Vol. 1, The techniques of resistance 1871–1948*. Keighley: The Moor Press, 1992.

Baker, Julie J. 'The Silent Crisis': Black Labour, Disease, and the Economics and Politics of Health on the South African Gold Mines, 1902–1930. PhD thesis, Queen's University, Ontario 1989.

Benatar, S.R. 'Tuberculosis in the 1980s, with particular reference to South Africa'. *South African Medical Journal,* Vol. 62, No. 11, 4th September 1982, pp. 359–364.

Brock, J.F. 'Malnutrition in South Africa'. *The South African Medical Journal,* Vol. 17, No. 14, 24th July 1943, pp. 219–222.

Feinstein, Charles H. *An Economic History of South Africa: Conquest, Discrimination and Development.* Cambridge: Cambridge University Press, 2005.

Fox, F.W. 'Diet and Health in South Africa: Malnutrition'. *The South African Medical Journal,* Vol. 10, No. 1, 11th January 1936, pp. 25–36.

Fox, F.W. 'Some Nutritional Problems Amongst the Bantu in South Africa'. *The South African Medical Journal,* Vol. 13, No. 3, 11th February 1939, pp. 87–95.

Gale, G.W. 'The Prevention of Tuberculosis in the Union, with Special Reference to Urbanised Natives'. *The South African Medical Journal,* Vol. 17, No. 20, 23rd October 1943, pp. 321–322.

Gear, H.S. 'The Position of General Hospitals in South African Health Organisation'. *The South African Medical Journal,* Vol. 12, No. 17, 10th September 1938, pp. 642–648.

Gear, H.S. 'A Plea for Improved South African Medical and Vital Statistics'. *The South African Medical Journal,* Vol. 11, No. 5, 13th March 1937, pp. 149–154.

Herbst, J.F. 'The Transkeian Territories', *The South African Medical Journal,* Vol. 18, No. 8, 22nd April 1944, p. 127.

'Historical Survey of Medical Research in South Africa'. *The South African Medical Journal,* Vol. 31, No. 27, 6th July 1957, pp. 677–679.

Jackson, J.H. 'Malnutrition in the Native in the Transkei'. *The South African Medical Journal,* Vol. 26, No. 5, 21st June 1952, pp. 501–504.

Marais, D.P. 'The Prevention of Consumption'. *The South African Medical Record*, Vol. 10, No. 3, 10th February 1912, pp. 40–43.
Melville, D. 'Tuberculosis in the Native Territories'. *The South African Medical Record*, Vol. 11, No. 18, 27th Sept 1913, pp. 393–395.
Mitchell, J.A. 'The Problem of Tuberculosis in South Africa'. *The South African Medical Record*, Vol. 19, No. 12, 25th June 1921, pp. 226–227.
Packard, R.M. *White Plague, Black Labour: Tuberculosis and the Political Economy of Health and Disease in South Africa.* Berkeley: University of California Press, 1989.
Paterson, R.L. 'Medical Missions in the Transkei'. *The South African Medical Journal*, Vol. 17, No. 10, 22nd May 1943, pp. 154–156.
Tonkin, A. 'A State Medical Service for Natives'. *The South African Medical Journal*, Vol. 15, No. 15, 9th August 1941, pp. 283–288.
Tonkin, A. 'The Control of Pulmonary Tuberculosis in the Transkei'. *The South African Medical Journal*, Vol. 18, No. 8, 22nd April 1944, p. 128.
Watkins-Pitchford, W. 'The Relationship of Dust and Pulmonary Tuberculosis'. *The Journal of the Medical Association of South* Africa, Vol. 3, No. 16, 24th August 1929, pp. 452–454.
Wylie, D. *Starving on a Full Stomach: Hunger and the Triumph of Cultural Racism in Modern South Africa.* Charlottesville and London: University Press of Virginia, 2001.

8

Tuberculosis and Migrant Labour in the High Commission Territories: Bechuanaland: 1885–1998

Britain acquired the High Commission Territories (HCTs or protectorates) of Bechuanaland (Botswana), Basutoland (Lesotho) and Swaziland (eSwatini) largely as a result of conflict with the Boer Republics around the turn of the twentieth century. The Territories were poor, had dispersed rural populations and few natural resources. Britain was determined to keep the costs of its empire to a minimum. It administered the HCTs on the principle that expenditure should not exceed the revenue obtained through taxation and made little investment in basic services and infrastructure. Generating sufficient revenue was a constant problem. In 1929, for example, administrative costs absorbed 79 per cent of Bechuanaland's total government expenditure.[1] The South Africa Act of 1909 provided for the eventual transfer of all three territories to South Africa so long as *native interests* were protected, and successive South African governments made every effort to promote that transfer. Constitutional advances were slow after the Second World War, and the

[1] J.E. Spence. 'British Policy Towards the High Commission Territories'. *The Journal of Modern African Studies*, Vol. 2, No. 2, 1964, p. 230.

J. McCulloch, P. Miller, *Mining Gold and Manufacturing Ignorance*,
https://doi.org/10.1007/978-981-19-8327-6_8

perceived legitimacy of South Africa's repeated claims that the HCTs were part of the Union diminished with the advent of apartheid.[2]

There was little coordination between the Territories, which were run as separate administrative units. The High Commissioner resident in South Africa had final responsibility for the HCTs. Resident Commissioners controlled the day-to-day administration in each Territory. As heads of government, the governors of the HCTs could initiate policies, share in their implementation and preside over the executive council. In practice those powers were circumscribed by the Colonial Secretary who had to approve the colonies' annual budgets and could veto legislation. In addition, not least because of the need for British policy to conform to ILO conventions, most labour policy was decided in London. Despite those constraints, the slowness of communications and lack of central capacity meant that most governors had a good deal of independence. As late as 1914 the Colonial Office had a staff of less than 140, and of necessity British colonial policy favoured decentralisation.[3]

The High Commissioner's role was politically sensitive as he had to oppose Pretoria's discriminatory policies, resist its demands for the incorporation of the HCTs into the South African Union and maintain smooth relations between Britain and South Africa. Reconciling those interests was made more difficult by the Territories' dependence on selling labour to the mines. In addition to the income from capitation fees paid by the WNLA and the NRC and the repatriation of wages, migrant labour simplified tax collection. Some men paid their tax out of the advance they received in signing on; the arrears of others were collected on pay days by HCT representatives in Johannesburg.[4] The permanent suspension of recruiting would have created serious political and economic problems for the High Commissioner's office. The negative impact of the mines on public health was referred to regularly by medical officers

[2] See David E. Torrance. 'Britain, South Africa and the High Commission Territories: An Old Controversy Revisited'. *The Historical Journal*, Vol. 41, No. 3, 1998, pp. 751–772.

[3] On the day-to-day workings of the empire see L.H. Gann and Peter Duignan. *The Rulers of British Africa 1870–1914*. London: Croom Helm, 1978.

[4] Spence, 'British Policy Towards the High Commission Territories', p. 229.

and district commissioners, but after the ban on Tropical recruiting was lifted in 1938, such references all but disappeared from the Resident Commissioners' correspondence.

To a large extent, the interests of the Resident Commissioners and the Chamber coincided. The HCTs were starved of funds for essential services, and they soon became dependent upon the revenue from contracting labour to the gold mines. Selling migrant labour, however, came at a cost. From as early as 1912, the annual medical reports from the three Territories suggested that the mines were spreading tuberculosis into vulnerable populations. The correspondence from the Departments of Health, Native Commissioners and the Resident Commissioners' Offices reveals that the WNLA and the NRC were doing so without making any effort to reduce the impact of this entrenched practice. In fact, the Chamber's secrecy regarding repatriations was characteristic of recruiting throughout the colonial period. Medical repatriations were one of the obvious costs of a system in which a physical elite travelled south and, having served their contracts, returned home seriously ill. Another point of dispute, discussed in the next chapter, was the WNLA's refusal to pay pensions in place of lump sums to that small number of men compensated for occupational disease.

Taxation and Mining

Bechuanaland (now Botswana), a large territory with low rainfall and much desert, was proclaimed a British Protectorate in 1885, as part of an effort to prevent Boer settlers establishing Afrikaner states in Tswana territory.[5] The Protectorate covered a vast land mass with a rudimentary road and rail system. On European maps, it is divided in half by Latitude 22° South. To help defray the costs of colonial administration, in 1899 a Hut Tax ranging from 5s to 10s per annum was imposed on all adult males, with substantial penalties for non-payment. Chiefs were made tax collectors and received a 10 per cent commission. Since there were few alternative sources of cash, they encouraged men to go to the mines. In

[5] Torrance, 'Britain, South Africa and the High Commission Territories', p. 753.

1904, the Protectorate's major revenue streams consisted of an Imperial Grant in Aid of £15,000 and a Hut Tax of £11,500. Over the years, the Hut Tax was raised. From 1919, Hut and other Native Taxes became the largest components of recurrent revenue.[6] As locals tried to cram more people into each hut to minimise their tax liability, overcrowding intensified the risk of TB infection.[7]

In 1899, in order to counter abuses of labour, the Bechuanaland administration issued a Proclamation to regulate recruiting. That was followed in 1907 by a law requiring the attestation of all labour contracts for working *outside* the territory before a government officer. That legislation, which was more progressive than the labour laws operating *within* the Protectorate, gave men a further incentive to go to the mines. In Bechuanaland itself, a version of the draconian Masters and Servants Act of the Cape of Good Hope (No. 15 of 1856) was still in force in 1949. Under that Act all (non-white) employees were required to give notice before leaving an employer, and the abandonment or desertion of employment without notice was an offence punishable by fine or imprisonment.[8] Whatever the legal framework, there were few opportunities in Bechuanaland for wage labour, which in any case offered far lower pay than the Johannesburg gold mines. In 1913, when the Union government imposed a ban on the recruitment of Tropical labour, recruiting to the gold mines was focused on the southern part of the Protectorate. For the following 24 years, Bechuanaland lay at the centre of the Chamber's efforts to have the ban lifted.

The end of the First World War saw an increase in the number of men leaving Bechuanaland in search of work. At the beginning of 1924, the mines extended their normal period of contract from six to nine months. Four years later, the Assisted Voluntary Scheme (AVS) was introduced. Recruits were given an advance to meet their rail fare and on arrival in Johannesburg were free to select an employer. The AVS also enabled men

[6] Dile Delarey Kote, Recruitment of Mine Labour in Botswana 1899–1945, B.A. (History) thesis, University of Botswana, Lesotho and Swaziland, 1976, pp. 20–22.

[7] Packard, R.M. *White Plague, Black Labour: Tuberculosis and the Political Economy of Health and Disease in South Africa.* Berkeley: University of California Press, 1989, p. 102.

[8] See Letter from G. Nettleton, Government Secretary, Mafeking, to the High Commissioner, Cape Town, 15th March 1949. BNA S117/1/3. Labour Native, 1949–53.

to work for a shorter period than the normal nine months' contract. Under the bonus system, men who returned to the mines within four months were taken on again at the same rate of pay as when they left, instead of starting again at the minimum.[9]

The 1921 Bechuanaland census showed a population of 150,185 blacks and 1743 whites. Population density was very low. The low rainfall meant that only one-tenth of the land was suitable for cultivation. Villages congregated where there was water, but it was difficult to cultivate vegetable or fruit gardens. Meat, which at one time was easily obtained by hunting, had become a luxury.[10] The cattle posts and farmlands where milk and fresh fruit were available were so far from the villages that most of the population was malnourished. By 1930, migrants' wages were the most important source of family incomes, and it could be catastrophic for a man to fail the entry medical.[11] In 1938 in the Kweneng District, where several decades later Steen and his colleagues conducted their path-breaking study of occupational lung disease, more than £20,000 in remittances flowed from the gold mines, and nearly half of that sum went to pay the Native Tax.[12] Such heavy reliance on oscillating migration—and the necessity to pay taxes to the colonial administration—transformed the whole country from a self-supporting agricultural economy into a labour reserve.

When, in September 1938, the British Secretary of State asked Colonial governments to consider setting up machinery for regulating and monitoring labour conditions, the Bechuanaland Resident Commissioner declined. Secretary of State MacDonald pointed out that the far-reaching economic and social changes sweeping the empire were bringing

[9] I. Schapera. 'Migrant Labour and Tribal Life: A Study of Conditions in the Bechuanaland Protectorate'. Report Presented to the B.P. Administration, Public Records Office, DO 119 1268 Schapera, 1944, pp. 71–72.

[10] *Annual Medical & Sanitary Report, 1930*. Bechuanaland Protectorate, p. 18. BNA S88/9 Annual Medical Report, 1930.

[11] See Julie Livingston. 'Physical Fitness and Economic Opportunity in the Bechuanaland Protectorate in the 1930s and 1940s'. *Journal of Southern African Studies*, Vol. 27, No. 4, 2001, pp. 793–811.

[12] M. Leepile. 'The Impact of Migrant Labour on the Economy of Kweneng 1940–1980', *Botswana Notes and Records*, Vol. 13, 1981, p. 34.

potential conflict between employers and workers.[13] Those changes made labour departments an important sector of government. In reply, the Resident Commissioner Forsyth Thompson noted that the amount of labour employed *within* Bechuanaland was too small to warrant a special labour department. Besides, the High Commission Territories were unique among the Colonial Dependencies in that practically all their wage labour was employed *outside* their borders in South Africa. Consequently, the general supervision of employment and worker welfare was in the hands of the Union Government rather than the Protectorate.[14]

The impact of increasing labour migration on food production and people's health soon became obvious. In his review of health services in the Protectorate in the mid-1930s, Sir Walter Johnson found the physique of the Bechuana poor compared with that of South Africans. The principal cause was malnutrition. The normal diet was inadequate: 'Besides the lack of protein of good biological value the native of Bechuanaland is living on the verge of vitamin deficiency which shows itself from time to time in outbreaks of scurvy'.[15] From the time a child was weaned, its diet during most of the year consisted almost entirely of mealie meal and millet porridge.[16] With the exception of chiefs and a few headmen, it was difficult or impossible for the bulk of the population to obtain milk.

Initially, transport costs were recovered from the miners, but from 1934 the ILO insisted that travelling expenses be paid by the recruiter or employer. In 1939 the WNLA and NRC agreed to bear the cost of rail fares *to* the mines. However, repatriation costs continued to be paid by the miners. Men from the north were also charged 30s for trousers, a vest and blankets, which were deemed necessary for Tropical recruits.[17] The

[13] Circular signed Malcolm MacDonald, The Dispatch, Downing Street, London, 5th September 1938. BNA, S116/3 Labour Supervision.

[14] Letter from A. Forsyth Thompson, Resident Commissioner, Mafeking, to The High Commissioner. Pretoria, 8th November 1938. BNA, S116/3 Labour Supervision.

[15] *Commission of Enquiry into Medical Matters, 1937*, p. 46. BNA S392/7/1.

[16] See *Annual Medical & Sanitary Report, 1930 Bechuanaland Protectorate*, p. 1. BNA S88/9 Annual Medical Report, 1930.

[17] Pauline Cuzon. The History of TEBA in Botswana. unpublished manuscript, July 1985, p. 72.

Native Labour Proclamation of 1941 No. 56 made written contracts and a medical certificate compulsory.[18] In the period from 1938 to 1942, doctors from the Scottish Livingstone Hospital examined 6829 mine recruits at Molepolole and Thamaga. Of that number, 793 were rejected with chest conditions.[19] From February 1949, all medical examinations of men recruited by the NRC were conducted at Mafeking by the Bechuanaland Government Medical Officer. The first 500 examinations in any year were paid for at the rate of 2s 6d per head, and examinations in excess of 500 at 1s.[20] In Bechuanaland deferred pay was voluntary, but the vast majority of men chose to accept it. In 1937, the proportion of men accepting deferred pay was 89 per cent; in 1943, when the average for the Union and High Commission Territories was just under a half, it reached 95 per cent.[21]

In 1937, the Union Government formally removed the prohibition on recruiting from North of latitude 22^0 South. The WNLA was confident that the 'Tropical Areas' would become its main source of additional labour. Northern Bechuanaland was to supply at least 2600 men per annum who, on their return home, would draw a total of £25,000 in deferred pay. The WNLA anticipated that around 10 per cent of men offering for employment would be rejected.[22] Soon after the ban was lifted, C.N.A. Clarke, who had replaced Charles Rey as the Bechuanaland Resident Commissioner, began receiving complaints from Chiefs that too many men were leaving the Reserves. Like his predecessors, Clarke was aware that his administration relied on revenue from migrant labour, but he was also concerned about the social costs of mining employment.

[18] Proclamation No. 56 of 1941, Bechuanaland Protectorate, pp. 14–15, 20. BNA S114-7-2 Labour, Native Recruitment, Bechuanaland, 1937–44. The Proclamation was based on the ILO's Recruiting of Indigenous Workers Convention, 1936 (No. 50) and the Contracts of Employment (Indigenous Workers) Convention, 1939 (No. 64). See also I. Schapera, 'Migrant Labour and Tribal Life, p. 137.

[19] Letter from Dr P.M. Shepherd, Superintending Missionary, Scottish Livingstone Hospital, Molepolole, to the Principal Medical Officer, Bechuanaland Protectorate, Mafeking, 8th March 1943. BNA S438-2-2 Tuberculosis in B.P. 1940–1953.

[20] Letter from District Superintendent, NRC Ltd, to the Director of Medical Services, Bechuanaland Government, Mafeking, 8th February 1949. BNA Med 15/8. Recruits for Mines, Farms, 1947.

[21] Schapera, 'Migrant Labour and Tribal Life', pp. 140–42.

[22] Notes from the WNLA on Tropical Areas Administration, 21st July 1945, W. Jones, District Manager, Francistown, Northern Bechuanaland. BNA S344/10/1, Recruiting north of Parallel 22.

A large proportion of teenage males were at the cattle posts during their school-going years. As soon as they left the posts, many went to work in the mines. As a result, the schools were filled with girls, and a large proportion of young men were illiterate. Clarke wanted a compulsory system of deferred pay to ensure men returned home at the end of their contracts. He also wanted educational facilities for workers at their place of employment and a formal agreement on the repatriation of medical rejects. He hoped to persuade the NRC and WNLA to concentrate on certain specified areas and thus make it easier to monitor the social impact of recruiting.[23]

Tuberculosis was the other subject which was often discussed by the Bechuanaland Native Advisory Council. At a meeting in March 1939, for example, Chief Bathoen (Bangwaketse) noted that many young men were repatriated from the mines with the disease. He pointed out that the number of recorded cases only referred to hospital patients rather than to all of those who were infected. People were suffering, but patients who could not be looked after at home were being rejected at the hospitals.[24] There was a marked increase in admissions during 1939, with Mission Hospitals bearing much of the burden. However, in the absence of isolation wards, tuberculosis cases were advised to live on the lands and not in the villages.[25] Bathoen asked if the government could provide special care at Lobatse and Serowe. S.J. Molema (Barolong) agreed that the tuberculosis figures were alarming. Like Chief Bathoen, he wanted isolation wards where cases could be treated. Lot Moswele (Batlokwa) commented that 'in our childhood days no such disease was known amongst us'. The Principal Medical Officer agreed that the official figures for tuberculosis did not give an accurate picture, as many patients were not reporting to local Medical Officers. While on the mines, migrant workers were fed well, but when they returned home there was the problem of malnutrition. He explained that the government wanted to conduct a survey

[23] Letter from C.N.A. Clarke, Resident Commissioner, Mafeking, to the High Commissioner, Cape Town, 8th December 1937, pp. 2; 5–6. BNA S114/7/1 Labour Recruitment, Bechuanaland.

[24] Extract from Minutes, 20th Session Native Advisory Council Held 6th to 10th March 1939, pp. 76–77. BNA S438/2/1 Tuberculosis in the B.P., 1938–39.

[25] *Annual Medical and Sanitary Report 1939*, Bechuanaland Protectorate, p. 4. BNA, S397/1/1 Annual Medical Reports 1939–1945.

before it invested in prevention and care. He promised to get more information from the WNLA on those repatriated with tuberculosis, so they could be traced.[26]

With the outbreak of the Second World War the competition for labour intensified. The WNLA was pressing for a higher quota, while the Resident Commissioner was anxious to preserve local food production. In order to restrict the flow of men to the mines, Clarke initiated Proclamation No. 56 of 1941, which placed a number of conditions on cash advances, the recruitment of agricultural workers and repatriations.[27] Before the Proclamation came into effect, Mr Gordon Turner of the NRC and Bechuanaland Resident and Assistant Resident Commissioners had met in Mafeking to discuss its likely impact. Clarke acknowledged that the Territory relied on revenue from miners and that it was vital that such revenue continue. The NRC was paying the government a capitation fee of 24s per head, and any curtailment of recruiting after the war would have to be gradual because of the impact on the Territory's economy.[28] Following a heated discussion, Clarke made a number of concessions. He was prepared to grant exemption from the requirement that recruits should not stay away from home for more than 18 months. In contradiction of ILO conventions, Clarke also agreed that the NRC and the WNLA would be exempt from paying the recruits' travel costs. By December 1942, it was estimated that 15,000 men were working on the Rand and probably another 20,000 on South African farms and in other South African industries. That represented well over a half of the Protectorate's 60,000 (male) taxpayers, and according to the Government Secretary, the labour exodus was disrupting family life.[29]

The one area in which the Bechuanaland administration refused to cooperate with the WNLA and the NRC was over their demands to

---

[26] Extract from Minutes, 20th Session Native Advisory Council, pp. 75–78.

[27] Proclamation No. 56 of 1941, Bechuanaland Protectorate, pp. 14–15, 20. BNA S114-7-2 Labour, Native Recruitment, Bechuanaland, 1937–44.

[28] Notes of Discussion at Mafeking on 10 October 1941, between the Resident Commissioner, the Assistant Resident Commissioner and Mr Gordon Turner of NRC to consider aspects of Proclamation No. 56 of 1941. BNA. S114-7-2, 1937–44, Labour, Native Recruitment, Bechuanaland.

[29] Memo from the Government Secretary to A.D. Forsyth Thompson, Resident Commissioner, Mafeking, 5th December 1942. BNA S387/5 Recruitment of Labour, 1942.

siphon off mine rejects to white farms in Southern Rhodesia. In April 1941, William Gemmill wrote to the Resident Commissioner requesting the WNLA be allowed to recruit mine rejects for Southern Rhodesian tobacco farms. In return, the WNLA would establish a system of deferred pay and compulsory repatriation.[30] The Bechuanaland government rejected Gemmill's request on the grounds that the Protectorate's manpower was already insufficient for local food production.[31] Soon after the war, the WNLA again requested permission to forward reject labour to Southern Rhodesia.[32] The WNLA pointed out that the local population was facing a famine and men either remained at home, aggravating the situation, or wandered in search of work. The WNLA's Francistown office had seen unprecedented numbers seeking recruitment, and there was the usual high proportion of rejects. The District Commissioner at Francistown supported the WNLA request, citing the drought and the prospect of crop and stock losses.[33] The Government Secretary was unsympathetic. Although there was a crisis, the wages offered in Southern Rhodesia were too low to support a family.[34] The Resident Commissioner agreed.[35] In a subsequent review, the Government Secretary wrote: 'In plain fact it is that we do not want to encourage the recruitment of labour

[30] Letter from William Gemmill, Witwatersrand Native Labour Association Ltd., Salisbury, to Lt.-Col. C.N.A. Clarke, Resident Commissioner, Mafeking, 15th April 1941. BNA S344-11. Recruiting of Natives, Suggested Scheme for Rejects, 1941–44. For a discussion of the complex political background see Alan Jeeves. 'Migrant Labour and South African Expansion, 1920–1950'. *South African Historical Journal*, No. 18, 1986, pp. 73–92.

[31] Letter from Government Secretary, B.P. Government, to W. Gemmill, General Manager, Witwatersrand Native Labour Association Ltd., Salisbury, 17th June 1942. BNA S344-11 Recruiting of Natives, Suggested Scheme for Rejects, 1941–44.

[32] Letter from District Manager, Witwatersrand Native Labour Association Ltd., Francistown, to The Government Secretary, B.P. Government, Mafeking, 8th April 1947. BNA S344-11. Recruiting of Natives, Suggested Scheme for Rejects, 1941–44.

[33] Letter from Acting District Commissioner, Francistown, to The Government Secretary, Mafeking, 7th May 1947. Subject: Recruiting of Agricultural Labour for Southern Rhodesia. Recruiting of Natives, BNA S344-11 Suggested Scheme for Rejects, 1941–44.

[34] Letter from District Commissioner to The Government Secretary, Mafeking, 30th May 1947. Subject: Recruiting of Agricultural Labour for Southern Rhodesia. BNA S344-11, Recruiting of Natives, Suggested Scheme for Rejects, 1941–44.

[35] Letter from G.E. Nettelton, Government Secretary, Resident Commissioner's Office, Mafeking, to The District Manager, Witwatersrand Native Labour Association Ltd, Francistown, 4th July 1947. BNA S344-11, Recruiting of Natives, Suggested Scheme for Rejects, 1941–44.

in the Bechuanaland Protectorate for Southern Rhodesia in any way'.[36] The government used the same reasoning to reject subsequent requests from the WNLA.[37] Clarke's resistance is easily explained. Employment on white-owned farms in Southern Rhodesia was the last resort for migrant workers whose health was in decline after extended periods underground. The pay was at best less than a third of that on the gold mines, and white farmers had a well-deserved reputation for violence and the non-payment of wages.[38]

High-Level Negotiations About Mine Labour After the Second World War

After the Second World War, the demand for mine labour increased again, fuelled in part by the opening of the new Free State Mines in South Africa. Low mine wages tied men and their families to a perpetual cycle of migrancy. For the administration, repatriated wages under the deferred pay scheme enabled gold miners to pay their tax and to meet their family obligations. In that sense at least, the interests of the administration and the mines coincided. From the perspective of local communities, the situation looked worse. In January 1947 Bathoen II, Paramount Chief of Bangwaketse, wrote to the Resident Commissioner about migrant labour. Bathoen pointed out that migration gave the mines access to a large pool of skilled men and made families dependent on their income. The system of voluntary deferred pay was supposed to induce men to return home, but the mines also encouraged miners to

[36] Letter from G.E. Nettelton, Government Secretary, Resident Commissioner's Office, Mafeking, to Mr Bent, District Commissioner, 20th December 1947. BNA S344-11, Recruiting of Natives, Suggested Scheme for Rejects, 1941–44.

[37] Letter from Acting District Manager, WNLA Ltd., Francistown, to the Government Secretary, B.P. Government, Mafeking, 1st July 1952. Subject: A.N.I. Engagements. BNA S344-11, Recruiting of Natives, Suggested Scheme for Rejects, 1941–44.

[38] See Jock McCulloch. *Black Peril, White Virtue: Sexual Crime in Southern Rhodesia, 1902 to 1935*. Bloomington: Indiana University Press, 2000, pp. 36–56.

return to Johannesburg within six months by offering the same rate of pay they received when they had left.[39]

Six months later the High Commissioner, Sir Evelyn Baring, met with Mr W. Lawrence, General Manager of the NRC and Mr G. Lovett, the Chamber's legal advisor, to discuss deferred pay. Men were tending to go to work in the Union at an earlier age, stay away longer and return home later. Baring wanted an arrangement whereby HCT men would not be permitted to work continuously.[40] While Lovett and Lawrence were sympathetic, they offered no solution. The issue of deferred pay was just as intractable. Baring pointed out that the main purpose of deferred pay was to encourage men to return home, thereby ensuring they spent their earnings in the Territories. While he appreciated that a compulsory deferred pay system was out of question, he wanted a limit placed on the withdrawal from deferred pay accounts while men were in the Union. Mr Lovett explained that neither the NRC nor the WNLA had the right to withhold wages.

Baring briefed London on the meeting later the same month. From previous negotiations with the Chamber, he was fully aware of his weak bargaining position. If he quarrelled with the mines' representatives, they could draw as much voluntary labour from the Territories as they wanted, and he would gain no concessions. As a result, his administration's fragile control over the workforce would diminish even further.[41] Baring believed he had only one card to play. The mining industry knew that Africans resident with their families in towns hardly ever worked on the mines. The industry was beginning to realise that, if nothing was done, tribal and family life in the HCTs and the Scheduled Areas of the Union would break down, with large numbers of men moving to the towns with their

[39] Letter from Bathoen II, Paramount Chief of Bangwaketse, Kanye, Bechuanaland Protectorate to 'My Friend', 6th January 1947. Subject: Migratory Labour. BNA S387/7/1. Recruitment of Native Labour in B.P. General, 1941–48.

[40] Notes of Meeting between the High Commissioner and Representatives of the Chamber of Mines on 8th August 1947, Regarding Certain Native Recruiting Matters, p. 4. BNA S387/7/1. Recruitment of Native Labour in B.P. General, 1941–48.

[41] Letter from Sir Evelyn Baring, High Commissioner, Pretoria, to A. Sillery, London, 23rd August 1947. BNA S387/7/1. Recruitment of Native Labour in B.P. General, 1941–48.

families.[42] Those men would then be lost to the mines.[43] The attraction of the towns was very great, and Baring shared the Chamber's desire to slow that process down.

In mid-1946, Baring held discussions with the NRC about recruiting in Bechuanaland. He wanted a maximum service period of 18 months, at the end of which miners would be repatriated and remain at home for at least six months. The NRC agreed in principle but insisted on the exclusion from such an agreement of highly skilled men such as drillers and shaft sinkers.[44] The District Commissioners were asked for their views on the proposed changes. Mr W.F. Mackenzie at Serowe felt that there was little the administration could do. The limit of 18 months could not be forced on the NRC as a very large number of men re-engaged, many within a week. In addition, the mines were 'likely to make use of any available loophole' to get the labour they wanted. He was sure the only effective means of securing the return of miners was compulsory deferred pay. Mackenzie was also sure that the NRC would oppose any such measure as it would probably divert labour from the mines.[45] As usual, the outcome favoured the mines. In theory re-engagements were allowed so long as the total period of continuous service did not exceed 18 months. The Resident Commissioner could, however, waive that clause following a request from the NRC or the WNLA.[46]

In July 1948, the Bechuanaland High Commissioner gave up. At a conference of Resident Commissioners he convened in Johannesburg, he explained that if they made deferred pay a condition of contract, it would

---

[42] Scheduled Areas were introduced in South Africa under the Native Land Act of 1913. The Act in effect enshrined racial segregation into law, reserving about 7 per cent of land for the country's 67 percent of black inhabitants, and making it illegal for them to purchase land elsewhere in the country.

[43] Letter from Sir Evelyn Baring, High Commissioner, Pretoria, to A. Sillery, London, 23rd August 1947. BNA S387/7/1. Recruitment of Native Labour in B.P. General, 1941–48.

[44] Circular Memorandum from V.P. Kilenberger, Acting Government Secretary, Mafeking, to All District Commissioners, 4th November 1946. Subject: Migratory Labour. BNA. DCS 30/6. Native Labour Recruiting, 1941–47.

[45] Letter from W.F. Mackenzie, District Commissioner, District Commissioner's Office, Serowe, 15th November 1946, to The Government Secretary, Mafeking, Subject: Migratory Labour. BNA, DCS 30/6. Native Labour Recruiting, 1941–47.

[46] Letter from W. Gemmill, General Manager (Tropical Areas), WNLA Ltd., at Johannesburg, to the Government Secretary, Mafeking, Bechuanaland Protectorate, 9th September 1944. BNA S387/7/1 Recruitment of Native Labour in B.P. General, 1941–48.

serve no useful purpose as the contract was between the labourer and the NRC and could be varied by mutual consent. In an admission that he was powerless, he suggested the Resident Commissioners should individually try their hand negotiating the issue directly with the Chamber.[47] Writing in the same year, a critic of the WNLA questioned its impact on Bechuanaland and its government. 'A single powerful trade or interest is generally an unhealthy, if not a sinister, thing, especially in a country which is naturally poor'. 'The South African gold-mining industry', he concluded, 'is permanently dependent upon being subsidised by a semi-bankrupt African peasant pastoral economy in the distant reserves beyond its own national borders'.[48]

By 1960, the critical chorus abated. In that year, the Resident Commissioner in Mafeking, R.H.M. Thompson, gave the migrant labour system his full support. Each year around 20,000 Bechuana went to the South African mines on nine-month contracts. According to Thompson, the recruiting system and the treatment of workers were excellent. The mine compounds, although modest, were clean, the medical care was meticulous and the men were well fed. To monitor their welfare, the mines were visited by officers of the High Commission Territories Agency. Neither District Officers nor Chiefs forced young men to go to the mines. Thompson did admit, however, that when an able-bodied man appeared before a Court charged with the non-payment of taxes, he would frequently agree to go to the mines to pay his arrears.[49]

Tuberculosis and the Medical System

During the nineteenth century, visiting physicians commented on the absence of tuberculosis in Bechuanaland, a fact they attributed in part to the region's dry climate. This view was endorsed by Dr Neil Macvicar in

[47] Summary of Minutes of Resident Commissioners' Conference, 19th–21st July 1948. BNA S387/7/1 Recruitment of Native Labour in B.P. General, 1941–48.

[48] A. Sandiland, review of I. Schapera, *Migrant Labour and Tribal Life*. In *Africa. Journal of the International African Institute,* Vol. 18, No. 2, 1948, p. 145.

[49] Letter from R.H.M. Thompson, Resident Commissioner, Mafeking, to the Rt Hon A. Creech Jones, MP, 5th October 1960. BNA S387/7/4 Recruitment of Labour 1960.

his pioneering medical history from 1908.[50] Subsequent surveys, including the South African Institute for Medical Research (SAIMR) 1932 study of tuberculosis and mine labour, reached the same conclusion.[51] Prior to 1920 tuberculosis remained uncommon, and it appears transmission only began in earnest with increasing migration of gold miners to South Africa. Whatever the case, there was a much higher incidence of TB in the southern half of the Protectorate, where mine recruitment was more common, and most of those diagnosed with TB were returning miners. The relationship between mine work and tuberculosis was widely recognised among the Tswana and the term 'maintisl' or 'minetisis' associated the disease with mine work.[52]

Bechuanaland's tiny Medical Department served a vulnerable and dispersed population prone to malnutrition. The dominant themes in the Annual Reports for the period 1915–1930 refer to the lack of staff; the threat posed by infectious diseases, especially malaria, smallpox and syphilis; and the widespread problem of a diet deficient in vitamins and protein. The Reports also acknowledge the lack of reliable data.[53] At the beginning of that period, tuberculosis was seldom mentioned and always in connection with men returning from the gold mines.[54] As late as 1926, the Annual Report notes that tuberculosis was 'not a disease of the Bechuanas or their country'. The returns for the following year show that most of the 142 tuberculosis cases were former miners.[55] Active tubercu-

---

[50] See Neil Macvicar. 'Tuberculosis among the South African Natives'. *South African Medical Record*, Vol. VI, No. 11, 10th June 1908, pp. 161–176.

[51] See *Tuberculosis in South African Natives with Special Reference to the Disease Amongst the Mine Labourers on the Witwatersrand*. South African Institute for Medical Research, Johannesburg, March 1932, pp. 31–45.

[52] Andrew S.C. Mushingeh. A History of Disease and Medicine in Botswana, 1820–1945. PhD Dissertation in History, St John's College, University of Cambridge, August 1984, pp. 189–190. See also Wazha G. Morapedi. 'Migrant Labour and the Peasantry in the Bechuanaland Protectorate, 1930–1965'. *Journal of Southern African Studies*, Vol. 25, No. 2, 1999, pp. 211–21; 189.

[53] Registers of births and deaths were only introduced from 1941. See I. Schapera. 'Migrant Labour and Tribal Life: A Study of Conditions in the Bechuanaland Protectorate'. Report Presented to the B.P. Administration, Public Records Office, DO 119 1268 Schapera, 1944, p. 28.

[54] The Annual Report for 1915, for example, makes no reference to tuberculosis. See *Annual Medical Report for 1915–16 Bechuanaland Protectorate*, p. 1. BNA S189/9 Health Section, League of Nations 1935–45.

[55] *Annual Medical & Sanitary Report 1927–29, Bechuanaland Protectorate*, p. 5. BNA S417 1927–29. Annual Medical & Sanitary Reports.

losis was noticed in several cases as a sequel to bronchial pneumonia in men who, some years earlier, had worked on the mines.[56]

There was a sustained expansion of medical services in the years after the First World War. Funding increased tenfold from £2260 in 1919 to £21,509 in 1936, with most of the money going to government hospitals in the population centres of Lobatse, Serowe and Francistown. In addition, grants from the Colonial Development Fund provided two travelling dispensaries and support for the building of mission hospitals at Maun and Sofala. In 1936, the Protectorate's European staff consisted of a Principal Medical Officer, eight Medical Officers, three matrons and six nurses. There were 164 hospital beds for 260,000 Africans, or one per 1585 persons.[57] There were, in addition, six medical missionaries whose salaries were subsidised by the government. While that extra funding was welcome, the Resident Commissioner, C.F. Rey, acknowledged that there was still an urgent need for additional staff. During 1933 the expenditure on Medical Services represented just 8 per cent of the Protectorate's total budget, well below the average of 10 per cent of revenue spent in territories under the Colonial Office.[58]

In 1934, about 389 cases of tuberculosis were reported throughout the territory. This was a dramatic increase on the 36 cases just seven years earlier. From that point, there was a steady rise in the numbers. As the NRC and the WNLA intensified their recruiting, several Medical Officers complained that the economic benefits of oscillating migration were outweighed by the spread of disease. There were no specialist beds, pathology testing was outsourced to the SAIMR in Johannesburg and the first X-ray plant in the territory was only installed at Lobatse in 1933. The official policy was to advise chiefs and headmen to send those with active TB to the cattle posts where they would die in relative isolation.[59]

---

[56] *Annual Medical & Sanitary Report 1930, Bechuanaland Protectorate*, p. 13. BNA S88/9 Annual Medical Report, 1930.

[57] *Commission of Enquiry into Medical Matters 1937*, pp. 8; 16. BNA S392/7/1.

[58] Memorandum on Bechuanaland Protectorate Health Proposals from Colonial Office, London, signed A.J.R. O'Brien and A.T. Stanton, 19th November 1934. BNA, Health Deputation MPS 1934 S392/5.

[59] Mushingeh, A History of Disease, pp. 190; 192–3.

There was little improvement over the next decade. In 1943, the Principal Medical Officer identified a number of flaws in the health service. There was a lack of medical facilities and personnel, the indigenous population was reluctant to accept new medical doctrines and the distances separating the centres of settlement presented a severe handicap.[60] According to the 1940 census, there was one hospital bed for each 1067 members of the population and just one X-ray plant for the whole population, but no specialist radiographer. Twelve years later the European staff had increased by just three. Low salaries and strenuous work conditions made it difficult to attract Medical Officers and European Sisters.[61]

In June 1933 Dr P.M. Shepherd, a mission medical officer at Molepolole, reported on the large numbers of men being rejected for mine work. Dr Shepherd noted that the NRC had good reason for instructing medical officers to examine 'with special care' the lungs of those who reported a previous mining experience. Out of a group of 500 men Shepherd reviewed between January and May 1933, only half were passed fit. Of the 207 men rejected outright, 125 had already been to the mines. Of that group, most had served five or more contracts and were no longer capable of underground work. Shepherd acknowledged that persistent droughts and malnutrition were a factor in the men's poor physiques. 'It raises the question as to how far the Bechuana generally at present are suited for mining'.[62] Shepherd, who was writing six months before the ban on Tropical recruitment was lifted, had in effect identified a cycle in which drought and malnutrition drove men into migrant work. After serving a number of contracts their health began to fail, and they were rejected at the next entry medical.

Even as the negative impact of mining on public health became more visible, observers noted that it was difficult to estimate its extent. In 1935 the Principal Medical Officer, Dr H.W. Dyke, noted that while there had been a threefold increase in tuberculosis in just eight years, at best

[60] Memorandum prepared by Principal Medical Officer on Health Services of the Bechuanaland Protectorate, 15th January 1943, p. 1. BNA S393/1/1 Health & Medical, 1942–50, S393/1/1.

[61] *Annual Medical & Sanitary Report, 1948. Bechuanaland Protectorate*. BNA S397/1/3, 1948 & 1949, Annual Medical, p. 1.

[62] Mine Recruits: Report of Dr P.M. Shepherd re poor physique of and main causes of rejection, Molepolole, B.P., June 1933, p. 3. Mine Recruits, 1934, S398/6 Botswana National Archive.

hospital admissions represented only a fraction of the actual cases.[63] Owing to a lack of accommodation, those admitted were kept for only a week and then returned to their villages, where Dyke believed they became 'reservoirs of infection'. He recommended segregation on the lines used with lepers. In the 1940 annual review, when a total of 289 new tuberculosis cases were reported, the Principal Medical Officer wrote: 'Facilities for the exact diagnosis of dormant disease and of disease resulting from work on the mines in the Union of South Africa do not exist in the Territory. I am therefore very doubtful about the value of the present figures on the incidence of the disease'.[64] In theory, the Chamber of Mines provided District Commissioners with a weekly list of men repatriated with tuberculosis so that treatment could be continued by local medical officers. In practice that did not happen. Dr J.W. Sterling lamented that there were no facilities for the diagnosis of dormant disease, nor for the treatment for returning miners.[65] Like the other HCTs, Bechuanaland had no specialist facilities.[66]

Two Health Studies into Mines and Disease in Bechuanaland

In response to persistent reports that miners were spreading disease, in 1937 the British Secretary of State appointed a Commission headed by Sir Walter Johnson, former Director of Medical and Sanitary Services in Nigeria. Johnson was asked to evaluate the threat posed by venereal disease and tuberculosis in the Bechuanaland Protectorate and to identify how best to monitor men invalided from the mines.[67] The date is signifi-

[63] Extract, *Annual Medical and Sanitary Report, 1934*. H.W. Dyke, Principal Medical Officer, 30th April 1935 p. 1. BNA S 438/2/1. Tuberculosis in the B.P. 1938–39.

[64] *Annual Medical and Sanitary Report 1940. Bechuanaland Protectorate*, p. 2. BNA, S397/1/1. Annual Medical Reports 1939–1945.

[65] Minute from Dr J.W. Stirling, Principal Medical Officer, Mafeking, to the Government Secretary, 15th November 1940. BNA S438-2-2 Tuberculosis in B.P. 1940–1953.

[66] Notes on Tuberculosis in the High Commission Territories, June 1940, p. 6. BNA S438-2-2 Tuberculosis in B.P. 1940–1953.

[67] *Commission of Enquiry into Medical Matters*, 1937, p. 1. BNA S392/7/1.

cant: in 1937 the South African government permanently lifted the ban on Tropical recruiting.

Johnson dismissed the official data on tuberculosis as unreliable, concluding that so little was known about the disease it was difficult for him to make recommendations. He was certain, however, that men repatriated with phthisis were a threat to public health. According to staff at the Free Church of Scotland Mission hospital in Molepolole, although tuberculosis accounted for only 2 per cent of out-patients, among those men examined at entry medicals for the NRC, roughly 10 per cent were tubercular. Johnson noted that the rise in the tuberculosis rate from 216 cases in 1934 to 332 in 1936 corresponded to the renewal of the WNLA's recruitment in the North. There was an obvious hazard from the mines, and Johnson recommended that repatriated cases be carefully monitored. He also found that the large numbers of men being rejected at entry medicals suggested that mine work was taking a heavy toll. In 1936, there were 9205 men recruited for the mines and 493 rejected. Of those 318 were experienced miners with defective lungs, and a further 68 had hearing loss acquired on the mines. Johnson pointed out that NRC and the WNLA had the capacity to establish a referral system, and he recommended that as a matter of urgency isolation wards be built at Lobatse. He also recommended that a survey be set up, using a travelling dispensary fitted with an X-ray plant, to establish the extent of the problem. He hoped the Chamber may help with funding it.[68]

In early December 1939, the Bechuanaland European Advisory Council complained that cattle production had fallen because of the absence of so many men on the mines.[69] The Resident Commissioner agreed, and commissioned Professor Schapera, a South African social anthropologist known for his work on the Tswana, to report on the causes, extent and impact of labour movements.[70] Of necessity, Schapera relied heavily on data and other assistance from the WNLA and the NRC. In a 300-page study, he devoted only 3 pages to the spread of

[68] *Commission of Enquiry into Medical Matters*, 1937, pp. 37–38; 57; 61–61; 40. BNA S392/7/1.

[69] Minutes of 27th Session of European Advisory Council, 27th November to 6th December 1939, pp. 60–61. BNA S387/5 Recruitment of Labour, 1942.

[70] The report was eventually published as I. Schapera. *Migrant Labour and Tribal Life*. London: Oxford University Press, 1947.

infectious disease from the mines and made no reference to tuberculosis. Schapera did, however, comment at length on the controversial topic of the conduct of mine medicals.

In his report, Schapera estimated that during the period 1928–1943, almost 13 per cent of men applying for the mines were rejected because of 'chest troubles and poor physique'. Those figures were all the more disturbing given that labour agents made a preliminary selection and unfit men were unlikely to offer themselves to the NRC. The high rejection rates were accompanied by longer periods of service of those passed fit. From 1930, the average length of time spent at home by migrant labourers fell, and the intervals between visits home became longer. Men went to the mines to support their families, but after serving three or four contracts they were no longer able to perform such work. Schapera estimated that nearly two-thirds had a career lasting less than five years.[71] His report, like the 1937 report by Sir Walter Johnson, offered a warning about the hazards of mining for men drawn from such a vulnerable population.

Prior to the introduction of chemotherapy in the early 1950s, patients with tuberculosis were encouraged to leave their villages and live at the cattle posts to lessen the spread of infection. For the small number admitted to hospital, the recommended treatment consisted of rest for several months with a diet rich in protein, with patients expected to drink at least two pints of milk a day.[72] In practice such diets were never provided. Chemotherapy dramatically changed the potential outcomes for patients. However, the new medicine brought its own problems. A large part of the health budget was spent on tuberculosis, but there was a lack of uniform treatment schedules. In theory, once chemotherapy came on stream, those in home care were to be brought to a hospital by ambulance and have a chest X-ray every six weeks.[73] None of that happened. There were

[71] Schapera, 'Migrant Labour and Tribal Life: A Study of Conditions in the Bechuanaland Protectorate', Report, pp. 22–23; 69; 74.

[72] See Diet for African Tubercular Patients, Athlone Hospital, Lobatsi. Undated (1944). BNA Med 7/1 TB General, 1944.

[73] Circular No. 13/55, 'Tuberculosis', from M.L. Freedman, Director of Medical Services, Medical Department, Mafeking, to all Government Medical Officers and Medical Missionaries, 3rd March 1955, p. 1. BNA S438/2/3. Tuberculosis in the Bechuanaland Protectorate, 1960–1963.

just 44 dedicated tuberculosis beds in the territory, there were no BCG or other vaccination programmes, no facilities for thoracic surgery, and no special clinics or dispensaries to treat or monitor out-patients.[74]

According to WNLA data, between 1953 and 1954 silicosis alone and silicosis with tuberculosis was diagnosed in a total of 269 cases, at a rate of 3.28 per 1000.[75] In theory, the NRC was to notify the Bechuanaland administration of all recruits repatriated on medical grounds, and to forward a report to the Medical Officer or Medical Missionary in the recruit's home district. This never apparently happened. The official correspondence shows that in 1957, the HCTs' office in Johannesburg believed that the WNLA kept tubercular miners in hospital until the disease was arrested or the patient had become non-infective.[76] That was not the case in the 1950s, and it was still not the case at majority rule in 1994. As the Oosthuizen report put it in 1954, 'Sick natives are all repatriated through the WNLA hospital and it appears that the criterion for deciding whether they are fit for repatriation is fitness to travel, the measurement of which is the ability to stand'.[77]

In 1957 Dr T.H. Davey, a Bechuanaland government medical officer, wrote a brief but highly critical review of tuberculosis management. The service was understaffed, and medical officers spent much of their time on outside commitments, such as conducting medicals for the WNLA to generate income. As a result, they had heavy workloads, but much of

[74] Memorandum, 'Particulars for Bechuanaland Protectorate', from Dr M. L. Freedman, Director of Medical Services, Mafeking, to The Secretary-General, The National Association for the Prevention of Tuberculosis, London, 2nd August 1956. BNA Med 7/5 1955. Medical Tuberculosis. BCG, or Bacillus Calmette-Guerin, was a vaccine to prevent tuberculosis. It was first administered to humans in 1921.

[75] Extract from *Bechuanaland Protectorate News-Letter* No. 24, January 1956, 'Recruiting', Dated 7th January 1956. BNA Med 15/8. Recruits for Mines, Farms, 1947.

[76] Memo from Agent for the High Commission Territories, Johannesburg, to Deputy High Commissioner for Basutoland, the Bechuanaland Protectorate and Swaziland, Cape Town, 10th July 1958. Subject: Treatment in South Africa of African Workers from other Territories. BNA, S387/7/3. Recruitment of Labour, Native in BP, 1957.

[77] See Report of the Departmental Committee of Enquiry into the Relationship Between Silicosis and Pulmonary Disability and the Relationship Between Pneumoconiosis and Tuberculosis. Part 2 The Relationship Between Pneumoconiosis and Tuberculosis, 1954. SANA, F33/671 Treasury, p. 133.

their effort was not devoted to government business.[78] Davey also noted that the lack of a laboratory service made the management of infectious disease all the more difficult. Treatment was often not completed, and this led to a high rate of drug resistance. At a Medical Officers' Conference at Gaborone in August 1966, one senior officer, Dr Thomas, spoke of the threat posed to school children by teachers with drug-resistant strains. Second-line drugs were expensive, and neither government nor mission hospitals could afford to treat the patients who needed them. More than 2000 teachers were employed in state and mission schools, and Thomas suggested that they should be X-rayed on appointment and periodically thereafter.[79]

The National Tuberculosis Programme (NTP) was launched in 1975, but many of problems that had plagued the service since the 1920s persisted. The lack of capacity meant that less than 30 per cent of tuberculosis patients completed treatment.[80] That was in part due to the high turnover of medical staff, as well as the failure of some doctors to follow the NTP's prescribed regimes. At the inception of the Programme, the reported incidence of tuberculosis was between 200 and 500 cases per 100,000 and the rate of drug resistance around 30 per cent. A generation later, with the arrival of HIV/Aids, the lack of biomedical capacity would prove equally devastating.

Two Unsuccessful Compensation Cases

The British High Commissioners' Office was aware that it was very difficult for men from the HCTs to apply for compensation once they had left employment. Most miners were not aware of their rights; once their disease developed, many were too ill to travel to Johannesburg for

[78] Dr T.H. Davey, 'Comments on Government Medical Activities and Projects in Bechuanaland, 1957. Prophylactic Campaigns Now in Being', to The Government Secretary, Mafeking, 18th June 1957, p. 10. BNA S393/1/3. Health & Medical Position in B.P. Health Services up to 1963.

[79] Minutes of the Medical Officers' Conference Held at Gaborone on the 6th and 7th of August 1966. BNA MED 12/16 Medical Officers' Conference, 1966.

[80] E.T. Maganu, *Botswana National Tuberculosis Programme: Tuberculosis in Botswana Results of an Epidemiological Survey 1981*, p. 4. Ministry of Health, Gaborone, Government Printer.

examination. Some of the obstacles to compensation were due to circumstances such as distance and the lack of medical capacity in the HCTs. Others seem to have been systematically created by employers. Two cases illustrate the difficulties faced by black miners and their surviving relatives in receiving compensation, the employers' efforts to shift costs from the mines onto the Bechuanaland medical service and the Chamber's determination to control health data. Chomati Letsepe was living at Mochudi when he was recruited by the Main Reef at Rustenburg in 1936. He worked at the mine for the next 21 months. Letsepe fell ill and was sent to hospital. He then asked to be allowed to return home. The mine doctors agreed to his request but did not tell him what their diagnosis was. Letsepe did not receive compensation and paid for his journey home himself. He was then examined by Dr Burger, the Medical Officer at Mochudi, who diagnosed tuberculosis.[81] The case eventually reached the Department of Native Affairs in Johannesburg, which sought to lodge a claim for compensation on Letsepe's behalf. As was required under the relevant regulations, the Department asked that a specimen of the patient's sputum and a chest X-ray be sent to the Miners' Phthisis Medical Bureau in Johannesburg.[82] Chomati Letsepe died at Mochudi in July before he could be examined. No claim was lodged by his surviving relatives, who were opposed to a post-mortem.[83]

The case of Seatuma Gopolang, a fellow miner from Mochudi who also contracted lung disease, provides another typical example of company practices. Gopolang had worked underground at Robinson Deep mine for only five months in 1937 when he became ill and was admitted to the mine hospital. From there he was sent to the WNLA Compound and four days later repatriated without compensation. Gopolang was ill

[81] Memo from Acting District Commissioner, District Commissioner's Office, Mochudi, to The Government Secretary, Mafeking, 14th March 1939. Subject: Phthisis Compensation Chomati Letsepe. BNA S332/5/1. Phthisis: Miners' Compensation for B.P. Native Mine Workers.

[82] Letter from Director of Native Labour, Department of Native Affairs, Johannesburg, to The Government Secretary, Resident Commissioner's Office, Mafeking, 7th June 1939. Subject: Claim for Compensation Native Chomati Letsepe. BNA, S332/5/1. Phthisis: Miners' Compensation for B.P. Native Mine Workers.

[83] Memo from Assistant District Commissioner, District Commissioner's Office, Mochudi, to The Government Secretary, Mafeking, 26th July 1939. Subject: Chomati Letsepe. BNA S332/5/1 Phthisis: Miners' Compensation for B.P. Native Mine Workers.

but had been given no diagnosis, so he went to the local doctor at Mochudi. The doctor concluded that Gopolang had tuberculosis. The District Commissioner wrote in protest to Government Secretary: 'The fact that he was discharged before the completion of his contract leads one to think that the mine authorities were aware of his condition'.[84] The Government Secretary wrote to the Robinson Deep asking that compensation be paid.[85] According to the mine's management, Gopolang was repatriated because he was 'mentally deficient'. The mine medical officer claimed there was nothing wrong with his health and that he could not have contracted phthisis in the short period he was employed.[86] That was the end of the matter.

The obstacles faced by men like Chomati Letsepe and Seatuma Gopolang were raised by the Agent for the HCTs, Mr A.G.T. Chaplin, at the Stratford Commission in March 1942. Chaplin pointed out that the applications for compensation had to be submitted to the Director of Native Labour in Johannesburg and include a record of the applicant's work history supported by a medical report, a sputum specimen and an X-ray. Based on those materials, the Medical Bureau would then decide on whether compensation was warranted. The problem of distance, as well as the lack of X-ray facilities in the Territories, made it impossible for miners to comply.[87] Chaplin also made a written submission in which he argued that the improvements in detection and monitoring used to exclude sick men from employment meant that tuberculosis must have been contracted on the mines. He wanted the Chamber to keep comprehensive work and medical records, and to use X-rays at entry and exit medicals. Chaplin suggested that on arrival in Johannesburg, recruits be

[84] Letter from Acting District Commissioner, District Commissioner's Office, Mochudi, to the Government Secretary, Mafeking, 13th June 1938. Subject: Phthisis Compensation—Seatuma Gopolang. BNA S332/5/1 Phthisis: Miners' Compensation for B.P. Native Mine Workers.

[85] Letter from G.H. Nettleton, for Government Secretary, to the Manager, Robinson Deep Limited, Johannesburg, 11th July 1938. BNA S332/5/1 Phthisis: Miners' Compensation for B.P. Native Mine Workers.

[86] Letter from Manager, Robinson Deep Limited, Johannesburg, to The Government Secretary, Mafeking, 17th August 1938. Subject: Native Seatuma Gopolang. BNA, Phthisis: Miners' Compensation for B.P. Native Mine Workers, S332/5/1.

[87] A.G.T. Chaplin, Draft Memorandum to be submitted by the Agent for the High Commission Territories in Johannesburg to the Miners' Phthisis Commission. Undated, circa March 1942, p. 2. BNA S332/5/1 Phthisis: Miners' Compensation for B.P. Native Mine Workers, S332/5/1.

issued with a card recording their medical and mining history, and that a full report of all miners suffering from miners' phthisis or tuberculosis be submitted to the Commissioner of his home district.[88]

Tuberculosis Surveys After the Second World War

The two major barriers to controlling the spread of tuberculosis were the lack of medical capacity and the lack of data. The first was due to imperial policy and the state of the economy and lay beyond the reach of the local administration. There was little industry in Bechuanaland, and it was not until the eve of independence that the first occupational health legislation was enacted.[89] The Moshaneng Asbestos Mine, with just 350 workers, was the largest employer, followed by the abattoir at Lobatse, which employed 300 men. The work conditions at Moshaneng were appalling, but there was no resident doctor, and the mine produced no health data.[90] The second barrier to controlling the spread of tuberculosis could have been easily resolved by the Chamber: the WNLA and the NCR had the expertise and capacity to collect whatever data they wished.

In 1937, the Commission headed by Sir Walter Johnson recommended that a survey be held to identify the extent of occupational lung disease.[91] The Principal Medical Officer, Dr D. Drew, suggested the administration approach the Chamber for assistance. The Chamber, he noted, had an interest in the control of tuberculosis in a Territory which supplied a great deal of the mine labour.[92] The Resident Commissioner also sought

[88] Memorandum of Further Evidence Submitted by the Agent for The High Commission Territories in Johannesburg to the Miners' Phthisis Commission, 5th August 1942, p. 1. BNA S332/5/1 Phthisis: Miners' Compensation for B.P. Native Mine Workers.

[89] Bechuanaland Protectorate: Answers to 'Policy Questionnaire' for 5th Inter-African Labour Conference, May 1957. BNA S504/1. CCTA Research in Africa South of the Sahara, 1957.

[90] See, for example, Memo from Acting Government Secretary, Mafeking, to the High Commissioner, Mafeking. 6th June 1950. BNA, Asbestos Deposits Moshaneng Mine, 1950. S62/3/3.

[91] *Commission of Enquiry into Medical Matters 1937*, p. 40. BNA S392/7/1.

[92] Letter from Dr D. Drew, Principal Medical Officer, Bechuanaland Protectorate, to Dr A.J. Orenstein, Johannesburg, 16th September 1937, p. 2. BNA S438/2/1 Tuberculosis in the B.P., 1938–39.

assistance from the British Colonial Development Fund.[93] While the Chamber did make a small donation towards the purchase of an X-ray machine, after much delay the request to the Development Fund was deferred indefinitely because of the war.[94]

During the Second World War the management of tuberculosis was complicated by soldiers returning from the African Auxiliary Pioneer Corps. Chiefs complained to the administration: 'When our men came back from the army we noticed that many of them were afflicted'.[95] The disease was alarmingly common and accounted for almost half of the deaths of Tswana soldiers who had served in the Middle East. In contrast to the management of repatriated miners, specialist tuberculosis shelters were built at various medical centres to contain the threat to public health.[96] The Deputy Director Medical Services, Dr Mackenzie, suggested that if those soldiers were kept in quarantine, the problem solved itself as most would soon die. 'This may seem an inhuman viewpoint, but it certainly compares favourably with that of allowing an African with tuberculosis to return and infect others—notably women and children of his own family group'.[97]

At a conference on post-war development held at Mafeking in January 1946, the delegates agreed on the urgent need for a tuberculosis survey. The nine years since Sir Harry Johnson's original recommendation for such survey had seen 'a menacing increase' in the incidence and distribution of the disease.[98] When the newly created World Health Organization

---

[93] Memorandum, Dominions Office, 12th July 1938. BNA S392-7-2 Health & Medical Position in Bechuanaland Protectorate, 1937.

[94] Letter from Resident Commissioner's Office, Mafeking, to His Excellency, The High Commissioner, Pretoria, 26th September 1939. BNA S438/2/1 Tuberculosis in the B.P. 1938–39.

[95] Mushingeh, A History of Disease and Medicine in Botswana, p. 194.

[96] Minute from G. Nettleton, Government Secretary, the Resident Commissioner's Office, Mafeking, to the Administrative Secretary to the High Commissioner, Pretoria, 9th November 1945. Subject: Tuberculosis in the African Pioneer Corps. BNA S438-2-2 Tuberculosis in B.P. 1940–1953.

[97] Memorandum from D.J.T.M. Mackenzie, Deputy Director Medical Services, Mafeking, to the Government Secretary, 3rd November 1945. BNA S438-2-2 Tuberculosis in B.P. 1940–1953.

[98] Memorandum from D.J.T.M. Mackenzie, Deputy Director Medical Services, Mafeking, to The Government Secretary, 4th August 1946. BNA S438-2-2 Tuberculosis in B.P. 1940–1953.

(WHO) Expert Committee on Tuberculosis met in Geneva in February 1948, it issued a set of policy guidelines for tuberculosis control. The Committee noted the importance of nutrition, housing and occupational health, and stressed that any prevention programme must include after-care and rehabilitation. Prevention was best done by identifying and isolating cases. Sample surveys using tuberculin testing and mass radiography were also recommended.[99] As always, the Committee emphasised the importance of education to inform patients and their families about hygiene and prevention.[100]

Under the Bechuanaland Public Health Act tuberculosis was a notifiable disease, and medical practitioners were required to immediately inform the local authority of new cases. Those authorities were in turn required to transmit weekly lists of cases to the Chief Health Officer.[101] In theory, the local authority was to take adequate measures to prevent the spread of the disease, including providing accommodation, maintenance, nursing and medical treatment. In practice the lack of funds and capacity meant none of that happened. During 1948, there were 932 tuberculosis cases reported in Bechuanaland, almost double the number from four years earlier. According to the Medical Director: 'There is little useful comment that can be made at this stage. The disease is apparently on the increase, but the true state of affairs is obscure'. Funding for a survey was finally approved in 1948.[102]

At a Cape Town meeting in February 1949, the South African Secretary for Health agreed to assist Bechuanaland by providing staff and a mobile X-ray unit.[103] In preparation, a pilot study using tuberculin testing, and

[99] Savingram [circular] from the Secretary of State for the Colonies, London, to the Officer Administering the Government of [Colonial Governments including B.P.], 20th December 1950, Annexure 1. Subject: Control of Tuberculosis-WHO. BNA Med 7/2 Tuberculosis, 1947.

[100] World Health Organization Supplementary Report, Tuberculosis, 22nd April 1948. Report on the Second Session of the Expert Committee on Tuberculosis held at Geneva, 17th–20th February 1948, p. 4. BNA Med 7/1. TB General, 1944.

[101] Memorandum, T.B. Control, May 1955. BNA Med 7/5 1955. Medical: Tuberculosis.

[102] *Annual Medical & Sanitary Report*, 1948. Bechuanaland Protectorate. BNA S397/1/3, 1948 & 1949, Annual Medical, pp. 2; 10.

[103] Memorandum from D.J.M. Mackenzie, Director Medical Services, Mafeking, to The Government Secretary, 11th February 1949. Subject: Yellow Fever and TB Surveys Scheme D.1037. BNA S438-2-2. Tuberculosis in B.P. 1940–1953.

mass radiography was carried out by the South African specialist Dr Dormer and a team of four. The pilot began well, with large numbers attending for examination at the Kanye Hospital. Unfortunately, public support collapsed with the deaths of a number of children following pneumonia and whooping cough vaccinations.[104] After several delays, in September 1952 a tuberculin testing and mass X-ray survey using a mobile X-ray unit was conducted in rural communities.[105] As most young male adults were on the mines, of the 21,270 examined 13,540 were females, an imbalance which compromised the results. A total of 273 abnormalities were diagnosed as 'active pulmonary tuberculosis', suggesting there was a significant problem.[106]

The Schechter Survey was an important step. However, it appears that its results underestimated the incidence of tuberculosis. In March 1953 Dr A.M. Merriweather, of the Scottish Mission Hospital at Molepolole, asked the government to help with the cost of medication. His hospital was treating patients with streptomycin and para-aminosalicylic acid (PAS) but lacked the resources to deal with the large numbers presenting. Most patients were asked to pay £10, a huge sum for a family, and this in any case did not cover the full cost of treatment.[107] The Director of Medical Services agreed that the government should pay for medication, providing the Mission bore the cost of hospital care.[108] Six months later, Dr J.A. Hay of the Seventh Day Adventist Mission at Kanye made an

[104] Dr D.J.M. Mackenzie, Director of Medical Services, Mafeking, 5th May 1949, Notes on a Discussion with Dr Dormer on 4th May 1949 Regarding a Tuberculosis Survey. BNA, Med. 8/1/1, TB Survey 1952.

[105] The survey was led by Dr M. Schechter from Durban. See M. Schechter. 'Mass X-Ray Survey: Bechuanaland Protectorate 1952'. *The South African Medical Journal,* Vol. 28, No. 4, 24th April, 1954, pp. 351–356.

[106] Schechter, 'Mass X-ray Survey: Bechuanaland Protectorate 1952', pp. 351–356.

[107] Letter from Dr A.M. Merriweather, Superintending Missionary, Scottish Livingstone Hospital, Molepolole to Dr Freedman, Director of Medical Services, B.P. Government, Mafeking, 17th March 1953. BNA MED 8/1/2. Medical TB Survey 1952.

[108] Letter from Dr Freedman, Director of Medical Services, Medical Department, Mafeking, to The Government Secretary, Mafeking, 10th April 1953. Subject: Tuberculosis, BNA MED 8/1/2. Medical TB Survey 1952.

urgent request for government assistance. The cases identified by the Survey had reported for treatment, but so many additional cases were presenting weekly with positive X-rays and sputum that the hospital was overwhelmed. Dr Hay had recently ordered a 12-year-old girl with open pulmonary tuberculosis to stay away from school. He was certain that many like her were a menace to their classmates.[109] Hay asked the government to consider building lying-in shelters to isolate infectious cases. The lack of capacity meant that many patients were turned away from hospitals because there were no beds.[110]

In response to a request from the government, the WHO carried out a Tuberculosis Survey in Bechuanaland in 1956. The survey confirmed the findings of a similar project undertaken with South African equipment in 1952. It showed that between 1 and 2 per cent of the population—almost double the official South African rate—had open tuberculosis.[111] In a territory with around 300,000 people, it was estimated that there were 3000 highly infectious cases. The Director of Medical Services considered that tuberculosis was the greatest public health problem. The vastness of the country posed serious difficulties. Most of the population lived in large villages which were several days' walk from their fields. Government policy was to isolate and treat as many infectious cases as possible, and it had spent more than £40,000 on specialist wards. While those 288 dedicated beds were essential for dealing with advanced patients, it was only possible to isolate a small proportion of the estimated pool of 3000 cases. Out-patient care using combined chemotherapy was provided by most hospitals, but patients often ceased treatment before they had recovered.[112]

---

[109] Letter from Dr J.A. Hay, Seventh Day Adventist Mission, Kanye, to J.C.E. Bowen, District Commissioner, Kanye, B.P. 29th September 1953. BNA S438-2-2. Tuberculosis in B.P. 1940–1953.

[110] Extract, Divisional Conference—Southern Protectorate, October 1955. BNA, Tuberculosis in the Bechuanaland Protectorate, 1960–1963, S438/2/3.

[111] Minutes, 31st July 1958. Subject: Tuberculosis Control. BNA S439/1/1 Tuberculosis in Bechuanaland Protectorate Assistance from the WHO 1955.

[112] Report on Visit to Bechuanaland by WHO Tuberculosis Consultant Anton Geser, 4th–7th June 1958, pp. 1–2. BNA Med 8/4 Medical: TB Surveys & Reports 1959.

Independence from Britain

Political independence in 1966 brought rapid economic and social change. The British Protectorate of Bechuanaland became Botswana, and the capital was moved from Mafikeng in South Africa to Gaberones (now Gaborone). The following year, diamonds were discovered in a remote region of central Botswana. While the country gradually developed a number of small enterprises, diamond mining became the dominant industry and accounted for most of the 30-fold increase in Gross Domestic Product, which occurred between 1967 and 1983.[113] Within 25 years, almost half of Botswana's population were urban dwellers.

However, South Africa's mines remained a major source of employment, and tuberculosis continued to pose profound challenges to the country's health system. In 1960, the 10,000 wage earning jobs in the Protectorate provided work for just 2 per cent of the resident population. In 1962, the total of remittances and cash inputs from migrant labour amounted to around £2 million per annum.[114] After independence the flow of labour continued. In 1970 there were at least 55,000 Batswana employed on the gold mines and in other sectors of South Africa's economy.[115] The flow of labour south peaked at 40,390 in 1976.[116] According to the 1991 Census, the vast majority of Batswana abroad were men working on contracts in South African mines, and tuberculosis was the major cause of morbidity and mortality in Botswana. A central Tuberculosis Register had been kept since 1962, but it did not provide reliable data. In the same year, a pilot BCG vaccination campaign began, but was not sustained. A comprehensive National Tuberculosis Programme based on BCG vaccination, case finding and treatment was finally launched in 1975 in collaboration with the World Health

[113] John Taylor. 'Some Consequences of Recent Reductions in Mine Labour Recruitment in Botswana'. *Geography,* Vol. 71, No. 1, 1986, pp. 34–46.

[114] Wazha G. Morapedi. 'Migrant South African and the Peasantry in the Bechuanaland Protectorate, 1930–1965'. *Journal of Southern African Studies,* Vol. 25, No. 2, June 1999, p. 203.

[115] Francis Wilson. *Migrant Labour in South Africa: Report to the South African Council of Churches Johannesburg.* The South African Council of Churches and SPRO-CAS, 1972, pp. 115–116.

[116] Taylor, 'Some Consequences of Recent Reductions in Mine Labour Recruitment in Botswana', p. 41.

Organization. In the early months of 1981, a tuberculosis prevalence survey was undertaken with the assistance of the South African Medical Research Council. There was a stigma attached to tuberculosis, and the results were compromised by the low numbers who presented for screening. In addition, the survey was conducted after people had migrated to their lands for agricultural work. Despite these problems, the survey uncovered an epidemic of tuberculosis. The Southern Health Regions of Lobatse, Gaborone and Molepolole had the highest reported incidence, usually around 500 cases per 100,000.[117] In the early 1980s, tuberculosis was joined by HIV/AIDS. The high volume of migrant labour, and the compounding effects of TB, silicosis and HIV/AIDS, resulted in Botswana having one of the highest HIV prevalence rates in the world. Botswana National Policy on HIV/AIDS (1998) provided a framework for a multi-sectoral response to the epidemic. However, it did not mention migrant or mobile populations.[118] The settlement of the miners' class action in 2019 gave many workers the first meaningful access to compensation for their occupational injury. As of 25 August 2022, about 1585 claims for compensation have been lodged, but none have so far been paid.[119]

References

Cuzon, Pauline. The History of TEBA in Botswana, unpublished manuscript, July 1985.

Gann, L.H. and P. Duignan. *The Rulers of British Africa 1870–1914.* London: Croom Helm, 1978.

Jeeves, Alan H. 'Migrant Labour and South African Expansion, 1920–1950'. *South African Historical Journal,* Vol. 18, No. 1, 1986, pp. 73–92.

Kote, D.D. Recruitment of Mine Labour in Botswana 1899–1945, B.A. (History) thesis, University of Botswana, Lesotho and Swaziland, 1976.

---

117 Maganu, 'Botswana National Tuberculosis Programme', p. 10.

118 *Briefing Note on HIV and Labour Migration in Botswana.* IOM International Organization for Migration, 2005, p. 5.

119 https://www.tshiamisotrust.com/information/progress-report/.

Leepile, M. 'The Impact of Migrant Labour on the Economy of Kweneng, 1940–1980'. *Botswana Notes and Records,* Vol. 13, No. 1, January 1981, pp. 33–43.

Livingston, Julie. 'Physical Fitness and Economic Opportunity in the Bechuanaland Protectorate in the 1930s and 1940s'. *Journal of Southern African Studies,* Vol. 27, No. 4, December 2001, pp. 793–811.

Macvicar, Neil. 'Tuberculosis among the South African Natives'. *South African Medical Record,* Vol. 6, No. 11, 10th June 1908, pp. 161–176.

Maganu, E.T. 'Botswana National Tuberculosis Programme. Tuberculosis in Botswana: Results of an Epidemiological Survey', 1981. Ministry of Health, Gaborone. Government Printer.

McCulloch, Jock. *Black Peril, White Virtue: Sexual Crime in Southern Rhodesia, 1902 to 1935.* Bloomington: Indiana University Press, 2000.

Morapedi, W.G. 'Migrant Labour and the Peasantry in the Bechuanaland Protectorate, 1930–1965'. *Journal of Southern African Studies,* Vol. 25, No. 2, June 1999, pp. 211–213.

Mushingeh, A.S.C. A History of Disease and Medicine in Botswana, 1820–1945, PhD Dissertation in History, St John's College, University of Cambridge, August 1984.

Packard, R.M. *White Plague, Black Labour: Tuberculosis and the Political Economy of Health and Disease in South Africa.* Berkeley: University of California Press, 1989.

Schapera, I. *Migrant Labour and Tribal Life.* London: Oxford University Press, 1947.

Schechter, M. 'Mass X-ray Survey: Bechuanaland Protectorate 1952'. *The South African Medical Journal,* Vol. 28, No. 4, 24th April 1954, pp. 351–356.

Spence, J.E. 'British Policy Towards the High Commission Territories'. *The Journal of Modern African Studies,* Vol. 2, No. 2, 1964, pp. 221–246.

Taylor, J. 'Some Consequences of Recent Reductions in Mine Labour Recruitment in Botswana'. *Geography,* Vol. 71, No. 1, January 1986, pp. 34–46.

Torrance, D. 'Britain, South Africa, and the High Commission Territories: An old controversy revisited'. *The Historical Journal,* Vol. 41, No. 3, 1998, pp. 751–772.

Wilson, Francis. *Migrant Labour in South Africa: Report to the South African Council of Churches.* Johannesburg: The South African Council of Churches and SPRO-CAS, 1972.

9

Tuberculosis and Migrant Labour in the High Commission Territories: Basutoland and Swaziland: 1912–2005

Basutoland (now Lesotho), a small and densely populated mountainous territory with a relatively cool climate, came under British rule in the late nineteenth century. By the 1930s, its transformation into a labour reserve for South Africa's mines had made it a net importer of maize. When the 1912 South African Tuberculosis Commission collected evidence on the spread of disease in the HCTs, the reports from Basutoland suggested that tuberculosis had been increasing, especially among those who had worked on the mines. Dr H.N. Macfarlane noted that during his 18 years as a Medical Officer, he had treated 322 cases, 80 per cent of them males. It was only after 1902, when significant numbers of men began going to the mines, that tuberculosis became prominent. Dr H.W. Dyke, a private practitioner at Butha Buthe, tended similar evidence. Men usually acquired tuberculosis on the mines and, on returning home, infected their families. Many of them also had fibrosis. In Dyke's opinion, if those men remained away from mining, they may recover, but if they went back, they usually returned with 'their lungs breaking down'. According to Dr G. Hertig, the medical officer at Morija, not a week passed that he did not see fresh cases, mostly from the gold mines.[1] Over time, the

[1] *Report Tuberculosis Commission, 1914.* U.G. 34-'14, Cape Town: Government Printer, 1914. Union of South Africa, pp. 78–80.

J. McCulloch, P. Miller, *Mining Gold and Manufacturing Ignorance*,
https://doi.org/10.1007/978-981-19-8327-6_9

numbers of men entering the mines rose, but there was no corresponding increase in the capacity of the local medical service. In 1924, the medical staff at the capital Maseru consisted of a Principal Medical Officer, one Medical Officer, one relieving Medical Officer and the superintendent of the Leper Settlement. There were also seven Medical Officers stationed at major settlements such as Mohale's Hoek.[2]

The Costs and Benefits of Migrant Labour in Basutoland

The mortality rates among Basuto miners were far higher than for South Africans. In June 1928, the High Commissioner wrote to the Secretary for Native Affairs in Johannesburg asking for an explanation. According to data on a cohort of 110 consecutive deaths among Basuto compiled by the NRC, the major causes were pneumonia and tuberculosis. The High Commissioner was greatly concerned, especially as Basutoland was sometimes called the Switzerland of South Africa, because so many Europeans with tuberculosis had been cured by living there.[3] In October that year, Dr E. Cluver from the Department of Health was asked to investigate.[4] He calculated that the death rates for Basuto on both the gold and coal mines were almost double those of South African workers. The data also showed that most deaths were due to pulmonary disease and accidents. The Basutos in general did the heaviest classes of work such as tramming and shovelling, but Cluver made no reference to the high dust exposures involved. Instead, he wondered if fatigue might account for their susceptibility to infection. He also thought it significant that a relatively large proportion of Basuto worked in B Mines, which had a depth of over 760

[2] 'Basutoland Medical and Sanitary Report', 1924, in *South African Medical Record,* 10th April 1925, p. 163.

[3] Letter from High Commissioner's Office, Pretoria, to The Secretary for Native Affairs, Pretoria, 26th June 1928. SANA, PM 1/2/63 18/46, High rate of mortality from disease among Basuto Natives employed on the Mines.

[4] Letter from H.D.J. Bodenstein, Secretary for External Affairs, Pretoria, to The Imperial Secretary, High Commissioner's Office, 2nd November 1928, 'Re. Mortality among Basutos on Witwatersrand Gold Mines'. SANA, PM 1/2/63 18/46, High rate of mortality from disease among Basuto Natives employed on the Mines.

metres. Importantly, when men returned to the mines, as most did to serve successive contracts, they started again as recruits, without any history of their previous service. The resulting lack of reliable data made it difficult for Cluver to make firm recommendations.[5]

Eight years later, statistics remained elusive. After he was appointed Principal Medical Officer of Basutoland in 1936, Hamilton Dyke wrote an overview of tuberculosis in the Territory. As there were no notifications of deaths, it was impossible to estimate the mortality rate. In addition, almost all diagnoses were made without an X-ray or laboratory aids. In Dyke's view, the chief factors which favoured the spread of the disease were improper diet and the lack of sufficient ventilation in sleeping rooms. Dyke opposed treating patients in hospitals: a good proportion of cases were acute and most died within a short period. More chronic cases would soon tire of hospital and leave prematurely, thereby exposing their families to infection. To protect other patients would require building specialist wards at a prohibitive cost.[6] Dyke favoured controlling the disease by improved hygiene and enhanced diet rather than institutional care.[7]

The effects of the gold mines on Basutoland extended far beyond lung disease. In April 1943, C.N.A. Clarke, of the Resident Commissioner's Office in Maseru, wrote an appraisal of the costs and benefits of migrant labour. In the 1940s, the population of Basutoland numbered some 650,000 people. Recruiting for the gold mines had expanded and by 1942, there were an estimated 70,000 Basuto employed in mines and industries. There were in addition 15,000 enlisted in the Pioneer Corps and a further 20,000 on farms. In all, over half of the adult male population of working age was away from the Territory at any given time.

---

[5] Letter from Dr E.H. Cluver, Assistant Health Officer, Department of Public Health, Union of South Africa, to The Secretary for Public Health, Pretoria, 5th November 1928, 'Health Conditions on Witwatersrand Gold Mines, 1928: Mortality among Basuto employed on Transvaal Mines', pp. 8; 3. SANA, PM 1/2/63 18/46, High rate of mortality from disease among Basuto Natives employed on the Mines.

[6] Memorandum on the Incidence of Tuberculosis in Basutoland and the Necessity for Hospital Treatment of Tubercular Patients, by H.W. Dyke, Principal Medical Officer, Maseru, 25th August 1936, pp. 2; 5. BNA S438/2/1 Tuberculosis in the B.P., 1938–39.

[7] Notes on Tuberculosis in the High Commission Territories, June 1940, p. 1. BNA S438–2-2 Tuberculosis in B.P. 1940–1953.

However, only a small proportion of a labourer's earnings found its way back to the Territories. Under the Remittances and Voluntary Deferred Pay scheme, the largest return was in 1941 when 50,000 men brought home approximately £300,000 out of estimated earnings of £1,500,000.[8]

The first wave of emigration to the Union, Clare wrote, provided a useful addition to men's incomes, enabling them to pay their taxes. Since then, the populations of the HCTs had doubled and the consequent pressure upon land had reduced the earning capacity of the peasantry. Families relied more and more on mine earnings. However, because wages had scarcely increased, the real income per head of population was lower than it had been 40 years earlier. By paying below-subsistence wages, the mining industry was retarding the Territories' development. Clarke warned: 'The payment, therefore, of an uneconomic wage by the Mines on the grounds that the labourer is only supplementing his income which is mainly derived from agriculture is unsound and will lead to disaster'. Migrant labour in Basutoland did not supplement farm incomes: there were usually no such incomes in a mountainous country where only 16 per cent of the land is arable.[9]

Clarke concluded that the Territories were responsible for maintaining the families of migrant workers and for producing medically fit men for the mines, with little contribution from the employers. While the Union Government received substantial revenue from taxation, the Territories benefitted only from the payment of a two-shilling capitation fee by the WNLA.[10] The health and physique of labour improved after one or two months on the mines, but that gain was offset by the ill-effects of the compound system. Although Clarke makes no mention of silicosis or

[8] Enquiry into remuneration and conditions of living of Natives employed therein with special reference to Natives of the High Commission Territories from C.N.A. Clarke, Resident Commissioner's Office, Maseru, Basutoland to Lt Col. A.D. Forsyth Thompson, Bechuanaland, marked 'Confidential', 29th April 1943, p. 14. TEBA Archives, Social Work Union Mines, S464/1.

[9] Kate B. Showers. *Imperial Gullies: Social Erosion and Conservation in Lesotho.* Athens: Ohio University Press, 2005.

[10] Enquiry into remuneration and conditions of living of Natives employed therein with special reference to Natives of the High Commission Territories from C.N.A. Clarke, Resident Commissioner's Office, Maseru, Basutoland to Lt Col. A.D. Forsyth Thompson, Bechuanaland, marked 'Confidential', 29th April 1943, pp. 7–8; 15. TEBA Archives, Social Work Union Mines, S464/1.

tuberculosis, his portrayal of the migrant labour system is damning. He also noted that it was imperial policy to encourage the formation of indigenous trade unions. The Colonial Development and Welfare Act, which set aside £50 million to be spent in the Colonies and Protectorates over the coming decade, stipulated that no assistance would be given to a colony which did not recognise unions. Trade Unions were already recognised in the three HCTs, but those rights were not available to men working in South Africa.

Clarke's confidential enquiry complemented the conclusions of a 1942 report on migrant labour prepared by the HCT's Office representative in Johannesburg. A.G.T. Chaplin found that it was difficult to arrive at reliable figures of the numbers of men, women and children from the Territories living on the Witwatersrand. He estimated that in 1936, half of Basuto men aged between 18 and 50 were working in the Union. Economic need was the major factor pushing men into the South African labour markets. However, there were other contributory factors, among them the dull life in the Territories, the breaking down of traditional authority, and the desire of young men to escape from the control of parents and Chiefs. Migrancy, Chaplin noted, came at a high social cost. Men living in mine compounds were divorced from their families and often formed illicit unions with local women. Once such a union had been established, men would often stop supporting their families at home. Large numbers of Basuto women too worked in South Africa. Their domestic conditions were unstable, with few living in a family unit. There were many cases where women went to Johannesburg to visit their husbands or for medical treatment but ended by staying permanently. According to Chaplin, the unauthorised influx of women was caused by the long absences from home of their menfolk, the uncontrolled and casual nature of male labour, the failure of men to support their families, and the lack of adequate medical care in Basutoland.[11]

[11] 'The Chaplin Report', Letter from A.G. Chaplin, dated 1st September 1942, No. 153, the Agency for the High Commission Territories, Johannesburg, pp. 2; 4–5. SNA GS 512/42. Income and Expenditure in Johannesburg – A Study of African Income (Conditions of Natives in Rand Mines).

Like Clarke, Chaplin was highly critical of the inequalities inherent in labour migration. Through taxation on gold mining, the Union Government collected revenue, some of which went to the native Reserves inside South Africa. There was no such benefit for the HCTs, which in effect were subsidising the gold mines. The mines used the Territories for recruiting in order to force down local wages in South Africa. Chaplin recommended that deferred wages schemes be made compulsory, and that the period of employment be prescribed to ensure that men returned home.[12] Given the allied costs and social disruption, Chaplin questioned the benefits of the entire system of labour migration.[13]

The HCTs were asked to respond to the Report. Although the scale and impact of migrant labour was different in each of the Territories, there was common ground. The Government Secretary of Bechuanaland was concerned about the large number of young men who left to work in the south. There were an estimated 15,000 men on the Reef and probably another 15,000 to 20,000 employed elsewhere in the Union. They represented a large proportion of the Protectorate's 60,000 taxpayers and their absence disrupted family life.[14] The Resident Commissioner's Office in Basutoland noted that the main cause of migration was economic pressure due to the high rates of local taxation. A simple form of income tax, based on the individual's capacity to pay in place of a flat-rate poll tax, would see fewer men seeking work in the Union. Emigration was not in the best interests of the country or the Basuto, as it disrupted family life and social organisation. The estimated 50,000 Basuto on the Rand played an important part in gold production, from which the Union Government received tax revenue. The only return to Basutoland was the 2 s capitation fee and

[12] Enquiry into remuneration and conditions of living of Natives employed therein with special reference to Natives of the High Commission Territories. Notes on Resident Commissioners' Conference, January 1943. Conditions of Living of High Commission Territories Natives on the Rand. TEBA Archives, Social Work Union Mines, S464/1.

[13] 'The Chaplin Report', pp. 6–7.

[14] Minutes, Government Secretary, Bechuanaland, Undated, 1942. Enquiry into remuneration and conditions of living of Natives employed therein with special reference to Natives of the High Commission Territories. TEBA Archives Social Work Union Mines, S464/1.

in some cases the payment of outstanding poll tax. The Resident Commissioner suggested an increase in the capitation fee to 10s or 15s. He also wanted contracts to require that two-thirds of a miner's wages be deferred, with payment being made in the Territory.[15] In contrast, so few men left Swaziland for the gold mines the Territory's Resident Commissioner was not much concerned. In September 1941, there were 5677 Swazi on the gold mines, 633 on coal mines and a further 1087 in other employment. The Resident Commissioner reminded the High Commissioner that all the HCTs received a great deal of taxation revenue from miners.[16]

During the Second World War, the regional medical services' lack of capacity to provide care or to collect data on the disease burden was raised at a number of forums. One of these was a Conference on the Medical and Health Services in the HCTs, held at the beginning of 1942. In part, the Conference was intended to justify the Territories' retention under British control once the war had ended, rather than their transfer to South Africa. The delegates agreed that the Territories should be an example to the Union of the advantages of liberal policies regarding race, and of the need to develop social and welfare services for the local population.[17] They also concluded that the provision of more hospital beds, health education and the creation of rural clinics should be a priority. As always, a lack of funding was a major hurdle.

The territories might well serve as an example of enlightened policy in the future, but at the time, they were far from ideal. The High Commissioner, Lord Harlech, was highly critical of the HTCs' medical services. In such small departments, there was little scope for specialised

[15] Enquiry into remuneration and conditions of living of Natives employed therein with special reference to Natives of the High Commission Territories. Memorandum from C.N.A. Clarke, Resident Commissioner's Office, Maseru, Basutoland, dated 3rd November 1942. TEBA Archives, Social Work Union Mines, S464/1.

[16] Enquiry into remuneration and conditions of living of Natives employed therein with special reference to Natives of the High Commission Territories. Letter from the Resident Commissioner, Swaziland, to the High Commissioner's Office, Pretoria, 2nd December 1942. TEBA Archives, Social Work Union Mines, S464/1.

[17] Conference on Development of Medical and Health Services in the High Commission Territories, 1942, p. 1. BNA S393/1/1 Health & Medical, 1942–50.

work or promotion, and the low salaries made it difficult to attract recruits. The annual rates of pay for medical officers varied between £600 in Bechuanaland and £500 in Swaziland. In contrast, medical officers in the South African health service were appointed at £800 per annum.[18] During a conference at Cape Town in January 1943, Harlech reiterated his concerns: 'Ever since I have been High Commissioner I have been dissatisfied with the existing provisions as regards public health in all these territories, and have so informed the Secretary of State'. He concluded: 'The urgency of need of the patient and not his or her capacity to pay fees must be the governing criteria in receiving skilled medical attention if we are to get on to any morally defensible basis'.[19]

From the mid-1950s, there were occasional exposés in the British press about men from the HCTs being forced to work on South Africa's mines to pay their taxes. The Labour politician, Arthur Creech Jones, took a particular interest in the issue: 'It seems to me that … the threat of work in the mines—a foreign state for all practical purposes—ought never to be exercised, that this form of recruitment is indefensible … The practice exposes the people to abuse and inflicts on them an evil. Have we the moral right to condemn people to labour where the policy of apartheid is practised?'[20] In response, the Commonwealth Relations Office in London pointed out that compulsion was never used by authorities, but that sometimes, particularly in cases where men owed several years' tax, it was suggested that by going to the gold mines they could earn sufficient to pay their arrears.[21]

[18] Memo: Sir W.B. Johnson, the Administrative Secretary to the High Commissioner, to Mr Priestman, 9th October 1942. BNA S393/1/1. Health & Medical, 1942–50.

[19] Report of Proceedings of the Resident Commissioners' Conference held at Cape Town on Wednesday, 20th January 1943, pp. 1; 6. BNA S393/1/1 Health & Medical, 1942–50, S393/1/1.

[20] Cited in Letter from N.V. Redman, Commonwealth Relations Office, Downing Street, London, to J. A. Steward Esq. CBE, Pretoria, 11th July 1960. BNA S387/7/4 Recruitment of Labour 1960.

[21] Letter from N.V. Redman, Commonwealth Relations Office, Downing Street, London, to J.A. Steward Esq. CBE, Pretoria, 11th July 1960. BNA S387/7/4 Recruitment of Labour 1960.

Occupational Disease and Injury After Independence

Basutoland gained independence from Britain in 1966 and became the Kingdom of Lesotho. The country's extreme dependence on mine wages made it a special case among the labour sending states. In 1976 there were just 27,500 Basotho wage labourers employed inside the country while 200,000 worked regularly in South Africa, half of these on the mines. In the late 1980s, remitted wages accounted for three quarters of Lesotho's GDP, with each absentee miner directly supporting ten dependents and six others downstream.[22] Lesotho's dependency on migrant labour came at a high price in terms of family breakdowns, female headed households, and the impoverishment of women.[23] By the early 1960s, the capital Maseru had a population of less than 10,000, with more than 90 per cent of people living in rural areas. The Black Laws Amendment Act of 1963 made it impossible for migrant workers to gain rights to continuous residence in South Africa. Men could only engage for work in the mines by a contract not exceeding a year made in Lesotho.[24]

According to the World Health Organization, a country with more than 200 tuberculosis cases per 100,000 of its population is experiencing an epidemic. In the period from 1991 to 2001, the recorded tuberculosis prevalence in Lesotho gradually increased from 159 to 506 per 100,000, with a heavy concentration of the disease among young men. The Ministry of Health and Social Welfare coordinates and manages the National Tuberculosis Program. The Maseru District has three hospitals, all of which treat tuberculosis. Treatment in government hospitals is free and since 2000, case reporting has improved. Because of limited resources, however, the health and welfare of migrant labourers, including mineworkers, had been neglected. In 2005, there were no reliable national

[22] J.S. Harington, N.D. McGlashan and E.Z. Chelkowska. 'A Century of Migrant Labour in the Gold Mines of South Africa'. *The Journal of the South African Institute of Mining and Metallurgy*, Vol. 104, No. 2, March 2004, p. 70.

[23] Colin Murray. *Families Divided: The Impact of Migrant Labour in Lesotho*. Cambridge: Cambridge University Press, 1981, p. 88.

[24] Murray, *Families Divided*, p. 22.

health data.[25] In addition to the miners repatriated with tuberculosis or silicosis, the mines produced such a steady stream of sick and injured that mine accidents constituted the largest single cause of disability amongst Lesotho men of working age.[26]

Due to the downsizing of the mining sector, the number of Basotho employed on South African mines fell from 95,913 in 1996 to 52,450 in 2005. Despite those retrenchments, Lesotho's economy remained dependant on remittances. Oscillating migration, which disrupts normal family life, has been one of the key drivers of the current HIV/AIDS epidemic in Lesotho. Single-sex hostels for young men and limited home leave lead to loneliness and a breakdown in social cohesion. The vulnerability of migrant workers has been compounded by the dearth of information and the lack of HIV programmes in rural areas.[27] The UN International Convention on the Protection of the Rights of Migrant Workers, ratified by Lesotho in 2005, states: 'migrant workers and members of their families shall have the right to receive medical care'. The lack of state capacity has meant that the care available to former miners has been rudimentary at best.

Swaziland

Swaziland (renamed eSwatini in 2018) was the smallest of the three protectorates. Land alienation to white settlers under British concessions meant that by the early 1930s, the territory produced only a fifth of its food needs. However, commercial agriculture and large deposits of asbestos generated local employment and foreign exchange and made the territory less dependent on migrant wages. The major private sector employer was the Havelock/Bulembu asbestos mine, which operated from 1939

[25] Lugemba Budiaki. Tuberculosis and Compensation: A Study of a Selection of Basotho Mineworkers from Maseru District. Masters of Public Health, University of the Witwatersrand, Johannesburg, 2005, pp. 1–8.

[26] R.S. Arkles, A.J. Weston, L.L. Malekela and M.H. Steinberg. *The Social Consequences of Industrial Accidents: Disabled Mine Workers in Lesotho.* National Centre for Occupational Health. NC0H Report Number 13/1990, p. 17.

[27] See *Briefing Note on HIV and Labour Migration in Lesotho.* International Organization for Migration (IOM), 2007.

until 2001. It was owned by the British conglomerate Turner & Newall, which also had asbestos mines in Southern Rhodesia. Havelock was the country's third largest settlement, and the largest earner of foreign currency. The work and living conditions were hazardous, but the pay was far higher than on white estates, and the mine provided family housing, medical care and a school.[28]

As in the other HCTs, tax collection and occupational lung disease posed serious problems to the territory's administration. In his 1932 report on Swaziland's economy, Sir Alan Pim suggested that special arrangements be made for collecting tax from men employed on the Rand Mines. A representative of all three HCTs should liaise with the South African mines. His duties should include assisting men disabled by accident. Such an appointment, Pim suggested, could be combined with the collection of tax.[29] Four months before Pim's report came out, Basutoland established an office in Johannesburg to assist Chiefs in collecting arrears and current tax, and to encourage miners to send remittances home. Collection was made at the various compounds on pay days. However, because of the large number of mines, the officer found it impossible to visit them all. In addition, some of the miners became aware of the collection and delayed drawing their pay to avoid paying tax.[30] In the following year, the number of staff was increased to three, and in line with Pim's recommendation, the office began representing Swaziland.[31] The agents also dealt with domestic matters and discouraged men settling permanently in Johannesburg. However, there is no evidence of them assisting miners gain access to compensation.

Pim's report also expressed concern about the high rates of occupational disease. When Swazi labourers left the mines, they did not pass through the compounds and undergo a medical examination. Pim was

---

[28] See Jock McCulloch. 'Dust, Disease and Labour at Havelock Asbestos Mine, Swaziland'. *The Journal of Southern African Studies*, Vol. 31, No. 2, 2005, pp. 251–266.

[29] Sir Alan Pim. *Financial and Economic Report on Swaziland Protectorate*, 1932, excerpt, p. 1, paragraph 52. BNA S299/2. Hut Tax Collection on Rand Mines, 1933.

[30] Letter from G.J. Armstrong, Sub-Inspector, B.M.P., Report to the Government Secretary, 'Recent Collection of Tax from Basutoland Natives on the Witwatersrand', Maseru, Basutoland, 4th August 1932, p. 2. BNA S299/2. Hut Tax Collection on Rand Mines, 1933.

[31] Undated Memo: Government Representative at Johannesburg, 1934? p. 1. BNA S387/5 Recruitment of Labour, 1942.

sure that many men were unaware of their rights regarding compensation.[32] The brief review of the Territories' medical service by the Principal Medical Officer at Mbabane, Dr R. Jamison, endorsed Sir Alan Pim's conclusion that the Administration had failed to provide adequate medical facilities, as well as his recommendation that such a small Territory should establish outposts run by trained black or Coloured nurses to treat women and children rather than build expensive central hospitals. In its first year of operation, the new hospital at Mbabane treated over 8000 African and 1095 European out-patients.[33] The staff workloads were heavy. On average, doctors saw 40 in-patients a day, examined another 35 out-patients, and carried out operations. Medical Officers also made home visits, did weekly court attendance and gaol inspections, had a small private practice and performed post-mortems. The service remained under-funded, even as the number of patients increased. During 1949, Swazi hospitals treated a total of 68 in-patients and 213 out-patients for tuberculosis, and it was proposed to begin localised tuberculin surveys. In the following year, the numbers of both in- and out-patients almost doubled, without an increase in capacity.[34]

After the war, there were occasional references from the Resident Commissioner's Office about lung disease in returning miners, but the medical service still lacked the staff and technology to treat tuberculosis. In 1950, the WHO Expert Committee on Tuberculosis' guidelines for disease control in developing countries was circulated by the Colonial Secretary. The Committee endorsed the conventional wisdom that a reduction in tuberculosis was best achieved by improvements in nutrition, housing, education and occupational health. A control programme should aim to prevent the spread of infection from known cases; it must also protect highly exposed groups and promote preventive and curative measures, including after-care and rehabilitation. The WHO endorsed

[32] Sir Alan Pim. *Financial and Economic Report on Swaziland Protectorate,* 1932, excerpt, p. 1, paragraph 52. BNA S299/2. Hut Tax Collection on Rand Mines, 1933.

[33] Letter on Pim's Report from R. Jamison, Principal Medical Officer, Mbabane, Swaziland, to Acting Government Secretary, 13th October 1932, pp. 1; 4. Swaziland NA RCS 777/12, Medical Comments on Pim Report, 1932.

[34] Swaziland Annual Medical and Sanitary Report for the Year 1950, pp. 6; 8. Swaziland NA RCS 3021 M –V, Tuberculosis.

sample surveys using tuberculin testing and mass radiography to identify infection and morbidity rates. Surveys should be supported by a central laboratory for diagnosis and a comprehensive record system augmented by epidemiology and clinical services. Dispensaries, preferably at existing hospitals or public health centres, should be established to isolate infectious cases. Home nursing or health visiting services were essential. A nurse or health visitor could teach the patient and their family how to prevent infection. To control the disease, intensive education of the public about tuberculosis was essential.[35] The WHO's 1950 prescriptions were based on public health principles used in Britain and Prussia from 1910 and this gives an indication of the inadequacy of the HCTs' response.

From the early 1950s, the British Red Cross Society took an interest in tuberculosis and the impact of migrant labour in Swaziland. The Society was concerned '[t]hat the incidence of TB among Swazi mine workers is the highest among any African tribe'.[36] The Society favoured treatment which catered for bed cases, convalescents and rehabilitation. On discharge, a system of Medical Officers and Outstation Clinics should monitor patients. The opening of a Hospital at Mahamba, with X-ray facilities and the proposed establishment of a rehabilitation settlement for tuberculotics, were major steps. Following the visit by a WHO specialist, a control scheme for the Territory was due to begin in 1960.[37]

After political independence in 1968, Swazi nationals continued to work on South Africa's gold, platinum and coal mines. The Swazi Labour Department attributed this to the lack of local employment and the better pay on offer across the border. If the working conditions in South Africa continued to improve, the Department expected the numbers

[35] World Health Organization 1950, 'Plan for Control Programmes: Suggestions for the Control of Tuberculosis in Countries with Undeveloped and Underdeveloped Programmes', p. 21. Swaziland NA RCS 2251, Secretariat, Native Labour Tuberculosis Control.

[36] Letter to Mrs M. Armstrong, Hon Secretary, British Red Cross Society, Mbabane, Swaziland, from Chairman, Goedgegun Division, Red Cross, 2nd June 1951. Swaziland NA RCS 2251, Secretariat, Native Labour Tuberculosis Control.

[37] Session of the Reconstituted European Advisory Council Held at Mbabane, 11th November 1959. Swaziland NA 3021 M. Tuberculosis.

would grow.[38] Nevertheless, of the more than 800,000 foreign migrants working in the economies of Southern Rhodesia and South Africa in 1972, just 4 per cent came from Swaziland. In contrast, Lesotho accounted for a quarter.[39] By 1975, there were 16,278 Swazi recruits on the gold mines and just 257 on the coal mines. In the following year those numbers rose to 20,334 and 451 respectively.[40] The recruitment of labour to South Africa was administered under the Employment Proclamation No.51 of 1962. Contracts could not exceed one year, with renewal for a further maximum period of nine months. The law stipulated that conditions of employment must be satisfactory. However, there was no machinery to ensure compliance by employers.[41]

In its annual report for 1979, the Swazi Labour Department presented a brief review of the Worker's Compensation Act of 1941, under which many Swazis were employed. Although the South African Government had introduced a common scale for occupational injuries, these were based on earnings and thus there was a considerable difference in the benefits payable to black and white workers. The Workmen's Compensation Commissioner made little effort to trace men entitled to compensation, and South African employers did not provide notification of accidents involving migrants. The Labour Commissioner noted that it would be difficult for any country in the region, acting individually, to change contractual conditions faced by migrant workers. There could be major benefits if the HCTs formed a cartel, enabling them to make joint approaches to the South African authorities. This was particularly so given that employers, particularly the Anglo-American Group, had consistently shown themselves to be ahead of the Government regarding the liberalisation of employment policies.[42]

---

[38] *Annual Report Swaziland Labour Department, 1976*, pp. 5–6.

[39] Francis Wilson. *Migrant Labour in South Africa: Report to the South African Council of Churches.*, Johannesburg: The South African Council of Churches and SPRO-CAS, 1972, p. 104.

[40] Appendix C in *Annual Report Swaziland Labour Department, 1976*, pp. 6; 26.

[41] *Annual Report Swaziland Labour Department, 1974*, p. 9. The Worker's Compensation Act of 1941 excluded mining-related lung disease but covered all non-lung occupational diseases and traumatic injury in miners.

[42] *Annual Report Labour Department, 1979*, pp. 52; 54.

Access to Compensation for HCT Miners

The HCTs were selling labour into a racialised state which denied migrant workers the most basic rights. One of the glaring inequalities concerned the progression of occupational lung disease. Under the Miners' Phthisis Acts from 1916, Europeans were entitled to pensions. Their health was monitored, and their awards were increased as the disease worsened. The small minority of black miners who received compensation were paid one-off lump sums. In theory at least, migrant workers had advocates in the British Colonial Office and the ILO. The Departments of Native Affairs and Native Labour repeatedly raised the issue of compensation with the Chamber. The issue was also raised in the South African Parliament, where Senator Ballinger pointed out the various inequalities inherent in the system. All this made the subject of pensions and disease progressions a sensitive political issue. For the Chamber, there was much at stake. Like all aspects of the Johannesburg mines, the racialised compensation system provided one of the pillars of the companies' profitability.[43]

The Chamber's usual response to critics was to reiterate its opposition to providing pensions to blacks, and to deny that there were large numbers of former miners in the rural areas with uncompensated disease. It estimated that each year, there were at most around 30 men with miners' phthisis unable to make the journey to Johannesburg for an examination.[44] How the Chamber arrived at that figure was not explained. By the late 1930s, the Chamber's legal advisors had come to believe that if the industry did not change its stance, the government was likely to intervene. In 1939, the Gold Producers' Committee agreed to a system of *ex gratia* payments to black beneficiaries whose silicosis had progressed, but excluded men from the High Commission Territories. In late 1941, it

[43] By the 1970s, based on a notion of the 'civilised wage ratio', the gap in compensation payments from the same category of disease reached 13 or 14:1 for silicosis.

[44] Letter from D. Smith, Joint Secretary, Transvaal Chamber of Mines, Johannesburg, upon Benefits to Native Labourers under Miners' Phthisis Act 1925 – Proposed Consolidated Fund, to Director of Native Labour, Johannesburg, 10th April 1940. SANA K105, Miners' Phthisis 1941–48, Vol.2 Correspondence: Transcripts, Stratford Commission.

approved an increase in payments and three years later, it extended benefits to recruits from the three HCTs. The payments were modest, and as most migrant workers were unaware of the scheme, there were few beneficiaries. In the four years from 1939 to 1943, only 342 grants were made.[45] In March 1942, the Chamber's Legal Adviser, G. Barry, wrote a position paper on the compensation system. He commented on the refusal of mine medical officers to refer men with compensable disease to the Bureau: 'Dr Girdwood also tells me that there is a tendency on the part of Mine Medical officers to reject Natives with considerable underground history as such natives may soon be certified to be silicotic and the responsibility for compensation rests with the employer who has last signed him for underground work'.[46] There is no evidence that Barry or the Chamber made any effort to remedy that injustice.

Migrant workers from the HCTs faced even greater barriers in accessing compensation than black South Africans did. In November 1940 A.G.T. Chaplin, the HCTs representative in Johannesburg, wrote to the High Commission Office expressing his concern that so few men received payments. No circulars or instructions on the subject had been issued, miners were unaware of their rights and District Officers did not understand the application process. Consequently, in some districts of Basutoland, no claims had ever been lodged. There was no X-ray plant in the Territory capable of taking films of sufficient clarity, forcing those applicants who were able to do so to travel to Johannesburg. Chaplin's own experience suggested that the majority of cases only came to light when a man was close to death. In half the cases he dealt with as a District Commissioner, the applicant died before his claim was assessed. Chaplin accepted that it would be difficult to conduct medicals for foreign

[45] Letter from Dr Peter Allan, Department of Health to Secretary, Chamber of Mines. Subject: Repatriation of Native Tuberculotics from the Mines to Native Territories. February 1944, Miners Phthisis Compensation for Natives 1942–1949. NRC 390 1&2 TEBA Archives.

[46] Memo, Miners' Phthisis Commission: Pooling of Native Compensation, from G. Barry, Legal Adviser to H. Wellbeloved, the Chamber, 26th March 1942. Appended to Memo to Member of the Gold Producers Committee 8th April 1942. Subject: NRC 390 1&2 Miners Phthisis Compensation for Natives 1942–1949 TEBA Archives.

workers, but was confident that could be largely overcome if up-to-date X-ray equipment was installed in the Territories.[47] However, it was war time, and the necessary equipment was not available.

Those applying for re-classification of their occupational injury were required to attend an examination in Johannesburg, something which was not feasible for men who were dying.[48] In addition, African families were strongly opposed to post-mortems on religious grounds, and this ruled out applications from dependents of most deceased miners. By November 1946, the ex-gratia scheme for the HCTs had been in operation for two years, but there had been no applications from living beneficiaries and only four from the relatives of deceased miners. In only one of those four cases was there a post-mortem, so no further compensation was paid to the other surviving relatives.[49] The Chamber was adamant that X-ray facilities ought to be available in the Territories, and it refused to pay the transport costs to Johannesburg for ex gratia cases.[50] As the scheme was deemed a form of charity which fell outside of the Act, the Department of Mines ruled that all costs of transport, maintenance, and medical examination must be borne by the HCTs.[51] The Assistant Secretary in Mafeking acknowledged that the ex gratia scheme was not working in the Bechuanaland Protectorate.[52] During 1948, 16 claims were processed.

[47] Letter from A.G.T. Chaplin, Representative of the High Commission Territories, Johannesburg to the Government Secretary, Maseru, 27th November 1940, pp. 1–2. SNA RCS 141/42. Tuberculosis in the High Commission Territories and Claims for Compensation under Miners' Phthisis Act.

[48] Memorandum from District Commissioner, Kanye, to The Government Secretary, Mafeking, 5th November 1946, Subject: Miners' phthisis patients. BNA S 332/5/21945–48. Phthisis: Miners' Compensation for B.P. Native Mine Workers, 1945–48.

[49] Memorandum: 'Miners' Phthisis Patients' from Mackenzie (?), District Commission to the Government Secretary, Mafeking, 7th November 1946. BNA S332–5-3. Phthisis: Miners' Comp. for B.P. Native Mine Workers, 1948–50.

[50] Letter from R.F. Thompson, Office of the Agent for the High Commission Territories, Johannesburg, to G.J. Armstrong, The Secretariat, Maseru, Lesotho, 8th September 1948. BNA S332–5-3 Phthisis: Miners' Comp. for B.P. Native Mine Workers, 1948–50.

[51] Minute, Ex-Gratia Payments, Government Secretary, Mafeking, undated (1948) BNA S332–5-3 Phthisis: Miners' Comp. for B.P. Native Mine Workers, 1948–50.

[52] Minutes, Reply to 303, from First Assistant Secretary, 8th November 1946. BNA S 332/5/2 Phthisis: Miners' Compensation for B.P. Native Mine Workers, 1945–48.

However, ten of those men had died without a post-mortem and so no award was made. Only three men were compensated.[53]

The 1942 Stratford Commission provided another platform for highlighting the shortcomings of the then compensation scheme, and for proposing improvements. The HTC representatives argued that the application process for miners who had returned home made access to compensation all but impossible. There were no X-ray facilities in the Territories capable of taking plates which met the Bureau's requirements, and most applicants were too ill or too poor to make the long journey back to Johannesburg. The Territories asked the Commission to consider whether some concessions could be made for applicants in remote areas.[54] A second issue raised at the Stratford Commission was disease progression. The Director of Native Labour, E.W. Lowe, told Stratford that the failure of the legislation to take account of disease progression in back miners was unjustifiable. For that reason, he favoured pensions over lump sums. However, he acknowledged that it may not be feasible to provide pensions for foreign labour.[55] The Gold Producers Committee (GPC), which made written and oral submissions to Stratford, was unsympathetic. It had introduced an *ex-gratia* scheme in lieu of lump sums because of the administrative difficulties. Many black miners lived in remote areas, and it would be impossible to guard against impersonation and fraud. For the same reasons, the provision of pensions for surviving dependants was impracticable. In any case, the life expectancy of a black miner with tuberculosis was too brief to justify a pension.[56]

The Stratford Commission was not convinced by the GPC's submissions and made several recommendations, including the creation of

---

[53] Summary of Replies to Circular Memorandum No.199 of 1948. BNA S332-5-3. Phthisis: Miners' Compensation for B.P. Native Mine Workers, 1948–50.

[54] Memorandum submitted for the High Commission Territories in Johannesburg to the Miners' Phthisis Commission, undated 1942. SANA K105, Miners' Phthisis 1941–48 [Stratford] Correspondence, Volume 2.

[55] Memorandum from E.W. Lowe, Director of Native Labour, Johannesburg, to Miners' Phthisis Commission, 9th January 1942. SANA K105 Miners' Phthisis 1941–48 [Stratford] Correspondence, Volume 1.

[56] Statement by the Gold Producers' Committee of the Transvaal Chamber of Mines upon Compensation for Native Labourers to the Miners' Phthisis (Stratford) Commission, undated 1942. SANA K105, Miners' Phthisis 1941–48 [Stratford] Correspondence, Volume 2.

sanatoria for black miners. It also proposed that a monthly pension replace lump sums, with a minimum payment of £1 12 s 6d for a married man with two children.[57] Three months after Stratford's Report was tabled, in what was to be the first in a series of meetings, the Miners' Phthisis Board met with senior officers from the Departments of Mines, Native Affairs and Native Labour, and the Miners' Phthisis Bureau to discuss pensions. It was a high-level meeting. Those present included the Secretary for Mines, J.F. Muller; the Chairman of the Miners' Phthisis Board, C.G. Southgate; the Chairman of the Miners' Phthisis Medical Bureau, Dr J.M. Smith; a Senior Law Advisor from the Department of Justice, Dr A. Schoch; the Under-Secretary for Native Affairs, W.J.G. Mears; and the Director of Native Labour, C. Alport. In line with the Chamber's consistent position, the Secretary for Mines doubted whether it was possible to pay pensions to black miners because of the many administrative obstacles. In particular, there was the question of identity, which made fraud more likely. Messrs. Smith and Southgate agreed. In contrast, the Director of Native Labour and the Under-Secretary for Native Affairs could see no difficulty in tracing men after they left the mines, as they would willingly come forward to receive entitlements. Native Affairs also favoured pensions, which it wanted to be set at around £2 a month. However, the Director of Native Labour was concerned that there would be an outcry from farmers that 'the Government is paying the native pension at the cost of labour supplies', and that the mining industry might say they were 'stopping Natives from coming back to the mines'.[58]

At a second meeting in November, the Minister of Native Affairs noted that the levels of benefits were very low, and he wanted the meeting to discuss the lack of compensation for disease progression. Mr Southgate reiterated that if a pension were paid, the first difficulty would be to trace men who had returned to the rural areas. How could the industry establish a scheme for 400,000 men? Mr Mears could see no such difficulties

[57] Memo from the Commission upon Compensation of Native Beneficiaries for The Minister, Dept. of Mines, 29/5/1946, p. 3. SANA K105 Miners' Phthisis, 1941–48 [Stratford] Correspondence, Volume 17.

[58] Miners' Phthisis Board, Resume of Discussion at the Board's Office on 5th October 1944, pp. 1–4. SANA K105, Miners' Phthisis 1941–48 [Stratford] Correspondence, Volume 4.

as the government was already paying military, old age and disability pensions to hundreds of thousands of recipients. The Board meetings produced only one notable outcome: the creation of a small pool of data on disease progression. The Chairman of the Miners' Phthisis Medical Bureau, Dr Smith, noted that according to data from the ex-gratia scheme, in the period 1939 to 1940 around half of the ante-primary and primary cases had progressed in the space of four years. That indicated a rate of progression of disease far higher than amongst white miners. Mr Southgate agreed that the information was significant and pointed out that if black miners were to be compensated for progression, it would require frequent medical examinations, at a high cost to the industry.[59]

At the beginning of 1945, the Chamber finally agreed that the *ex-gratia* scheme would be extended to the HCTs. Miners were required to submit an X-ray, a medical report and a sputum specimen to the Bureau. In the case of an applicant's death, his dependents could submit an X-ray, or a post-mortem examination could be performed. If the examining physician considered silicosis or tuberculosis to be present, the lungs were to be sent with a report to the Bureau in Johannesburg.[60] The small number of patients from the HCTs required to travel to Johannesburg for a medical were to be housed at the WNLA Compound, and the cost charged to their Territory of origin.[61] Arrangements were being made for the installation of an X-ray plant, capable of making suitable films, at Mbabane in Swaziland. Until those facilities were available, the Swazi government agreed to pay for the transport costs to Johannesburg.

The third and final pensions meeting was held in November 1946. As previously, there was heated discussion, but this time the differences between industry and Native Affairs and Native Labour became more apparent. The legislation recognised that black miners with silicosis had a marked tendency to contract tuberculosis. For that reason, they were

[59] Miners' Phthisis Board, Minutes of Meeting on Wednesday, 29th November 1944, pp. 3–6. SANA K105, Miners' Phthisis 1941–48 [Stratford] Correspondence, Volume 4.

[60] G.J. Armstrong, Government Secretary, Mbabane, Swaziland, Letter to All District Commissioners et al., 30th April 1945, p. 1. SANA RCS 141/42 Tuberculosis in the High Commission Territories and Claims for Compensation under Miners' Phthisis Act.

[61] Letter from E.C. Butler, Agent for the High Commission Territories, Johannesburg, to the Government Secretary, Maseru, 25th September 1944. SANA RCS 141/42 Tuberculosis in the High Commission Territories and Claims for Compensation under Miners' Phthisis Act.

barred from further employment in a dusty occupation. Despite acknowledging that principle, the Chairman of the Miners' Phthisis Board, G.G. Southgate, remained opposed to pensions. Under the Mozambique Convention, compensation to recruits from the Portuguese Territory was required to be the same as that payable to black South African miners. Therefore, it was not possible to provide pensions for men domiciled inside the Union unless the same provision was made for Portuguese recruits. He argued that until facilities for radiographic examination existed in the labour sending states, it was not possible to provide graduated compensation.[62] There were also administrative problems regarding identification, proof of payment and questions of dependency, which he believed were insurmountable.

According to the Secretary for Native Affairs, Mr Mears, both his department and Native Labour had always opposed lump sums. Pensions were already provided in the Union for the aged, the blind and for black soldiers under the War Pensions Act of 1942, and he could see no major obstacles. His department was willing to provide the necessary administrative support. Mr Barry from the Chamber acknowledged the principle that a workman seriously incapacitated by industrial disease or accident should receive a pension. The gold mines employed roughly 222,000 black miners, of whom approximately 85,000 were from the Union. The Chamber had introduced ex-gratia payments to South African miners in cases where the disease had progressed. That scheme required applicants to be examined and certified by the Bureau. Although few cases had come forward, the scheme was working satisfactorily. However, Barry could not visualise the extension of a pension scheme to recruits from outside of South Africa.

The Minister for Mines, Mr S.F. Waterson, pointed out that a chain of fully equipped X-ray depots staffed by radiographers and doctors would be required if examinations were to be carried out in rural areas. If a pension scheme was adopted for South African labour, the HCTs would demand that their workers be given the same benefits. Neither his department nor Native Affairs had data on disease progression, and the

[62] Memorandum from G.G. Southgate, Chairman, Silicosis Board, Johannesburg, 30th November 1946. SANA, K105, Miners' Phthisis 1941–1948, Silicosis Vol. 7, Act 1946, Natives Files.

Government Actuaries were unable to estimate the costs. Before the committee saddled the mines with a liability, it must know what that liability would be. The Bureau Director, Dr Smith, reminded the meeting that the black miner certified with silicosis progressed more rapidly than the European did and was also more liable to contract tuberculosis. The Bureau depended on the X-ray examinations and sputum samples made by District Surgeons. In most cases, those reports were unreliable and the X-rays useless. Sputum analyses were vital in diagnosing tuberculosis. In Johannesburg, the Bureau may take up to 30 specimens for one patient, but only single samples were submitted from the rural areas. Physical examinations were another difficulty, as former miners were usually too ill to travel long distances. Dr Smith had written to the Magistrates in all labour sending districts within the Union asking for details of the X-ray apparatus in use. The reports indicated that most machines were unsuitable for the Bureau's purposes.[63]

The final and most contentious issue was the building of isolation hospitals as recommended by the Stratford Commission. The Director of Native Labour, J.M. Brink, suggested that sanatoria be provided for black miners. The Chamber had much to gain, as sanatoria would have isolated infective cases, thereby protecting the labour pool and to some extent dampening criticism about repatriations from the Department of Native Affairs. Despite such potential benefits, Mr Barry from the Chamber strongly disagreed. For many years, Barry had been involved with the silicosis question and he was certain that under the existing scheme, the black miner was better off than if he was treated on the same basis as the European.[64] The Chamber's opposition to sanatoria suggests that other factors were at play. Perhaps it feared that such facilities would soon be filled with infected men and the extent of the tuberculosis problem would become visible.

Following the meeting, Brink traced a cohort of compensated tuberculosis cases. Of the 309 he investigated, 197 were reported dead, 26 not traced and 86 still under investigation. The 197 beneficiaries who had

[63] Minutes, Johannesburg, December 1946 of Conference re Pensions for Natives -Silicosis Act held in Pretoria on Thursday, 5th December 1946, pp. 5–6; 12. SANA K105, Miners' Phthisis 1941–1948, Silicosis Vol. 7, Act 1946, Natives Files.

[64] Ibid., p. 17.

died did so within a short period after being certified. In not one case was a post-mortem held, with the result that compensation could not be claimed by the dependents.[65] For that reason, Brink wanted the Silicosis Act amended as a matter of urgency. There was no response from the Chamber, and there was no attempt to replicate Brink's study. Almost a year later, the Silicosis Medical Board acknowledged that in the absence of any follow-up of cases, there was no data on the mortality from silicosis or tuberculosis in black miners. Nor was there any data on disease progression.[66]

There was no improvement in access to compensation during the immediate post-war period. In early 1948, the Government Secretary in Swaziland wrote to the High Commissioner that the failure to help phthisis sufferers was causing great hardship. On being told that they must attend Mbabane Hospital for an X-ray examination, several former miners at Mankalana said that they had no money for the bus and were too sick to walk. The Bureau in Johannesburg was not prepared to issue travel warrants.[67] The problems of gaining compensation for miners' occupational disease were also frequently raised by the Bechuanaland African Advisory Council. At a meeting at Mafeking in April 1949, for example, Kgosi Bathoen II commented that tuberculosis was rife amongst young men who had returned from the mines. When miners were recruited, care was not taken with the spelling of Setswana names, and often the wrong name was recorded. Problems also rose with the interpretation of dependence. The meaning of the word in Setswana was altogether different—and much broader—than the English idea of a dependant. Mr A.M. Bome from Bangwaketse pointed out that few Setswana men consulted European doctors, and asked that if a miner

[65] Letter from J. M. Brink, Director of Native Labour, Department of Native Affairs, Johannesburg, to The Secretary for Native Affairs, Pretoria, 6th September 1948. Subject: Proposed Amendment of the Silicosis Act, No. 47 of 1946. SANA K105, Miners' Phthisis 1941–1948, Silicosis Vol. 7, Act 1946, Natives Files.

[66] Summary of Discussion which took place on Tuesday 28th June 1949, Regarding Statistical Data to be Collecting Relative to Native Labourers. SANA K105, Miners' Phthisis 1941–1948, Silicosis Vol. 7, Act 1946, Natives Files.

[67] Minute from C.J. Armstrong, Government Secretary, Mbabane, Swaziland to the Administrative Secretary to the High Commissioner, Cape Town, 7th February 1948. 'Union Silicosis Act, 1946'. BNA S332–5-3. Phthisis: Miners' Comp. for B.P. Native Mine Workers, 1948–50.

contracted disease while at the mines, was he given a certificate as proof. Mr G.F. Nettelton, the Deputy Resident Commissioner who chaired the meeting, explained there were no certificates.[68] In 1950, the question of compensation was again reviewed by the HCTs, with all three territories agreeing that awards to deceased miners should be paid in full to their heirs.[69] The amounts involved were trivial: during the previous year, 15 awards, totalling £1753 -12d, were made to Bechuanaland miners and their dependents.[70]

At the request of regional medical services, during 1963 a number of compensation examinations were conducted in decolonising territories outside of South Africa. There were 262 in Lesotho, 54 in Botswana, 9 in Swaziland and 21 in the Central African Federation (Southern and Northern Rhodesia and Nyasaland).[71] In the following year, 245 benefit exams were carried out, with 180 in Lesotho, 40 in Swaziland and 25 in the Rhodesias. The numbers were minuscule in terms of the volume of labour flowing to the gold mines. From that point, the volume of examinations fell sharply. During 1966, a total of 24 benefit exams were performed at the request of the Director of African territories. There were 16 in Lesotho, three in Swaziland, four in Botswana and one in Malawi.[72]

Over the following decades, the Chamber maintained its opposition to pensions. In 1960, the Director of Medical Services in Mafeking pointed out that many former miners who had received compensation were never reviewed for an increase in their awards. That state of affairs had come about partly because District Officers and Medical Officers had such heavy workloads, and partly because former miners were reluctant to

---

[68] Minutes of the African Advisory Council held at Mafeking April 21st to 29th 1949 Thirtieth Session, pp. 56, 59. BNA S332–5-3 Phthisis: Miners' Comp. for BP Native Mine Workers, 1948–50.

[69] Savingram [circular] from The Chief Secretary to The Government Secretary, Maseru/ Mafeking/ Mbabane, 3rd June 1950. BNA S332–5-4, Phthisis: Miners' Compensation for B.P. Native Mine Workers 1950.

[70] Letter from the District Commissioner, Botswana, to The Government Secretary, Mafeking, 26th September 1950. Subject: Payment of Silicosis Awards to Dependents of Deceased Beneficiaries. BNA S332–5-4 Phthisis: Miners' Compensation for B.P. Native Mine Workers, 1950.

[71] *Report of the Miners' Medical Bureau for the Period 1st October 1962 to 31st March 1963*. Pretoria: Government Printer, 1963, p. 21.

[72] *Report of the Miners' Medical Bureau for the Period 1st April 1965 to 31st March 1966*. Pretoria: Government Printer, 1966, pp. 11–12.

present themselves for examination. In any case, there was no qualified radiologist or radiographer in the Protectorate, and the Director suggested that the Chamber should accept more responsibility for re-examinations, which should take place in Johannesburg.[73] The Bechuanaland Government Secretary supported the Director's request. He asked the HCT's Office to raise the matter with the Chamber.[74] The NRC replied that it was impracticable to conduct re-examinations in Mafeking, as the necessary equipment was not available. Furthermore, it would be 'unsound to refer distant cases' for examination in Johannesburg.[75] The Chamber's opposition to pensions is curious: according to its own estimates, the life expectancy of repatriated men was so brief that paying pensions instead of lump sums would have saved the industry money.

Conclusion

The WNLA's senior management was aware that many families of repatriated miners lived in dire poverty. It was common for men to be in debt before they left for Johannesburg. Mine wages, they hoped, would solve their problems. When they were repatriated, their debts increased, and the men fell into poverty; some families lost their cattle.[76] The Chamber acknowledged that paying pensions rather than lump sums was a core principle of compensation schemes, including those available to white miners. However, it remained adamant that pensions could not be paid to black miners. Despite evidence to the contrary, it claimed that there

[73] The Director of Medical Services, Mafeking, to the Government Secretary, Mafeking, 28th November 1960. Subject: Miners' Phthisis Patients and other Injured Mine Workers. BNA S332/5/5, Compensation for B.P. Native Mine Workers 1961–62.

[74] The Government Secretary, Mafeking, to The Agent for High Commission Territories, 5th January 1961. BNA S332/5/51961–62, Compensation for B.P. Native Mine Workers.

[75] Letter from District Superintendent, Native Recruiting Corporation, to The Government Secretary, Mafeking, 17th January 1961, Subject: Medical Examinations - Pneumoconiosis Act. BNA S332/5/5 Compensation for B.P. Native Mine Workers 1961–62.

[76] Memo, Payments to Beneficiaries - Silicosis Act, 1946, 10th December 1948, from General Manager WNLA to The Legal Consultant, Chamber of Mines NRC 390 1&2 Miners Phthisis Compensation for Natives 1942–1949 TEBA Archives.

was no administrative grid to reach men who lived in remote regions, and hence that fraud would be inevitable. The Chamber also argued that since compensated miners died so quickly, their families received more money from a lump sum than they would from pensions. This rare example of the industry refusing to minimise a production cost suggests there were other factors at play.

The Chamber's refusal to pay pensions attracted criticism from the Departments of Health, Native Affairs and Native Labour. This was not in itself unusual, as these departments were often at odds with Chamber policies. During the 1920s and 1930s, the Department of Native Affairs had to a degree championed the rights of miners to pensions. That changed as apartheid was established after the war. In November 1947, the South African Institute of Race Relations asked the Department of Native Affairs if it would consider making several changes to the compensation system. It wanted black miners to receive pensions rather than lump sums, and for the Bureau to collect mortality data on the fate of repatriated men so the progress of disease could be gauged. The Department replied that a pension scheme presented major problems; neither was it possible for the Silicosis Board to compile mortality rates for repatriated miners. There was no provision for the periodic re-examination of compensated miners as many beneficiaries simply disappeared 'into the blue. The monitoring of extra-Union Natives would be impossible'.[77] Such difficulties, in turn, made the extent of the risk invisible and compensation for occupational injury to black miners de facto impossible. It was against this system that the recent miners' class action was fought.

References

Arkles, R.S., A.J. Weston, L.L. Malekela and M.H. Steinberg. *The Social Consequences of Industrial Accidents: Disabled Mine Workers in Lesotho*. National Centre for Occupational Health. NC0H Report Number 13/1990.

[77] Letter from Secretary for Native Affairs to Director, South African Institute of Race Relations, Johannesburg, 11th November 1947. SANA NTS 33/315, Correspondence of Director of Native Labour, NTS, Subject: Silicosis Act 1946.

Budiaki, Lugemba. Tuberculosis and Compensation: A Study of a Selection of Basotho Mineworkers from Maseru District. Master of Public Health thesis, University of the Witwatersrand, Johannesburg, 2005.

Harington, J.S.; N.D. McGlashan and E.Z. Chelkowska. 'A Century of Migrant Labour in the Gold Mines of South Africa'. *The Journal of the South African Institute of Mining and Metallurgy*, Vol. 104, No. 2, March 2004, pp. 65–71.

McCulloch, Jock. 'Dust, Disease and Labour at Havelock Asbestos Mine, Swaziland'. *The Journal of Southern African Studies*, Vol. 31, No. 2, 2005, pp. 251–266.

Murray, Colin. *Families Divided: The Impact of Migrant Labour in Lesotho*. Cambridge: Cambridge University Press, 1981.

Showers, K.B. *Imperial Gullies: Social Erosion and Conservation in Lesotho*. Athens: Ohio University Press, 2005.

Wilson, Francis. *Migrant Labour in South Africa: Report to the South African Council of Churches*. Johannesburg: The South African Council of Churches and SPRO-CAS, 1972.

10

Contests over Labour in British Central African Colonies: 1935–1953

As in South Africa and the HCTs, labour recruitment was the focus of smouldering tensions in Central African colonies. The interests and policies of the WNLA, the ILO, the government of South Africa, the British Colonial Office and the colonial administrations of Nyasaland (Malawi), Southern and Northern Rhodesia (Zimbabwe and Zambia) at times coincided, but frequently clashed. In charting a path through their complex negotiations, three issues stand out. There were constant references to tuberculosis and the lack of reliable statistics in low-level correspondence about returning miners. However, the risks to miners' health from silicosis and tuberculosis, or the reliability of official morbidity and mortality rates, were seldom mentioned in high-level meetings and documents involving the Resident Commissioners or the Colonial Office. Finally, the underlying assumption that employers had the right to pay below-subsistence wages, and colonial governments the right to impose taxes on indigenous populations regardless of their ability to pay, was rarely challenged.

Although specific individuals and changing political and economic conditions made a difference to the ongoing contests, there were some constants. The WNLA wanted an expanded recruiting zone to guarantee

J. McCulloch, P. Miller, *Mining Gold and Manufacturing Ignorance*,
https://doi.org/10.1007/978-981-19-8327-6_10

a supply of workers without increasing wages. The British Colonial Office and its local administrations wanted to collect the revenue that came from contracting labour to the WNLA. They also hoped to protect native interests and be seen to observe international treaties. The ILO was keen to promote labour rights and work safety. The South African government supported the WNLA and in addition wanted to channel mine rejects onto farms in the Transvaal. The Nyasaland (Malawi) government often sided with the WNLA. However, it did not want its migrants to work on South African farms because the pay and conditions were poor and there were no capitation fees. Nor did it want men to seek work on Southern Rhodesian mines outside of the quota system. The administration of Northern Rhodesia (Zambia) often seemed to agree with Nyasaland and oppose Southern Rhodesia. The Southern Rhodesian (Zimbabwe) government, in contrast, wanted secure access to cheap labour for its own mines and white farms.

The men contracted by the WNLA represented only a fraction of those moving south. A large proportion were assisted voluntary labourers (AVS) who found their own way to Johannesburg to work on scheduled mines.[1] There were others who, because of injury or age, were forced to seek work on small gold or asbestos mines. The least able ended up on white farms. The control of so-called clandestine labour added to the conflict between South Africa and the northern territories. From the mid-1930s, it was South African government policy to increase the supply of farm labour in the Transvaal.[2] White farmers were not only given access to those rejected for work in the mines but also hired illegals from the north.[3] The recruitment of such men, who were willing to work for very low pay, had been declared unlawful by South African government Notice No. 2091 of 1924. However, the Native Affairs Department had not been able to exclude these workers from the Union.

[1] In 1937 more than half of black miners were assisted voluntary labourers. Statement by the Native Recruiting Corporation Ltd. before the Native Labour Committee at Pietersburg, undated, December 1937, pp. 1–32. SANA. K 356 Native Labour Commission 1937.

[2] Notes of Interview with Representatives of Northern Governments regarding the Influx of Native Labour into the Union, Cape Town 2nd September 1938, p. 1. PRO CO 525/173/10 Nyasaland, 1938.

[3] Statement by the Native Recruiting Corporation Ltd. before the Native Labour Committee at Pietersburg, undated, December 1937, p. 2. SANA. K 356 Native Labour Commission 1937.

With the end of the ban on the recruitment of tropical labour approaching, the mining houses' most powerful negotiator, William Gemmill, stepped up his lobbying. In April 1937, he wrote to the Southern Rhodesian Secretary for Mines pointing out that the conditions of employment on the Rand were superior to those in Northern Rhodesia and Nyasaland. 'If, however', he added, 'the maximum number [of recruits] offered by each of the two Northern Governments concerned should be, in the opinion of the Chamber, too small, the object of the ... industry would be better served by the engagement of suitable Natives who travel voluntarily to the Union in search of mining employment'.[4] He repeated that threat to the Governor, Sir Herbert Stanley, with the proviso that if Southern Rhodesia agreed to an increase in recruiting from Nyasaland, the gold mines would not employ voluntary workers.[5]

All three Northern governments were keen to halt the unregulated flow of labour. A meeting of the Standing Committee on Migrant Labour in Salisbury in December 1937, attended by representatives from the Rhodesias and Nyasaland, noted that clandestine migration to South Africa had reached alarming proportions. Those present also agreed that the WNLA had not honoured its agreement regarding recruiting, and that it was instead encouraging voluntary labour to move south. The Committee recommended that the three governments approach South Africa to stop illicit migration.[6] Despite that unified approach, the three territories had their own conflicts. Southern Rhodesia wanted 95,000 Nyasa for its own mines and farms and questioned the Nyasaland estimates of the number of workers available.[7]

[4] Extract from a Confidential letter 13th April 1937, from William Gemmill, the Gold Producers Committee, to the Secretary for Mines. PRO CO 525/173/1 Native Labour Employment of Natives on Witwatersrand Gold Mines.

[5] Letter from W. Gemmill, Gold Producers Committee to Sir Herbert Stanley, the Governor of S. Rhodesia, Salisbury, 15th December 1937. PRO CO 525/173/1 Native Labour Employment of Natives on Witwatersrand Gold Mines.

[6] Annexure: A Memo of Proceedings of Meeting of the Standing Committee, Salisbury, 3rd and 4th December 1937. PRO CO 525/173/1 Native Labour Employment of Natives on Witwatersrand Gold Mines.

[7] Memo Migrant Labour Agreement, from G. Bullock to the Prime Minister, Salisbury, Southern Rhodesia, 5th January 1938. PRO CO 525/173/1 Native Labour Employment of Natives on Witwatersrand Gold Mines.

Acting under the terms of the Salisbury Agreement (discussed in Chap. 5, pp. 7–9), in late August 1938 the three Northern governments filed a set of proposals with Pretoria. They wanted South Africa to exclude migrant workers who did not have identity certificates. In return, the Northern governments would not hinder the movement of legitimate workers to the Union. They also requested that after two years' absence, men should be compelled to return home.[8] The Secretary for Native Affairs in Cape Town, Mr D.L. Smit, replied that as all farmers and many mines in the Northern and Eastern Transvaal relied upon clandestine labour, there would be trouble if the flow was stopped. He suggested that the Union government could import contracted migrant labour from adjoining territories for employment on farms in the same way as the gold mines did. The Northern governments replied that under such an arrangement they would insist on conditions akin to those guaranteed by the WNLA, namely the fixing of wages, proper housing, rations and medical care.[9] They knew well that white farmers would never agree to such high costs.

Following protracted negotiations, Mr Smit met with William Gemmill to discuss the proposal. Gemmill protested that the Northern governments were seeking to control the flow of labour to the Union and therefore to the mines. If the Chamber agreed, the mines would get the residue after Southern Rhodesia had been served. Smit was unsympathetic and noted that the mines could always, if they wished, get clandestine labour to make up their needs.[10] Eventually, following representations from the white farming community, the South African government decided to issue clandestine workers temporary permits for employment on farms.[11] Southern Rhodesia was aware that labour was badly needed

[8] Memorandum from Mr. D.L. Smit, Secretary for Native Affairs, Cape Town, to The Minister for Native Affairs, 5th September 1938. Subject: Influx of Natives from the Northern Territories into the Union, p. 2. SANA, K356 1937, Native Labour Committee.

[9] Notes of Interview with Representatives of Northern Governments regarding the Influx of Native Labour into the Union, Cape Town 3rd September 1938, p. 11. PRO CO 525/173/10 Nyasaland, 1938.

[10] Note by D.L. Smit, Secretary for Native Affairs, Cape Town, Regarding the Influx of Native Labour into the Union, 2nd September 1938. SANA, K356 1937, Native Labour Committee.

[11] Memorandum from Mr D.L. Smit, Secretary for Native Affairs, Cape Town, to The Minister for Native Affairs, 5th September 1938. Subject: Influx of Natives from the Northern Territories into the Union, p. 1. SANA, K356 1937, Native Labour Committee.

in the Transvaal and that restrictive measures would have political repercussions. White farmers would probably engage workers who had no permits, and even though that would be illegal, the South African government would hesitate to prosecute them.[12]

Southern Rhodesian Mines

Southern Rhodesia was unique among the Northern governments as both an exporter and an importer of labour. As an exporter Southern Rhodesia saw local men, attracted by the higher wages on offer, migrate to the Union mines. As an importer of labour from the north, the colony was limited by the relatively low wages it offered, and after 1936 by the Salisbury Agreement, which was designed to close the frontiers to the north. The Southern Rhodesian economy was reliant on gold mining, which by 1924 accounted for three quarters of its export revenue. Although rock mining was dangerous and work conditions oppressive in both countries, there were significant differences between Southern Rhodesian and South African mines. The Rand mines were much deeper and larger in scale and the host ore had higher silica content. They were also subject to state regulation and a compensation system that produced masses of data, factors which appealed greatly to the British Colonial Office and the ILO. In Southern Rhodesia, the ore bodies were poor, and the average output was far lower than on the Rand. The scale of mining and the composition of the workforce were also different. By 1920, there were 295 mines but only 8 were substantial.[13] There were few white miners, and no powerful white trade union. State regulation was minimal, and flagrant abuses such as the non-payment of wages and the use of

[12] Letter from C. Bullock, Secretary for Native Affairs, Salisbury, Southern Rhodesia, to The Secretary to the Prime Minister (Native Affairs), 13th September 1938. PRO CO 525/173/10 Nyasaland, 1938.

[13] Charles van Onselen. 'Worker Consciousness in Black Miners: Southern Rhodesia, 1900–1920'. *The Journal of African History*, Vol. 14, No. 2, 1973, p. 239.

credit to defraud workers were common.[14] The low wages and poor conditions on Southern Rhodesian mines were often cited by the WNLA to bolster its own claims to migrant workers. Those criticisms were justified. Between 1900 and 1933, over 30,000 miners lost their lives in Southern Rhodesia; of those, 27,000 died from disease. As in South Africa, the mortality figures did not include those who died home from disease contracted on the mines. Mine profitability, as Charles van Onselen and Ian Phimister have shown, was achieved through low wages and cost cutting on safety, compounds, medical care, food and compensation.[15]

Southern Rhodesia's mines had perennial difficulty in securing labour; by 1906, they were almost a third below their requirements.[16] Attracted by higher wages across the border, about 8000 men went to the Union each year. Their places were often taken by Nyasa.[17] The labour shortages were resolved for a time by *chibaro* or forced labour, which van Onselen likens to slavery. Once *chibaro* ended in 1933, workers from Nyasaland and Northern Rhodesia filled the gap. The migration of men from the north was accompanied by heavy loss of life from pneumonia and scurvy. Workers from Nyasaland often arrived weakened from their journey and fell ill when exposed to sudden chills at higher altitudes. Crowded compounds, combined with poor ventilation, provided the ideal conditions for the spread of pneumonia.[18] In South Africa, the Departments of Public Health and Native Labour were large and complex organisations; many of their officials frequently sided with black workers. With the exception of the Director of Public Health, Dr Fleming, that was not the case in Southern Rhodesia. In official correspondence, black miners were

---

[14] Jock McCulloch. *Black Peril, White Virtue: Sexual Crime in Southern Rhodesia, 1902 to 1935*. Bloomington: Indiana University Press, 2000, pp. 36–54.

[15] Charles van Onselen. *Chibaro: African Mine Labour in Southern Rhodesia*. Johannesburg: Ravan Press, 1980, p. 56; Ian R. Phimister. 'African Labour Conditions and Health in the Southern Rhodesian Mining Industry, 1898–1953'. In Ian R. Phimister and Charles van Onselen (eds). *Studies in the History of African Mine Labour in Colonial Zimbabwe*. Gwelo: Mambo Press, 1978, pp. 102–150.

[16] Michael Gelfand. 'Migration of African Labourers in Rhodesia and Nyasaland, 1890–1914'. *Central African Journal of Medicine*, Vol. 7, August 1961, p. 296.

[17] Letter from K.L. Hall, acting Governor, Zomba, Nyasaland to W. Ormsby-Core, Secretary of State for the Colonies, 13th March 1937, pp. 3–4. PRO CO 525/167/1 Native Labour.

[18] *Report of the Committee Appointed to Enquire into the Prevalence and Prevention of Scurvy and Pneumonia Amongst Native Labourers*. Salisbury: Government Printer, 1910.

referred to as 'labour devices', 'labour units' or 'tax-paying units'. After 1923, Southern Rhodesia became a self-governing Crown Colony, and the Colonial Office acquired the ability to influence mining practices through regulating the migrant labour system. However, it chose not to do so.[19]

Conflicts Between Nyasaland and Southern Rhodesia

Mining practices were, however, a source of conflict between Salisbury and Zomba. On some pits, eleven- and twelve-hour shifts were common, and migrant workers were agitating for fixed hours of work. In December 1939, the Southern Rhodesian Minister for Mines complained to the Acting Governor of Nyasaland, K.F. Hall, about the protests. He was sure that neither the mining industry nor white farmers would ever consider such a proposal. Hall was unimpressed. He wrote to London: 'There are now on the Nyasaland statue book various ordinances ameliorating conditions of native labour and if Southern Rhodesia continues to shun similar legislation it will be difficult to reconcile such conflicting policies under a scheme of amalgamation [into the Central African Federation]'. Hall noted that the recent improvements in the labour conditions in Southern Rhodesia were due to pressure from Nyasaland.[20]

The return of sick and injured men from Southern Rhodesia was another contentious issue. In the first eight months of 1936, many of the more than 100 men medically repatriated to Blantyre were tuberculosis cases requiring isolation and prolonged care. The Senior Medical Officer, Dr Calleja, warned the Director of Medical Services that the lack of beds at the Blantyre Hospital meant those men were housed in general wards, which put other patients at risk.[21] There was also a steady stream of men

[19] van Onselen, *Chibaro*, pp. 112–113; 64.

[20] Letter from K.F. Hall, Acting Governor, Zomba, Nyasaland, 22nd December 1939 to Malcolm MacDonald, Secretary for State for the Colonies, London, p. 5. PRO CO 525/185/1 Native Labour: Recruitment for the Union of S. Africa and S. Rhodesia.

[21] Memorandum, Subject: Repatriates, from Dr Calleja, The Senior Medical Officer, Blantyre, to The Director of Medical Services, Zomba, 27th August 1936. MNA. Repatriation, Provincial Commissioner, Southern Province Migrant Labour NS 1/13/7.

with traumatic injuries who were given neither medical care nor compensation. The repatriations provoked criticism from the Secretariat in Blantyre, the Department of Health and the Missions. The Rev W.P. Young of the Ekwendeni Mission in Mzimba held a public meeting to discuss the problem. Many men from the district had been injured on the Rhodesian mines but few had received adequate treatment or compensation. One villager, James Nyanguru, had lost three fingers and was dismissed without an award. Samson Zimba had his leg amputated after working only three days on a mine at Que Que. He was awarded £25 for total disability but given no crutches.[22]

The Abraham (1937) and Burden (1938) Reports (discussed in Chap. 5, pp.9–14) brought no change to working conditions in Southern Rhodesia. The criticisms continued—as did the repatriations. In February 1941, eight repatriates from Salisbury were admitted to the Zomba African Hospital. Their injuries included paralysis of the upper and lower limbs, total blindness and the loss of a right arm. Four had tuberculosis. The Senior Medical Officer questioned why the tubercular cases had been sent on from Blantyre. Zomba had no specialist facilities and the most his staff could do was confirm an existing diagnosis. He complained to the Director of Medical Services that the repatriation system was chaotic and requested that the Southern Rhodesia authorities issue certificates of disability and verify what compensation, if any, had been paid.[23]

Complaints from the Missions and the Zomba hospital finally provoked Mr Juxon Barton, a member of the Secretariat, to set out a list of recommendations for improving the repatriation system.[24] There was an urgent need for periodic examinations and for the treatment of men with lung disease. Men repatriated with miners' phthisis should be compensated, and notification sent to the Nyasaland authorities so they could be monitored. Barton noted there were no sanatoria in Nyasaland, and it

[22] Letter from Duncan Campbell, Missionary, Ekwendeni Mission, to the District Commissioner, Mzimba, 3rd August 1935. MNA. Enquiry on Emigrant Labour. NNM 1/9/2.

[23] Letter from The Medical Officer, Zomba African Hospital, Nyasaland to The Hon. The Director of Medical Services, Zomba, 22nd February 1941. MNA. M2/5/53 TB and Sanatoria Medical Report.

[24] Memo from Juxon Barton, the Secretariat, Zomba, Nyasaland to the Secretary, Nyasaland, Northern and Southern Rhodesia Inter-Territorial Conference, Salisbury, 9th December 1941. MNA. Emigrant Labour Governors Permits. M 2/3/19.

was undesirable that men with tuberculosis be admitted to general hospitals. The Secretariat in Blantyre wanted those men treated before repatriation. However, if the Southern Rhodesian government preferred not to assume that responsibility, perhaps it might agree to contribute to the creation and maintenance of a sanatorium in Nyasaland?[25] A month later, the Governor of Nyasaland, Donald Mackenzie-Kennedy, granted the Southern Rhodesian Chamber of Mines a permit to engage 2000 men from the Northern Province. In return, the Chamber paid a capitation fee of ten shillings per recruit and agreed to remit any outstanding government taxes from the miners' wages. The contract made no reference to exit examinations, medical referrals or access to compensation.[26] By 1944, the estimated 21,000 Nyasa employed in Southern Rhodesia accounted for 30 per cent of the mines' workforce.[27] Each year, a significant number of Nyasa died on the mines or were repatriated. Migrant workers were often discharged while seriously ill, and many perished before reaching home. Their surviving dependents rarely received compensation. Few mine owners complied with the Mines and Mineral Act of 1935, which required that all tuberculosis cases be reported to a District Magistrate.[28]

The Framing of Public Knowledge

By the mid-1930s, a significant number of silicosis cases began to appear among white miners in Southern Rhodesia. Their plight led the Director of Public Health, Dr Fleming, to set up an enquiry.[29] The 1935 Report on Tuberculosis, Phthisis and Asbestosis found that black miners were

[25] Letter from The Secretary, Nyasaland, N. & S. Rhodesia Inter-Territorial Conference, Salisbury to Mr K. Lambert Hall, Chief Secretary, Zomba, Nyasaland, 10th December 1941. MNA. Emigrant Labour Governors Permits. M 2/3/19.

[26] Governor's Permit to engage Natives for work outside the Protectorate. Nyasaland Protectorate Reference, 19th January 1942. MNA. Emigrant Labour Governors Permits. M 2/3/19.

[27] Minutes of a Meeting of the Central Labour Advisory Board held at Blantyre on Friday, 14th April 1944. MNA. M2/24/30.

[28] In 1940 more than 60 men were repatriated because of lung disease. Transcripts, *Southern Rhodesian Committee of Enquiry into the Mining Industry*, 1944, pp. 5–6. MNA. LB 3/14/1.

[29] *Report on Tuberculosis, Phthisis and Asbestosis*, 1935. ZNA. Public Health Department S 2104/5.

often repatriated without a medical examination and that the lack of data made a review of lung disease all but impossible.[30] Many white silicotics were married men with dependent children, and the loss of employment plunged their families into poverty. In response, the government established a Silicosis Commission chaired by Dr Louis Irvine, Chair of the Bureau in Johannesburg.[31]

The Irvine Commission relied upon the mining companies and their medical officers for opinion and data; the only original research it cited was provided by the Salisbury Chamber of Mines. The Commission ignored black miners and found there was no serious hazard facing white miners, who as in South Africa only served in supervisory roles.[32] The Commission report was tabled in 1938, but it took a further six years before a new Miners Phthisis Act was passed. The war was one reason for the delay. The other was the refusal of the mining companies to improve conditions or provide compensation. In the Commission's report, as in the official correspondence, there was no suggestion that dust exposure was a problem.[33] Such omission was not due to lack of an alternative perspective. In 1940, a Northern Rhodesian government enquiry into silicosis on the local copper mines took a very different view. It found that silica dust exposure was a serious hazard and recommended regular X-rays and improved dust control measures.[34]

The 1944 correspondence between Chief Secretary of Nyasaland, Lambert-Hall, and the Prime Minister of Southern Rhodesia encapsulates the differences between the two administrations regarding miners.[35]

---

[30] Letter from the Minister for Internal Affairs to the Prime Minister, 11th December 1939. ZNA. Prime Minister's Office S 482 337/39.

[31] *Report of the Commission to Study Silicosis and other Industrial Pneumoconioses.* Salisbury: Government Printer, 1938.

[32] Appendix 1 'Silicosis in Southern Rhodesia: Report by the sub-committee appointed by the Rhodesia and Salisbury Chambers of Mines' in *Report on Tuberculosis, Phthisis and Asbestosis*, 1935, pp. 91–92. ZNA. Public Health Department S 2104/5.

[33] Letter from the Minister for Internal Affairs to the Prime Minister, 11th December 1939. ZNA. Prime Minister's Office S 482 337/39.

[34] *Silicosis Investigation: Statement by the Northern Rhodesian Government.* Lusaka: Government Printer, 1942.

[35] Letter from Lambert-Hall, Secretary Nyasaland, Northern and Southern Rhodesia Inter-Territorial Conference, Salisbury, to Secretary of the Prime Minister, 2nd June 1944. ZNA. Prime Minister's Office S 482 337/39.

Lambert-Hall was concerned that healthy workers went south, and sick men returned home. He did not want invalid workers repatriated without treatment because they placed pressure on local medical services. To reduce the numbers, Lambert-Hall proposed compulsory examinations prior to employment and at discharge.[36] He also wanted pamphlets informing workers of the dangers of mining and of their right to compensation, prepared in various languages. The Chief Secretary was unimpressed by the proposed Southern Rhodesian Miners' Phthisis Act, which offered lump sums well below those payable in South Africa for the same injuries. He was also concerned that although blacks constituted almost the entire mine labour force, there was to be no native representative on the Miners Phthisis Board which would decide compensation.[37] The Governor of Northern Rhodesia, E.J. Waddington, shared Lambert-Hall's concerns. In 1941, a total of 36 Northern Rhodesians who had mined in Southern Rhodesia died; in 1944 that rose to 54. In only one case was compensation paid.[38] Waddington pointed out that the proposed Act compared poorly with what was available in Australia and South Africa. For example, the criteria for compensation were total disability, and proof of injury or liability was not open to appeal or independent review.[39]

The Nyasaland government was represented at a Southern Rhodesian Enquiry into the Mining Industry held at the end of 1945. Its delegates referred to 'the deplorable effect unrestricted migration was having on the social and economic system of the Protectorate and the misery and hardship endured by abandoned wives and children'.[40] During the previous

[36] Letter from the Secretariat, Zomba, Nyasaland to the Secretary, Nyasaland, Northern and Southern Rhodesia Inter-Territorial Conference, Salisbury, 10th March 1942. ZNA. Prime Minister's Office S 482 337/39.

[37] Letter from Lambert-Hall, Secretary Nyasaland, Northern and Southern Rhodesia Inter-Territorial Conference, Salisbury, to Secretary of the Prime Minister, Salisbury, 2nd June 1944. ZNA. Prime Minister's Office S 482 337/39.

[38] Letter from E.J. Waddington, Governor of Northern Rhodesia to the Governor, Southern Rhodesia, 18th August 1945. ZNA. Prime Minister's Office S 482 337/39.

[39] Letter from E.J. Waddington, Governor of Northern Rhodesia to the Governor, Southern Rhodesia, 14th October 1944. ZNA. Prime Minister's Office S 482 337/39.

[40] Transcript. 19th April 1945, *Southern Rhodesian Committee of Enquiry into the Mining Industry*, pp. 2–3. MNA. LB 3/14/1.

year, approximately 40 per cent of able-bodied men were absent from the territory. The treatment of migrant labour on Southern Rhodesian mines, the delegates noted, was far worse than in South Africa. They wanted contracts of service limited to specified periods, exit medical examinations and a compulsory scheme of deferred pay. Those measures were in accord with the ILO Conventions to which Nyasaland subscribed but which were rarely honoured by Southern Rhodesian employers. Nyasaland's presentations to the Enquiry brought no change. The oppressive conditions on Southern Rhodesian mines and the refusal of London to force improvements did, however, ensure that the WNLA remained an attractive employer to the Colonial Office.

In negotiating quotas, British Colonial Office never specified what it considered an acceptable annual number of deaths per thousand recruits, nor did it ask for detailed morbidity and mortality statistics. The WNLA data was usually crude, with categories changing from year to year. There were no details about the men concerned, the type of work they performed or the types of mines on which they were employed. During what were often protracted negotiations, neither did the Colonial Office ever request data on exit medicals, repatriations or compensation.

Labour Negotiations After the Ban on the Recruitment of Tropical Labour Was Lifted in 1937

As soon as the ban on the recruitment of tropical labour was fully lifted in 1937, the WNLA initiated new negotiations with northern administrations. Sir Harold Kittermaster, the Governor of Nyasaland, was aware of the Protectorate's weak bargaining position. He also understood what his government stood to gain by contracting labour to the WNLA. Thirty thousand Nyasa going to the Rand would mean an extra income of £200,000 a year, most of which would be spent at home. At that time, the Protectorate's total annual revenue was less than £450,000. Kittermaster was concerned by the efforts of the Southern Rhodesian government to compel Nyasa to accept lower wages, and in January 1938 he wrote in protest to its Governor, Sir Herbert Stanley. Kittermaster acknowledged that Nyasa labour was essential to small mines and white farms in Southern Rhodesia, but he was aware that the work conditions and the pay offered were inferior to those on the Rand. He was also concerned about the potential impact of uncontrolled recruiting by South Africa's mines. 'I am filled to the brim with terror at the prospect of a free flow of uncontrolled labour going to the Rand. Can the Prime Minister show me any effective way of opposing the Witwatersrand Native Labour Association's economic dictation except by making terms with them? It has not done so yet.'[41]

Two months later, there was a meeting of the Governors of Southern Rhodesia, Northern Rhodesia and Nyasaland and the Chamber. They agreed that, as an experiment, 1000 Barotse and 4000 Nyasa be recruited by the WNLA. If the experiment proved satisfactory to those governments and the Secretary of State for the Colonies, further agreements could be reached subject to the Salisbury Agreement. The Governors of the Rhodesias and Nyasaland were, however, alarmed by the high death rates and wrote to Gemmill: 'The figures submitted by the Witwatersrand Native Labour Association show a disquietingly high mortality rate, which is borne out by the result of enquires made from individual mines by the representative of the Northern Rhodesian Government'. As a result, the three governments would not agree to an extension until the experiment was 'further developed', presumably meaning until the mines became safer. In Northern Rhodesia, half of able-bodied men were already absent and there was no labour surplus as 'the margin of safety [for preserving social life] had already been exceeded'. The three governors were adamant that as a condition for future recruiting, the WNLA must undertake not to encourage the flow of voluntary or clandestine labour. No concessions were made, but the prospect of financial benefit soon eclipsed the governors' initial concerns about the death rates, which suddenly disappeared from the correspondence.[42]

[41] Letter from Sir Harold Kittermaster, Governor of Nyasaland to Sir Herbert Stanley, Governor of Southern Rhodesia, 15th January 1938, p. 6. PRO CO 525/173/1 Native Labour Employment of Natives on Witwatersrand Gold Mines.

[42] Aide Memoir from the Governors of Southern Rhodesia, Northern Rhodesia and Nyasaland to Mr. Gemmill. Undated, March 1938, p. 2. PRO CO 525/173/1 Native Labour Employment of Natives on Witwatersrand Gold Mines.

While the negotiations were taking place, Lord Dufferin toured the Johannesburg mines as a guest of the Chamber. He was greatly impressed: 'The hospitals, even the old ones, are not over-crowded and the new one at the Sub-Nigel mine is the finest native hospital by far that I have seen'.[43] Dufferin noted that the death rate of Tropicals from pneumonia was double that of other miners, and he recommended that the Colonial Office bring pressure to bear on the Chamber to make the mines safer. He made no mention of dust, tuberculosis or repatriations.

Northern Rhodesia Negotiates with WNLA

In late October 1938, William Gemmill met with the Governor of Northern Rhodesia, J.A. Maybin, to discuss recruitment. The two men agreed that initial medical examinations would be conducted by a Northern Rhodesian government medical officer at a fee and that the WNLA would pay a commission of 1s for each month served by each recruit. The period of contracts was to be twelve calendar months. Finally, no man who had previously worked on the gold mines was to be recruited again until he had spent at least six months at home.[44] Having established that much common ground, the negotiations broke down over two issues. The first was what Maybin viewed as the unacceptably high mortality rate. As of 31 January 1937, the experiment with Barotse from Northern Rhodesia was 13 months old. According to William Gemmill, the death rate from disease for the Barotse was 14.83 per 1000 per annum: that compared unfavourably with the rate of 6.87 for other classes of recruits. However, as he pointed out correctly, the mortality amongst 'tropical natives' from Northern Rhodesia and Nyasaland employed on Southern Rhodesian mines in 1935 was 15.90 and 18.22 respectively.

[43] Extract of Letter from Lord Dufferin to Mr Ormsby Core, Under Secretary for State. Undated, 1938. PRO CO 525/173/1 Native Labour Employment of Natives on Witwatersrand Gold Mines.

[44] Note of an Interview between J.A. Maybin, the Governor of Northern Rhodesia and W. Gemmill, the General Manager the Transvaal Chamber of Mines, held at Lusaka, Northern Rhodesia, on 27th October 1938. PRO CO 525/173/2 Labour: Employment of Natives on the Witwatersrand Gold Mines. Johannesburg Agreement, 1938.

During the same period, the annual death rate among all Tropical Natives employed on the South African mines was 14.14.[45]

The second issue was who should pay for repatriation. While the WNLA had agreed to provide free transport *to* the mines, Gemmill was unwilling to pay return fares. He argued that the International Convention did not require such a payment. To provide free repatriation would involve all those territories from which the gold mines drew labour and would cost the mines £500,000 per annum. According to Gemmill, the wage rates were fixed deliberately high to enable recruited labour to pay their own return fares. Furthermore, the Union government would not agree to free repatriation as it would adversely affect white farmers. The meeting ended in a stalemate. A week later the Governor of Northern Rhodesia wrote to the Secretary of State in London explaining there had been a delay in negotiations with the WNLA. The *cumulative* mortality figures for Nyasaland recruits had shown a steady improvement since November 1937, but the figures for Northern Rhodesia remained high at around 22.75 per 1000. That was almost a third higher than for other Tropicals and more than three times higher than for other classes of labour. With surprising optimism, Maybin noted that once that issue had 'been resolved', he was willing for the gold mines to recruit a maximum of 10,000 Barotse per annum under the Salisbury Agreement. Maybin felt he had no alternative: 'In the latter event, of negotiations having broken down, our only weapon for preventing the Witwatersrand enticing our labour without any conditions at all would be organised propaganda through the Paramount Chief to advertise the mortality rates and discourage emigration. This might or might not be successful.'[46] A decade later, Maybin's retreat on the death rates was to be repeated by the Governor of Nyasaland.

[45] Memoranda from W. Gemmill, General Manager, Chamber of Mines. Subject: Native Labour: Mortality of Tropical Natives, 2nd March 1937. TEBA Archives, WNLA. 17/1. Statistical data and prepared for special purposes. From December 1922 to March 1957.

[46] Letter from the Governor of Northern Rhodesia, J.A. Maybin, to the Secretary of State for the Colonies, London, 5th November 1938, p. 7. PRO CO 525/173/2 Labour: Employment of Natives on the Witwatersrand Gold Mines. Johannesburg Agreement, 1938.

Negotiating a Big Nyasaland Quota

After the ban on tropical recruiting was lifted, Nyasaland offered the WNLA an initial quota of 8500 men. In December 1938, Harold Kittermaster received a request from William Gemmill for the quota to be increased to 15,000.[47] According to Gemmill, among Northern Rhodesians, nearly all of whom were Barotse, the total death rate for the 33 months ending 30 September 1938 was 26.4 per thousand. This compared unfavourably with the rate for Nyasa recruits for the 57 months ending September 1938, which was 16.08 per 1000. However, like the Colonial Office, Kittermaster was swayed by Dr A.J. Orenstein's assurance that the incidence of fatal disease among men employed for the first time in industrial undertakings was always heavy during the first decade but would soon decline.[48] The WNLA never conducted a study to test that supposition.

In accordance with the Salisbury Agreement, Kittermaster visited Salisbury to gain the support of the Southern Rhodesian government. The colony's Prime Minister, G.M. Huggins, was however alarmed by the number of voluntaries moving south and insisted that the WNLA quota form part of a scheme for the overall regulation of migration to South Africa.[49] As always, the negotiations dragged on. In March 1940, there was a further meeting between Huggins and the Governors of the Rhodesias and Nyasaland in Salisbury. The Acting Governor of Nyasaland, Mr Hall, who supported the WNLA application, pointed out that it had invested a good deal of capital and wanted a guarantee it would receive some return. Huggins would not agree to an increase. The discussion

[47] Notes by F.J. Pedler, Colonial Office, 23rd November 1938. PRO CO 525/173/2 Labour: Employment of Natives on the Witwatersrand Gold Mines. Johannesburg Agreement, 1938.

[48] Note for Lord Dufferin, Under Secretary of State, on the Disparity between the Mortality of Natives of Barotseland and Nyasaland on the Rand Mines. Undated, December 1938. PRO CO 525/173/2 Labour: Employment of Natives on the Witwatersrand Gold Mines. Johannesburg Agreement, 1938.

[49] Letter from Harold Kittermaster, Governor, Government House, Zomba, Nyasaland, 9th December 1938 to Malcolm MacDonald, Secretary of State for the Colonies, London, pp. 1–2. PRO CO 525/173/2 Labour: Employment of Natives on the Witwatersrand Gold Mines. Johannesburg Agreement. 1938. [Part Two].

then turned to illicit migration. The war had just begun and the clandestine movement of labour to South Africa was assuming greater importance.[50]

Following the Salisbury meeting, the new Governor of Nyasaland, D.M. Kennedy, wrote to London. Frustrated that Southern Rhodesia would not endorse the new quota, he framed his argument around competing white visions of desirable African futures. Gemmill claimed that oscillating migration helped preserve traditional African society while at the same time modernising it. In contrast, white farmers in the labour-sending states argued it helped destroy it. Kennedy told the Secretary of State that the rights of the African to sell his labour in the best market were being challenged by white farmers in Nyasaland. These wanted to pay low wages and were claiming that migration to the mines was destroying African society. The farmers were supported by District Commissioners as well as the Missions and Native Authorities, who were all disturbed by the drain on the adult male population. Kennedy warned London that any attempt to halt the flow of labour to the gold mines would meet strong political opposition. The South African mines paid well and provided good rations and accommodation. The men who returned from the mines were healthy.[51] At 6.06 per 1000 per annum, the death rate from all diseases was negligible.[52]

With the outbreak of war, the WNLA wanted a dramatic increase in the quota, from 4500 to 45,000. In February 1940 K.F. Hall, the Acting Governor of Nyasaland, informed London that he would raise the matter at the up-coming Inter-Territorial Conference in Salisbury. Hall had made little progress with South Africa over reducing clandestine migration and warned the Colonial Office that until that was resolved there

[50] Record of the Proceedings of a Conference held at Salisbury on the 6th and 7th March 1940. PRO Nyasaland 1940, pp. 1–2; 8. PRO CO 525-185-2, Native Labour, Recruitment of for the Union of S. Africa and S. Rhodesia.

[51] Letter from D.M. Kennedy, Governor of Nyasaland, Zomba, 27th April 1940, to Malcolm MacDonald, Secretary of State for the Colonies, London, pp. 2; 6. PRO Nyasaland 1940, CO 525-185-2, Native Labour, Recruitment of for the Union of S. Africa and S. Rhodesia.

[52] Statistical Statement attached to Letter from D.M. Kennedy, Governor of Nyasaland, Zomba, 19th June 1940, to Rt Hon Lord Lloyd, Secretary of State for the Colonies, London. PRO Nyasaland 1940. CO 525-185-2, Native Labour, Recruitment for the Union of S. Africa and S. Rhodesia.

could be no effective agreement among the Northern governments. Hall believed that unless South Africa accepted a fixed annual quota and refused entry to men without certificates, Southern Rhodesian employers were faced with having to raise wages. Hall did not want to halt the flow of workers. 'We have no bargaining factor with the southern territories except our labour and the Witwatersrand Native Labour Association offers us the best terms.' He was unhappy with the stance taken by Southern Rhodesia, and he had considered renouncing the Salisbury Agreement in favour of a new agreement with South Africa.

Hall reminded London that each year, large numbers of men from Northern Rhodesia and Nyasaland went south in search of work. Some went under contract with the WNLA, but most went on their own to find work on farms, on mines and in domestic service. The Northern Rhodesia and Nyasaland governments accepted that migration was inevitable and welcomed higher WNLA quotas, as they guaranteed repatriated wages and the payment of taxes. In contrast, the Southern Rhodesian Prime Minister worried that the labour shortages in Southern Rhodesia would become acute and wanted South Africa to control clandestine migration. Until that was done, he opposed any increase in the quotas. The matter was referred to the Secretary of State in London, who sided with Southern Rhodesia.[53]

Increased Tensions After the Second World War

During the immediate post-war period, the struggle between the WNLA, the ILO and the British territories over the flow of labour intensified. The beginning of apartheid coincided with the creation of the UN. It also coincided with the development of the Orange Free State mines, which wanted recruits from Nyasaland and the HCTs to offset their inability to attract enough South African workers for the wages they offered. In negotiating higher quotas, the WNLA used ILO labour conventions to expose

[53] Letter from K.F. Hall, Acting Governor, Zomba, Nyasaland 22nd February 1940 to Mr. Boyd, the Colonial Office, p. 1. PRO CO 525/185/1 Native Labour: Recruitment for the Union of S. Africa and S. Rhodesia.

the contradictory policies of the HCTs and the Colonial Office. In his presentations to the ILO, Gemmill also invoked the UN Charter on Human Rights to berate the British delegation. Silicosis and tuberculosis were rarely mentioned in these high-level debates.

Britain's ILO commitments were part of a wider set of obligations regarding risk management in the workplace. In 1930, the Colonial Office Labour Committee was set up to review workmen's compensation law in the territories, and in particular to ensure that such legislation conformed with the ILO Conventions to which Britain was a signatory.[54] In 1935, the Secretary of State, Malcolm MacDonald, asked the colonial governments to create the machinery to inspect and examine workplaces. MacDonald pointed out that the far-reaching economic and social changes sweeping the empire were bringing potential conflict between employers and workers.[55] Between 1937 and 1943, protective legislation and the appointment of dedicated staff saw some progress in government supervision of labour conditions.[56] In those colonies where trade union movements were rudimentary, however, the Colonial Office tended not to insist on legislation based on metropolitan law.

At its Thirty-Second Session held at Geneva in 1949, the ILO debated a revised text on Migration for Employment. Before departure from their home territory, migrant workers should be examined for their occupational and medical fitness by a local physician.[57] Medical selection should be thorough and, if necessary, include an X-ray. Article 17 prescribed that 'equality of treatment in the workplace should apply, without discrimination in regard to nationality, race, religion or sex'. Finally, all labour should have the right to membership of a trade union and the benefits of collective bargaining. South Africa's gold mines met virtually none of those requirements.

[54] Labour Supervision in the Colonial Empire 1937–43, London, May 1943, p. 21. BNAs, S116/3 Labour Supervision.

[55] Circular signed Malcolm MacDonald, *The Dispatch*, Downing Street, London, 5th September 1938. BNAs, S116/3 Labour Supervision.

[56] Labour Supervision in the Colonial Empire 1937–43, London, May 1943, p. 4. BNAs, S116/3 Labour Supervision.

[57] International Labour Conference, No. 37 (1949), Provisional Record, Thirty-Second Session, Geneva, pp.VI–VIII. SNA 3068A, Secretariat, Migration for Employment, International Labour Office.

The ILO prescriptions for labour rights were incompatible with minority rule, and this exposed the ambiguities of British colonial states selling labour to the WNLA. In addition, when negotiating labour quotas, the Colonial Office operated under a major constraint. It was difficult for London to criticise the WNLA when the South African silicosis and tuberculosis legislation was enacted earlier, was more comprehensive and far more generous in its ambit than comparable laws in Britain. This problem was compounded by the fact that neither the ILO nor the Colonial Office understood the extent to which the South African laws were racialised.

In the early 1950s, the Colonial Office issued a number of memos on silicosis and occupational disease to assist administrations in framing legislation. It noted that compensating for such diseases was especially complex, and some territories had created problems for themselves by adding specific lung diseases to existing schedules. In the United Kingdom, the problem was dealt with under the Industrial Injuries Act, which treated pneumoconiosis separately from other prescribed diseases. To secure benefits, a workman had to prove to a Pneumoconiosis Medical Board that he had been employed in a prescribed occupation for a period amounting to at least two years. Tuberculosis was compensable for certain classes of health workers but not for miners.[58]

In 1952 and 1953, the ILO prepared position papers on migrant labour for the UN.[59] The organisation noted that in Africa, rapid population growth after the war had forced more men and women into wage labour. Most people lived in self-subsistence economies and so were not bound to employers as were workers in industrial states. Migrant workers often reacted to a wage increase by leaving employment, because they were able to satisfy their needs more rapidly. The Colonial Office believed that unskilled labourers' tendency to work for only brief periods was a

[58] Draft Circular Despatch to all Territories from Colonial Office, London, 23rd November 1952, p. 2. PRO CO859/259. Lung Diseases Workmen's Compensation 1952.

[59] United Nations General Assembly, Committee of Information from Non-Self-Governing Territories (1950–52), Third Session, 1952. *Social Problems Involved in the Economic Development of Non-Self-Governing Territories*, Prepared by the International Labour Office. PRO, CO 859 304 U.N. 1952, Labour Productivity.

major barrier to development.[60] A free and strong trade union movement was essential to ensure workers received a share of the benefits of economic development and to encourage permanent employment.[61]

Big Politics

Both the UN and the ILO assumed that migrant labour was transitory and would disappear as colonial economies matured. William Gemmill and the Chamber, the architects of oscillating migration, had a different view. As the winds of change swept the British empire, the Chamber remained determined that the HCTs and other British Protectorates continue to act as its labour agents. In 1943, the High Commissioner's Office in Cape Town submitted a memorandum to the Lansdown Commission into mine wages.[62] It proposed compulsory repatriation at the end of contracts, the compulsory deferment of pay, and for the mines to provide married quarters. Lansdown rejected those proposals, ostensibly because they would interfere with the African's freedom to sell his labour as he wished. According to High Commissioner's Office, the explanation lay in new challenges facing the mines. Production costs were rising while the gold price remained static. The mines also faced increasing competition in the South African labour market, with factory owners offering higher wages.

The High Commissioner, Sir Evelyn Baring, viewed migrant labour as one of the major challenges facing the HCTs. However, he also wanted to avoid what he termed 'a head-on collision' with the Chamber. The higher wages on offer in the urban areas were attractive and a journey to the

[60] Labour Productivity in Africa, paper prepared in the Colonial Office, London, in connection with the Third Inter-African Labour Conference, held at Bamako between 27th January and 5th February 1953. PRO, CO 859 304 UN 1952, Labour Productivity.

[61] United Nations Economic and Social Council, 'Economic Development of Under-Developed Countries: Question of Methods to Increase World Productivity'. Working paper prepared by the International Labour Office on the role of labour in programmes for increasing productivity, and measures needed to safeguard the interests of workers, 22nd May 1953, pp. 1–28. PRO, CO 859 304 UN 1952, Labour Productivity.

[62] *Report of the Witwatersrand Mine Native Wages (Lansdown) Commission on the Remuneration and Conditions of Employment of Natives on the Witwatersrand Gold Mines.* Pretoria: Government Printer 1943.

towns, and particularly to the South African mines, had become customary for young men. Baring was aware of the negative impact on indigenous farming and family life as economic forces encouraged men to start mine work at an earlier age, to spend longer periods away, and shorter periods home.[63] The mines were the largest employer of labour from the HCTs, and the WNLA and the Native Recruiting Corporation were in a strong bargaining position. If legal recruitment within the HCTs was forbidden, Baring was sure the NRC would simply place its agents along the borders and recruit men as they crossed. Gold mining was the foundation of South Africa's economy, and the Union government would always side with the industry.

In October 1950, representatives from the South African government, the Chamber and the Standing Committee on African Labour of the Central African Council met in Salisbury. William Gemmill and his son James, who represented the WNLA, made it clear that if they were not given permission to recruit it would be impossible to refuse entry to Northern men who entered the Union 'without properly endorsed identity certificates'. The Gemmills were keen to reach an agreement: 'In order to be able to say to the world that the South African government was co-operative and anxious to safeguard the interests of Africans entering the Union'.[64] The South African government supported the WNLA. If the Northern governments refused to meet the WNLA's quotas, it threatened to allow Northern labour to enter the Union without identity certificates.[65]

The government of Nyasaland shared Baring's concerns about the negative impact of WNLA recruiting. In 1950, it agreed that the WNLA could recruit 8000 men over the coming year: well below the 10,000 it wanted. The government also refused the WNLA's request that the contract period be increased from twelve to eighteen months. Owing to the development of the tobacco industry, recruiting in the Kasungu District

[63] Letter from Sir Evelyn Baring, High Commissioner's Office, Cape Town, 25th May 1949, to Philip Noel-Baker, Secretary of State, p. 1. PRO DO35/4110. Native Labour Conditions. Control of the Flow of Labour to Industry in the Union.

[64] Notes of Meetings in Salisbury on 5th and 6th October 1950, Commissioner for Labour and Mines, p. 3. PRO CO 525/220/12 Migration of Labour.

[65] Note prepared by G.J. Law to the Commissioner for Labour and Mines, Northern Rhodesia, on the discussions regarding migrant labour to South Africa at Salisbury 5th and 6th October 1950, dated 9th October 1950, pp. 1–3. PRO Nyasaland, CO 525/220/12, Migration of Labour, 1950.

was prohibited.[66] Gemmill wrote immediately in protest to the Colonial Office. The Nyasaland government was forcing Africans to work for low wages for the benefit of European tea planters. Gemmill went on: 'I shall regretfully be compelled to raise the matter in public at the forthcoming International Labour Conference at Geneva in June and to indict the Nyasaland Government, and through it the British Government, before the full Conference, of which I shall probably be Vice-President. It is only right to let you know the position in advance in order that your people may be prepared.'[67]

The Colonial Office was annoyed that Nyasaland adopted such a hard line at a time when negotiations were underway with South Africa over controlling illicit migration. It doubted that Nyasaland's decision to reduce the WNLA quota would increase the labour supply within the territory unless it was able to restrain the flow of voluntary migrants. London was sure, however, that the decision would worsen relations with the WNLA.[68]

Behind-the-Scenes Negotiations Before the 1951 ILO Meeting in Geneva

While the Secretary of State thought it disingenuous of the WNLA to present itself as a champion of African labour, he was concerned about Gemmill's warning. 'It might well be embarrassing if this matter were raised at an international meeting, particularly one constituted on a tri-partite basis, with workers' representation, as is the case with the

[66] Letter from A.C. Talbot Edwards, Chief Secretary to the Government, Zomba, Nyasaland, to Mr W. Gemmill, General Manager (Tropical Areas), WNLA Salisbury 23rd December 1950, p. 1. PRO CO859/197/2, International Labour Organization Representations by Mr Gemmill Regarding Restrictions on Migrant Labour in Africa, 1951. Talbot and Gemmill *do* refer to tobacco industry and tea planters respectively.

[67] Letter from William Gemmill, Salisbury, Southern Rhodesia to E. W. Barltrop Esq, Labour Adviser to the Secretary of State for the Colonies, London, 20th February 1951. PRO CO859/197/2, International Labour Organization Representations by Mr Gemmill (WNLA) Regarding Restrictions on Migrant Labour in Africa, 1951.

[68] Memo from E. W. Barltrop to Mr Watson and Mr Lambert 1st March 1951. PRO Co 525 220 13 Nyasaland Migration of Labour 1951.

International Labour Conference.' Should the matter be raised, the Secretary suggested a justification for refusing the WNLA demands. Under the provisions of ILO Convention No.50, the Nyasaland government was entitled to consider the effects on social life, health, welfare and development, particularly regarding food security. 'Despite this argument, however, it might well be difficult to present a convincing case regarding the present wage levels in agricultural employment in Nyasaland.'[69]

In preparation for the Geneva ILO meeting, Gemmill forwarded a brief history of the WNLA's agreements to the Nyasaland government. In 1938, he noted, the Rhodesias and Nyasaland agreed to allow the WNLA to establish recruiting stations and to recruit up to 9000 men. In return, the WNLA and the NRC undertook not to engage workers from those territories already resident in South Africa. According to Gemmill, while the mines had been scrupulous in observing their part of the agreement, the Governor's Permit of 1951 had reduced the quota to 8000. A still more serious breach was the prohibition on recruitment in the Southern Province. Men from that Province had always worked on the mines and the ban involved the exploitation of African labour for the benefit of white farmers.

Gemmill argued, with some justification, that any labour shortage in Nyasaland was due to the wages which ranged from 12s 6d to 30s a month: the minimum wage on the gold mines was £4. Gold miners also received a bonus on completion of 18 months' service and liberal re-engagement benefits. Free accommodation and ample food were provided, together with medical care. Those benefits attracted the best Nyasa workers. The official estimate of the number of men absent from the country was 150,000. Gemmill warned that he was merely stating the Chamber's position before the matter 'became a question for those international bodies interested in the welfare of indigenous populations'.[70]

[69] Confidential Letter from the Secretary of State for the Colonies to the Officer Administering the Government of Nyasaland, 11th April 1951. Subject: Recruitment of Native Labour. PRO CO859/197/2, International Labour Organization. Representations by Mr Gemmill (WNLA) Regarding Restrictions on Migrant Labour in Africa, 1951.

[70] Letter from W. Gemmill, General Manager (Tropical Areas), WNLA, Salisbury to the Chief Secretary to the Government, The Secretariat, Zomba, Nyasaland. Subject: Governor's Permit, undated 1951, pp. 1; 3; 5. PRO CO859/197/2, International Labour Organization Representations by Mr Gemmill (WNLA) Regarding Restrictions on Migrant Labour in Africa, 1951.

Nyasaland had no significant mineral resources. To encourage development, Governor Geoffrey Colby, who served from 1948 to 1956, sought to shift the economy from subsistence to cash cropping. Colby was successful in expanding social services, and during his term the civil service doubled in size, as did the education and health budgets. Despite that extra spending, the medical service was hampered by a lack of qualified staff.[71] Colby found that while Nyasaland was a major exporter of labour, an internal shortage of manpower was stifling the agricultural development on which the colony's future depended.

In response to the issues raised in Gemmill's memo, Colby set out his own position for the Secretary of State for the Colonies. Oscillating migration, he stated, was the most serious barrier to the territory's development. Those areas from which the greatest migration took place had the worst farming practices and it was there that the drought and famine of 1948–1949 had its most serious effects. Because of the absence of so many able-bodied men, large numbers of women, children and elderly had died of starvation. Colby had seen villages where there were no men between the ages of 20 and 45. Gold mining was diverting manpower from the production of food, which was left to women and old men. For that reason, members of African Congress had long been opposed to WNLA's recruiting. The government's restrictions on the quota, to which Gemmill so strenuously objected, were part of its efforts to reduce the threats that chronic shortages of labour posed to the Protectorate's development. Colby's initiatives to halt the flow of labour included a general improvement in the salaries, wages and conditions for government employees, a large-scale housing programme, an increase in the minimum wage for non-government workers and encouragement for employers to improve work conditions. Training schemes for artisans had also been established and the production of commercial crops had begun. Colby reminded the Colonial Office that William Gemmill would not hesitate to cause Nyasaland embarrassment at the ILO. He was employed by one of the most powerful financial groups in the world, whose profits depended on securing adequate supplies of labour. The mines were

[71] Colin Baker. *Development Governor: A Biography of Sir Geoffrey Colby.* London: British Academic Press, 1994, pp. 84–89.

already short of men, and a greatly increased workforce would be needed to develop the Orange Free State fields. 'The treatment [of migrant workers] however is more akin to the treatment of prize fat stock than of human beings since each of them represents a cash profit: it is sometimes forgotten that these men are herded in compounds in the Transvaal where they have no family life and are exposed to the most vicious influences.'[72]

William Gemmill at the ILO

When he spoke at the ILO's conference in Geneva in June 1951, William Gemmill did so as an 'Employers' Delegate from South Africa' rather than as a WNLA official. Gemmill reminded his audience that he had been a member of the ILO's governing body since 1922, during which time he had promoted African welfare. He had, for example, played a part in drawing up the Recruiting of Indigenous Workers Convention of 1936, and the Contracts of Employment (Indigenous Workers) Convention of 1939. Prior to the opening of the Conference, Gemmill promised the British delegation that he would give them a draft copy of his speech. At the request of the South African Ministry of Labour, who saw the speech only minutes before it was delivered, Gemmill left out the final paragraph, which accused Nyasaland of violating ILO Conventions.[73]

Gemmill explained that it was common in African territories for a tiny proportion of the population to be engaged in wage labour. Such backwardness could only be overcome by inculcating the habit of work and the desire for a higher standard of living. One of the most effective ways of promoting change was employment on South Africa's mines. Gemmill wanted to draw the Conference's attention to an oppressive and growing practice in some African territories, which prevented workers selling their labour in the best market. Those restrictions were designed to compel Africans to accept employment inside their own territory at lower wages

[72] Memorandum from the Governor of Nyasaland, Geoffrey Colby, to the Secretary of State for the Colonies, 2nd May 1951, pp. 1–2; 5. Subject: Recruitment of Native Labour. PRO CO859/197/2, International Labour Organization Representations by Mr Gemmill (WNLA) Regarding Restrictions on Migrant Labour in Africa, 1951.

[73] Note by G.M. Evans, British delegation, 24th July 1951. PRO, CO 525/220/13, Nyasa 1951.

and under inferior conditions than were available elsewhere. 'The British Colonial Office is itself a party to the practice, an instance being Nyasaland, a British Colonial Office territory.'

The wages on the gold mines were relatively high, Gemmill noted, and the conditions of employment excellent. In fact, the conditions were so good that the Witwatersrand had become the Mecca of the Central and South African labourer. In Nyasaland, the wages paid by employers ranged from 17s 6d to 30s per month, plus rations. On the Rand, the average pay was more than four times higher. Gemmill objected to the Nyasaland government's policies: 'I characterise restrictions such as these as constituting indirect compulsion and interference with the liberty of the subject and as coming very near to exploitation of the African for the benefit of low wage scale employers'.[74] Article 13 of the United Nations Declaration of Human Rights, adopted in December 1948, provided that 'everyone has the right to leave any country including his own and to return to his country'. Article 23 states: 'everyone has the right to work, to free choice of employment'. Forty-eight nations, including the United Kingdom, voted in favour of the Declaration. Gemmill concluded that it would be strange for a country to support the Declaration and at the same time compel a man to remain in his own country to work at very low rates of pay. He called on the ILO to end an abuse of power. Gemmill failed to remind his audience that of the eight countries which abstained from supporting the Human Rights Declaration in 1948 only two, namely South Africa and Saudi Arabia, were not members of the Soviet bloc.

In response, Alfred Robens, the British Minister of Labour, questioned Gemmill's right to raise the issue of WNLA's recruiting at such a forum. The fixing of the quota was entirely within the discretion of the Nyasaland government, which was entitled, and indeed obliged, under Article 5 of the ILOs Convention No.50, to consider the social impact of the withdrawal of adult males from the population. The comparison Gemmill

[74] International Labour Conference Provisional Record, 34th Session, Geneva, No. 7 (XXX1V-1951). Fifth Sitting, Monday, 11th June 1951, 10:15 a.m. Mr Gemmill, pp. 30–31. PRO CO859/197/2, International Labour Organization Representations by Mr Gemmill (WNLA) Regarding Restrictions on Migrant Labour in Africa, 1951.

drew between the wages a worker might earn in employment in Nyasaland and elsewhere was irrelevant.[75]

In a closing speech to the Conference, Gemmill attacked the ILO itself and reminded the audience of what he took to be endorsement of his critique of the British. He referred to 'certain unfavourable changes developing in the internal operations of the International Labour Organization', one of which was the method of electing the governing body. Gemmill also emphasised that early in the debate on the Director General's Report, he had drawn attention to an abuse of power prevalent in some African territories. Gemmill was pleased that the matter was to be considered by the ILO Committee of Experts on Non-Metropolitan Territories.[76] In his report on the Conference to the South African Minister of Labour, Gemmill claimed that he 'broke fresh ground' by drawing attention to the pernicious practice of some territories in restricting Africans from selling their labour. According to Gemmill, his own views were not seriously challenged, although Robens, the British Minister of Labour, had defended Nyasaland's policies.[77]

Since Gemmill had not directly accused Nyasaland of breaching ILO Conventions, the Colonial Office did not object to the matter being referred to the ILO's Committee of Experts. It assumed that the ILO would not consider it necessary to examine such criticisms in detail but rather to take the WNLA's views into account in so far as they were relevant to the general problem of migrant labour. Gemmill's speech was, however, an attack on British policy in Southern Africa, and the Colonial Office produced a review. The WNLA's notion of the free movement of labour to the best market implied that the social and economic impact on the sending-community was to be disregarded so long as the flow of

[75] Mr Alfred Robens, Minister of Labour and National Service, United Kingdom, in Provisional Record, International Labour Conference, Thirty-Fourth Session, Geneva, 21st June 1951. PRO, CO 525/220/13, Nyasa 1951.

[76] International Labour Conference, Provisional Record, 34th Session, Geneva, 26th Sitting, Friday, 29th June 1951, 3 p.m. No. 36, p. 487. PRO CO859/197/2, International Labour Organization Representations by Mr Gemmill (WNLA) Regarding Restrictions on Migrant Labour in Africa, 1951.

[77] Extract from a Report made by Mr Gemmill to the Minister of Labour, South Africa. PRO CO859/197/2, International Labour Organization Representations by Mr Gemmill (WNLA) Regarding Restrictions on Migrant Labour in Africa, 1951.

migrants was voluntary and spontaneous. Article 5 of ILO Convention No.50, however, recognised the need for the protection of labour-sending communities.[78] Nyasaland's manpower was constantly drained through migration, most of it voluntary and individual movement, and the local chiefs and other African representatives were opposed to WNLA's operations.[79]

Sir Geoffrey Colby Changes his Mind

While the Colonial Office was concerned about Gemmill's participation in the next ILO Conference, it was determined to adhere to its existing general principles regarding migrant labour. The problem, it believed, had two aspects: the preservation of the economic and social life of indigenous communities, and the protection of the individual migrant. A predominantly agricultural community which was adjacent to an industrialised territory would inevitably suffer a drain of its manpower. Should that increase to a level where agriculture suffered and social problems arose, then government had to intervene: a principle enshrined in the ILO Convention on Recruiting. The protection of migrant workers required the supervision of recruiting and contracts, and some supervision over the actual conditions of employment.[80] Since the Nyasaland government was unable to prevent spontaneous emigration to the Union, and South Africa seemed unwilling to control the movement of informal migrants, the Colonial Office concluded it could best *reduce* spontaneous migration by allowing the WNLA to *increase* its recruitment quota.

This contentious logic was now endorsed by the Governor of Nyasaland, Sir Geoffrey Colby. On his appointment, Colby had strongly opposed

---

[78] Colonial Office Responses, Undated, pp. 1–2. PRO CO859/197/2, International Labour Organization Representations by Mr Gemmill (WNLA) Regarding Restrictions on Migrant Labour in Africa, 1951.

[79] Memorandum from Mr. R. Turner, Colonial Office on the Gemmill Presentation, undated, 1951, p. 2. PRO CO859/197/2, International Labour Organization Representations by Mr Gemmill (WNLA) Regarding Restrictions on Migrant Labour in Africa, 1951.

[80] Memorandum from G. Foggon, Colonial Office, 6th December 1951 on CAA 148/233/01, pp. 1–2. PRO CO859/197/2, International Labour Organization Representations by Mr Gemmill (WNLA) Regarding Restrictions on Migrant Labour in Africa, 1951.

WNLA recruiting because of its harmful effect on food production. His administration was, however, starved of revenue at a time when there was a desperate need for investment in health, education and infrastructure. The economy's major export was labour, and the WNLA provided the best market. Within three years, Colby was keen for Nyasaland to sell labour to the WNLA. In January 1953, he participated in an important meeting held at the Colonial Office in London. Also present were Mr Barltrop, P. Nield and N.D. Watson from the Colonial Office. Providing that the annual quota be raised, the meeting was told, the WNLA had offered to pay the Nyasaland government £5 for each man who enlisted. That payment was in addition to a 10s attestation fee per recruit. Colby explained that for many years, Nyasaland had lacked public revenue and had few options to increase it. Many men earned a living outside the territory, and it was only reasonable that the administration gain some benefit. Colby acknowledged that the WNLA offer would be open to criticism, but that could be overcome if the money was paid into a Native Development Welfare Fund rather than General Revenue.

N.D. Watson agreed that there would be criticism if the government increased the quota because of monetary inducement from the WNLA. Since Colby's current proposal was at odds with his government's previous position, it was all the more important to avoid the ILO seeing any rise in the quota as being due to WNLA payments. The only justification for altering the quota must be the same as that used previously, namely to protect the social and economic welfare of the indigenous population. Mr Barltrop pointed out that on both political and moral grounds, the ILO opposed per capita payments to recruiting agencies or governments. To avoid conflict, it was suggested that Nyasaland should not enter into such an agreement. If, however, the WNLA later made a lump sum payment which was not formally associated with the quota, the money could be justified as a donation. After a long discussion, the meeting agreed to endorse a rise in the quota, and for Nyasaland to accept payment from the WNLA. To avoid criticism that the government was acting as a labour agent, the Committee decided that the rise in the quota should be announced immediately. Then, after a six-month interval, and as a completely separate arrangement, the WNLA should donate £5 per head to the government for labour supplied to the mines.[81]

That deception had its origins with Harold Kittermaster. In 1938, the then Governor of Nyasaland wrote to W. Ormsby Gore at the Colonial Office about the political implications of contracting labour to the WNLA. For some time, Kittermaster had wanted to charge a capitation fee for men working abroad. Such fees had been charged on a batch of men engaged to work in Tanganyika and several groups going to Southern Rhodesia. The WNLA had agreed to pay a monthly fee and it was Kittermaster's intention to create a Native Welfare Fund into which all such monies would be deposited. Kittermaster believed that was necessary if the government was to escape the accusation that it was selling labour.[82] Initially, the fees had no statutory authority, but were imposed as one of the conditions of receiving the Governor's Permit for agents engaging labour for work abroad. Soon after Kittermaster's sudden death from appendicitis in 1939, the Employment of Natives Ordinance was amended by Ordinance No. 25, which empowered the Governor to prescribe fees.[83] The way the Ordinance was written enabled the Colonial Office to hide from the ILO the fact that, from 1953, British territories were selling labour to an apartheid state.

William Gemmill's Vision of the WNLA

After the 1951 ILO Conference, William Gemmill presented the organisation with a review of the WNLA's operations in Southern Africa. Migrant labourers, he noted, were strongly attached to their tribal lives and their aim was consolidating a traditional way of life by earning the necessary money. In responding to those complex needs, the WNLA had

[81] Notes of Meeting held on 21st January 1953 on subject of Migrant Labour. PRO CO 859 306 Migrant Labour Conference.

[82] Letter 'Confidential', Harold Kittermaster, from Government House, Zomba, Nyasaland to the Right Hon W. Ormsby Gore, PC MP, Colonial Office, London, 5th March 1938. PRO CO 525/173/7. Nyasaland, Native Labour Capitation Fee 1938.

[83] A history of capitation fees is found in Letter from The Labour Department, The Secretariat, Zomba, Nyasaland 20th April 1947. M2/24/30 National Archives Malawi.

devised the controlled migrant system. Gemmill explained that the WNLA was a non-profit organisation which, since its foundation in 1902, had engaged almost four million men for employment. Under its auspices, men were employed on the condition that they must return to their homes at the end of a stipulated period. The wages paid were usually much higher than in their countries of origin, and the conditions of engagement conformed to those laid down by the ILO. The system had the approval of the South African, Portuguese and British governments.

The migrant labour system, Gemmill emphasised, preserved family life. Better still, the educative influence of men on returning to their homes was an important factor in social progress. The evils associated with migrant work were confined almost entirely to uncontrolled migration. The annual death rate amongst black miners was under 5.00 per 1000 from disease and under 2.00 per 1000 from accidents. 'Employment on the Witwatersrand Gold Mines is one of the greatest civilising factors in the whole field of employment of African labour. If the standard of living of the African peoples is to be raised, their productive capacity and efficiency must increase, and nothing is more conducive to this than employment on the Gold Mines.' Government restrictions on the free movement of labour, Gemmill reiterated, promoted the exploitation of the African for the benefit of local low-wage-paying employers.[84] Gemmill's account of WNLA's recruiting was accurate about wage rates and the support the system received from the British government. In most other respects, it was disingenuous. Borrowing from the ideology of South Africa's apartheid state, Gemmill claimed that migrant labour preserved traditional societies while modernising subsistence economies. Yet the Chamber's Northern strategy did just the opposite. It was designed to avoid the play of market forces within South Africa which would have driven up mine wages. In the long run, it retarded the economic development not only of South Africa but also of the whole region.[85]

---

[84] William Gemmill, Statement Presented to the International Labour Office Dealing with the Restrictions Placed on the Spontaneous Flow of African Migrant Labour, pp. 2; 3; 5. PRO CO859/197/2, International Labour Organization Representations by Mr Gemmill (WNLA) Regarding Restrictions on Migrant Labour in Africa, 1951.

[85] Charles H. Feinstein. *An Economic History of South Africa: Conquest, Discrimination and Development.* Cambridge: Cambridge University Press, 2005.

Conclusion

On the eve of the Second World War, the ILO drew up a schedule of rights to reduce the hardships of migrancy. The organisation wanted to prevent the separation or desertion of families, and to enable migrants to remit their savings to their country of origin. It also lobbied for pension rights under old age, invalidity and survivors' insurance schemes. The ILO was determined that labour be informed of the risks of work before entering a contract.[86]

In the protracted and often heated high-level contests over the labour of black miners, few of these conditions were met. The single clause honoured by the WNLA was the remission of wages. It was not until after the Leon Commission in 1994 that miners were warned by the industry of the dangers of silica dust. Britain's negotiations with the WNLA did little to help. Three failures were particularly important. The first was the failure to pressure the gold mines over repatriations, accurate mortality rates and the spread of infectious disease. For example, the Colonial Office never questioned Gemmill's mortality data, even though the figures were always far lower than the Bureau's own annual returns. The second was London's refusal to urge the Southern Rhodesian governments to improve wages and work conditions on their mines and farms. That in turn made the WNLA a more attractive employer. Finally, there was the willingness of the Colonial Office to violate the ILO Conventions to which Britain was a signatory. The ambiguity of the HCTs selling labour to the WNLA was worsened by apartheid. That encouraged the WNLA and London to emphasise the health benefits of migrancy, and to conceal the payments British administrations received for contracting labour. On their part, the HCTs were dependent upon the WNLA for revenue; that dependence grew in proportion with their perceived need to invest in housing, health and education.

[86] International Labour Conference, Draft Conventions and Recommendations Adopted by the Conference at its 25th Session 8th–28th June 1939, pp. 54; 48–50. PRO, Labour, International Labour Office, Migration of Workers, CO/859/9/7.

Over time, several key underlying assumptions were consolidated in negotiations between the WNLA, the Southern African governments and the Colonial Office. By the mid-1930s, the mines were no longer perceived as a phthisis-producing industry. As a result, tuberculosis spread by migrant workers disappeared from view as a factor in the WNLA's recruiting. After the war, negotiations assumed an even higher degree of dissembling. The core concealments were the missing health costs of contracting men to work in an industry which made them, their families and their communities sick.

References

Baker, Colin. *Development Governor: A Biography of Sir Geoffrey Colby.* London: British Academic Press, 1994.

Feinstein, Charles H. *An Economic History of South Africa: Conquest, Discrimination and Development.* Cambridge: Cambridge University Press, 2005.

Gelfand, M. 'Migration of African Labourers in Rhodesia and Nyasaland, 1890–1914'. *Central African Journal of Medicine,* Vol. 7, August 1961, pp. 293–300.

McCulloch, Jock. *Black Peril, White Virtue: Sexual Crime in Southern Rhodesia, 1902 to 1935.* Bloomington: Indiana University Press, 2000.

Phimister, Ian R. 'African Labour Conditions and Health in the Southern Rhodesian Mining Industry, 1898–1953'. In Ian R. Phimister and Charles van Onselen (Eds). *Studies in the History of African Mine Labour in Colonial Zimbabwe.* Gwelo: Mambo Press, 1978, pp. 102–150.

van Onselen, Charles. 'Worker Consciousness in Black Miners: Southern Rhodesia, 1900–1920'. *Journal of African History,* Vol. 14, No. 2, 1973, pp. 237–255.

van Onselen, Charles. *Chibaro: African Mine Labour in Southern Rhodesia.* Johannesburg: Ravan Press, 1980.

Part IV

Manufacturing and Contesting Ignorance

11

Dissenting voices: 1902–1956

Neil Macvicar in the Eastern Cape[1]

Neil Macvicar was born in Scotland in 1871 and graduated from Edinburgh University in 1894. He joined the Blantyre Mission of the Church of Scotland in Nyasaland as a medical officer and served there between 1896 and 1900. During that period, a smallpox epidemic swept Blantyre, and Macvicar began training female African nurses, an innovation he continued in South Africa despite strong opposition from state authorities.[2] Accompanied by his wife Jessie who was a nursing sister, Dr Macvicar was appointed to the Victoria Hospital at the Lovedale Mission in the Eastern Cape in 1902. The hospital was the first mission hospital to be established in South Africa and served a rural population of about 90,000. Macvicar also began health classes and with his colleagues

[1] See Martin J. Lunde. An Approach to Medical Missions: Dr Neil Macvicar and the Victoria Hospital, Lovedale, South Africa circa 1900–1950. PhD thesis, University of Edinburgh, 2009, pp. 240–253.

[2] For an account of Macvicar's life see Obituary, 'A Great South African Figure: Neil Macvicar MD LLD born 1st August 1871, died 2nd December, 1949'. *South African Outlook*, January 1950. Rhodes University Archives, 3(ii) and 'Neil Macvicar. In Memoriam'. *South African Medical Journal*, Vol. 24, 21st January 1950, p. 45.

J. McCulloch, P. Miller, *Mining Gold and Manufacturing Ignorance*,
https://doi.org/10.1007/978-981-19-8327-6_11

promoted the establishment of the South African Native College, later Fort Hare University. The region was a labour reserve for the gold mines, and Macvicar treated returning miners and lobbied provincial and national governments to halt the spread of tuberculosis. Macvicar was a persistent critic of the migrant labour system, and over a long career lobbied governments to make it less oppressive. The dominant themes in his representations and correspondence were the hardships of migrant life and the immense human cost and unsustainable economic and social consequences of exporting labour for the labour-sending regions and communities.

One of Macvicar's most important initiatives was the South African Health Society, which was founded at Lovedale in 1909. The Society promoted health education, especially about tuberculosis. From 1914, it published a quarterly magazine in English, *Xhosa and Sotho*, which was distributed to schools in the Transkei, the Ciskei, Basutoland, the Orange Free State and Bechuanaland. The Society presidents included Dr N. Macvicar in 1911, Dr Peter Allan in 1929, Dr A.J. Orenstein in 1933 and Sir Edward Thornton from 1937.[3] Macvicar retired in 1937 but continued to lobby governments over migrant labour and the prevention of tuberculosis. He died in 1949.

In addition to treating patients, Macvicar published a number of papers in medical journals. He also completed an MD thesis at Edinburgh University in 1907. The thesis dealt with the history of tuberculosis in southern Africa and was serialised in the *South African Medical Record*. Macvicar was greatly respected by his contemporaries. The British expert, Professor Lyle Cummins, believed that the Victoria Hospital should be used as a model for treating and preventing tuberculosis, while Dr H.W. Dyke, the principal medical officer for Basutoland, referred to Macvicar's work at Lovedale as 'outstanding'.[4] Despite his expertise,

---

[3] Dr Neil Macvicar, Editor-in-chief of Health Society's Magazine, Notes on the South African Health Society, undated, 1943. Macvicar Papers, Cory Library, Rhodes University PR 3095–3153, (One-ii).

[4] Lunde, An Approach to Medical Missions, p. 242.

Macvicar was not called before the Tuberculosis Commission of 1912. Neither did he appear as an expert witness at the 1932 SAIMR enquiry.[5]

During the 1920s, admissions at the Victoria Hospital averaged 700 per annum. The staff consisted of a resident physician, a matron and lady dispenser, a black staff-nurse, twelve student nurses and one orderly.[6] In 1932, the Board established separate Tuberculosis and Venereal Disease wards. By the mid-1930s, there were more than thirty mission hospitals in South Africa, supported by a still larger number of small clinics staffed by missionary nurses. Those hospitals provided a service which, according to Macvicar, was rightfully the responsibility of the state. The black population contributed more than its share to the national economy through taxation and through their low wages. It was therefore the state's duty to provide social services. Mission hospitals obtained over half of their revenue from their patients, with the rest coming from mission sources, Provincial Councils and the Public Health Department.[7] Lovedale was unusual in receiving financial support from the Cape government, and Macvicar lobbied that all missions should receive grants for capital works.

In most towns, Macvicar noted, there were public hospitals which treated African patients. However, most people lived in rural areas. Their only access to care was a district surgeon who was typically responsible for 30,000 to 40,000 patients. The Victoria Hospital provided general medical, surgical and maternity care, as well as a ward for crippled children. Miners and their families, among whom tuberculosis was common, were

---

[5] Macvicar was thanked by the SAIMR for his advice but had no formal role and his important essays on tuberculosis are not included in the Report's appendices. There are references to Macvicar on pp. 262–263 by Lyle Cummins. *See Tuberculosis in South African Natives with Special Reference to the Disease Amongst the Mine Labourers on the Witwatersrand.* South African Institute for Medical Research, Johannesburg, March 1932.

[6] Memorandum to the Rockefeller Foundation on Proposals for the Training of Native Nurses in connection with Victoria Hospital, Lovedale, South Africa, undated, 1930. Macvicar Papers, Cory Library, Rhodes University. PR 3095–3153, (One- ii).

[7] Memorandum on the Financing of Mission Hospitals, Submitted to the Provincial Financial Resources Committee by Dr Neil Macvicar MD, Chairman, Medical Work Committee of the Christian Council of South Africa, July 1941. Macvicar Papers, Cory Library, Rhodes University. PR 3095–3153, (One-I), p. 1.

served by mission hospitals. However, these could not survive on patient fees. In the Cape, the Transkei and Natal, people were becoming poorer and patients could not afford to pay.[8]

Soon after arriving at Lovedale, Macvicar recognised that migrant labour, and in particular the employment of men on the gold mines, was spreading tuberculosis in the Eastern Cape. During the four years to December 1906, 6 per cent of the 8000 patients treated at Lovedale had the disease. Macvicar refers to the 'fearful havoc' of tuberculosis among Coloured and black communities, with the average death rate being four times that of white South Africans.[9] In 1909, Macvicar wrote a Memorandum for the Minister of the Interior, in which he identified land hunger, low wages and migrant labour as drivers of infection. The disease was rare in whites but prevalent in Coloureds and blacks. Among urban populations, the death rate for Europeans was 15.3 per 10,000, the same as the corresponding figures for Western Europe and North America. Among the Coloured and black communities living in the same towns, the death rate was 65.8 per 10,000. The disease was becoming more widespread and because it killed breadwinners it was worsening the existing poverty. 'Tuberculosis in South Africa', Macvicar wrote, 'has two chief modes of spread. It spread from man to man, especially at the Rand mines, and within families at peoples' homes'. In terms of the Coloured and black populations, South Africa was probably the most consumptive country in the world and the mines were 'the chief distributing agencies'.

According to Macvicar, the Union government was the only authority with the resources to take effective action and yet it had done little. He recommended the appointment of government medical officers to conduct mine medicals in place of company doctors, the creation of sanatoria for early cases, the provision of hospitals for advanced cases, special dispensaries and sanitary inspectors at locations and, above all, a sustained

[8] Neil Macvicar, Statement to the South African Medical Mission of the Church of Scotland, Lovedale, May 1933. Macvicar Papers, Cory Library, Rhodes University, PR 3095–3153, (One-I), pp. 6; 8.

[9] Neil Macvicar. 'Tuberculosis Among the South African Natives'. *South African Medical Record*, Vol. VI, No. 11, 10th June 1908, p. 173.

public education programmes.[10] The whole country should be mapped out into sanitary areas, each under the control of a full-time medical officer who would advise local authorities and co-ordinate the response by hospitals, missions and charities. In 1910, Macvicar published an important article, based in part on his MD thesis, on the demography of tuberculosis. He explained that in South Africa tuberculosis was not a white people's disease because most whites lived in social settings above what he refers to as 'the tuberculosis line'. In fact, there were few white populations in the world in which tuberculosis was so uncommon. Tuberculosis was a *native* problem, with the disease spreading from labour centres and especially from the gold mines to rural areas. Living conditions rather than race determined the rates of infection and death.[11]

From his work at the Victoria hospital, Macvicar was sure that the official numbers of those who died on the Rand represented only a fraction of mine deaths. Some men went to the mines with incipient tuberculosis while others acquired the infection in the compounds. When a sick man returned home he infected others. After his death, his widow developed tuberculosis as did her baby and by the following year they were both dead. Her children became orphans and were scattered among relatives. Some developed tuberculosis and infected their new families. Macvicar concluded that 'the stream of consumptives which the [gold] mines are annually pouring out upon the country is still so large that, unless it is greatly reduced, one almost despairs of being able to counteract its effects whether by sanatoria or any other agency'. It was essential that every doctor have a hospital to which he could send patients for education and, if necessary, isolation. Every patient should receive a sanatorium education for a month: in that way, a facility with 50 beds could educate 500 patients a year. There should also be hospitals for advanced cases. Macvicar estimated that out of the 150,000 black children currently in school, between 5000 and 10,000 were doomed to die from

[10] Dr Neil Macvicar, Memorandum for submission to the Hon. the Minister of the Interior, prepared at his request by the deputation representing the South African Medical Congress, 'Tuberculosis among the South African Natives', undated but likely 1909. Macvicar Papers, Copy Library, Rhodes University. PR 3095–3153, (One- ii), pp. 1–3.

[11] Neil Macvicar. 'Tuberculosis Amongst the Coloured Population of South Africa'. *The South African Medical Record,* 26th February 1910, pp. 39–45.

tuberculosis before reaching the age of 35. It was vital that teachers be trained in prevention. Citing the recommendations of the International Congress on Tuberculosis from 1909, Macvicar wanted compulsory notification and registration, government-funded hospitals, the isolation of patients in sanatoria, the provision of dispensaries, improved housing, the regulation of factories and instruction in personal and public hygiene.[12]

Twenty years later, little had changed. Malnutrition remained a major cause of tuberculosis infection. The Ciskei Reserve was overcrowded, and scurvy and vitamin A deficiency were common. The rainfall was unreliable and when crops failed, cattle died and there was no milk. That environment produced a steady stream of tuberculosis cases, with peaks during the severe droughts of 1919–1920 and 1929–1930.[13]

Advocacy

By 1910 Macvicar was arguably the foremost South African expert on tuberculosis in labour-sending communities. He appeared as an expert witness before a number of Commissions into the mines and migrant labour, including the Economic Commission of 1925. He told the commissioners that men in the Eastern Cape were forced by necessity into migrancy. After two or three contracts, their capacity for intense labour would diminish and they would seek work on the coal or asbestos mines, the railways or with municipalities. A young man was usually about thirty before he could afford to marry, and he would be approaching fifty by the time his sons went to Johannesburg or his daughters into domestic service.[14] It was an oppressive system that made family life all but impossible.

In 1917, Macvicar submitted a report on migration in the Eastern Cape to the Native Economic Commission. He conceded that migrant labour was beneficial in the sense that many men could not marry or

[12] Macvicar, 'Tuberculosis Amongst the Coloured Population of South Africa', p. 43.

[13] N. Macvicar. 'The Prevalence of Certain Diseases among the Natives of the Ciskei'. *The South African Medical Journal,* 26th November 1932, pp. 721–724.

[14] Dr N. Macvicar, Lovedale, evidence to the Economic Commission, 30th October 1925. Macvicar Papers, Cory Library, Rhodes University. PR 3095–3153, (One- ii), pp. 4–5.

maintain families without such income. Once married, most continued of necessity to work on the mines, thereby placing massive strain on their families. Migrant workers lived in crowded quarters, and it was common for men to economise on food in order to save. As a result, many contracted tuberculosis. One man, who returned from Cape Town with consumption, described to Macvicar his life at Langa where the rent of his room and his railway fares to work took most of his wage, leaving him seriously undernourished. Such sacrifices were futile as migrant wages, including those from the mines, were too low to support a family.

Many of the outpatients at Victoria Hospital suffered from malnutrition. This was particularly common among small children. In a survey Macvicar conducted in 1930, of fifty consecutive cases where a mother brought in a sick baby, not a single one was able to pay for medication. Out of forty-four fathers who were alive and able-bodied, thirty-five were working at distant labour centres. Macvicar argued that men should be able to live with and support their families, whether in towns, on farms or on their own holdings. The one exception was the gold mines, where it would be disastrous for the health of miners and their families to have large communities living permanently.[15] The risk of tuberculosis was too great.

Neil Macvicar presented similar arguments in a memorandum on silicosis and tuberculosis to the Stratford Commission in 1942. The rainfall in the Eastern Cape was unreliable, and droughts were common. A miner's wage alone was insufficient to support a family. However, since a large number of families in the Reserves had no land, they were increasingly dependent upon migrant wages as their only source of income. Store-bought food was expensive, and diets were so poor that pellagra and scurvy were rife. Macvicar explained that it was a common for young men to work two or three contracts and then return home with pulmonary tuberculosis. Asked why he had given up such work, a man would commonly answer was that he had a slight cough, or perhaps a pain in the side, and was afraid of the chest diseases common on the mines. Many did not report sick as their contracts were nearly up and they did

[15] Report submitted by Dr N. Macvicar, Victoria Hospital, Lovedale, to the Native Economic Commission 1930. Macvicar Papers, Cory Library Rhodes University, 3(I), pp. 1–3.

not think it worthwhile. Most men were married with children, and those who were single usually supported their parents. By the time Dr Macvicar saw those patients they were dying, and it was too late to make compensation claims.

Macvicar was certain that the only way to ensure that tuberculosis cases did not escape detection was for every departing miner to be X-rayed. At the time, it was almost impossible for repatriated miners to apply for compensation. Few men had records of their service and often such men were too ill to make the journey to the Bureau in Johannesburg. Macvicar wanted the Chamber and the state to establish clinics in labour-sending areas to examine former miners. Such a scheme should include the High Commission Territories, and Portuguese Africa.[16] Macvicar believed there was a moral obligation on employers to provide for sick miners and that such costs should not be shifted onto rural communities. There was, for example, no justification for black miners with silicosis not being awarded further compensation as the disease progressed.

In 1943, Dr Macvicar submitted evidence to the National Health Services Commission. He noted that the South African population had doubled over the previous forty years but food production in the Reserves had declined. The plots were too small to support a family. In the Transkei, the absence of 60 per cent of able-bodied men hampered cultivation. Milk had become scarce and among the very poor pellagra was widespread. The decisive factor in vitamin A deficiency, so common among the patients at Lovedale, was the total absence of milk, fruit and vegetables at certain times of the year. During one severe drought in 1927–1928, over 200 cases of scurvy were admitted to the Lovedale Hospital.[17]

In addition to his appearances before commissions, Macvicar lobbied governments to lessen the negative impact of the migrant labour system on family life and public health. One of the principal issues he pursued was children's nutrition. In 1936, he wrote to the Minister of Native Affairs suggesting that the department provide free meals for black school

[16] Memorandum by Dr Neil Macvicar on Miners' Phthisis and Tuberculosis in Mine Labourers, January 1942. Cory Library, Rhodes University. PR 3095–3153, Macvicar Papers, (One-I), pp. 2–5.

[17] Dr Neil Macvicar, Evidence Submitted to the National Health Services Commission on behalf of the Christian Council of South Africa, undated [1943?]. Macvicar Papers, Cory Library, Rhodes University, PR 3095–3153, (Two-ii), pp. 2–3.

children to combat malnutrition and thereby reduce the risk of tuberculosis. The minister replied that while he was sympathetic, he could not help without 'running the risk of pauperising large sections of the Native people'.[18] Nine months later Macvicar wrote to the Minister of Finance pointing out that the wages for black labour were so low that malnutrition was almost universal among children in the Eastern Cape. Macvicar requested that the government extend to the Reserves the free supply of milk given to whites and Coloureds.[19]

Having failed in those approaches, Macvicar wrote to the leader of the opposition, the future prime minister Dr D.F. Malan, in the hope that the Parliament may intervene. Macvicar pointed out that the South African Health Society scrupulously avoided political matters, but a milk ration for black children was of great importance. 'We are aware that it is the policy to discourage Natives from going to towns, but whatever other measures might be legitimately employed to enforce this policy, a deprivation affecting the health of the children could surely not be regarded as a legitimate weapon.'[20] The appeal again failed; there was no extension of welfare programmes across racial lines.

In his public pronouncements, Macvicar usually appealed to Christian notions of charity and stewardship rather than to principles of social justice or political rights. One exception comes from 1945, when the possibility of Britain handing over the High Commission Territories to South Africa was being debated. In October 1945, Macvicar cautioned the Under Secretary for the Colonies, A. Creech Jones, about the looming transfer. Macvicar reminded Creech Jones that almost all heavy labour on the mines, factories and farms was done by African men. The majority had homes in the Reserves, and they depended upon their wages to support their families. Migrant labour depressed wages and stifled

---

[18] Letter from Department of Native Affairs, to Neil Macvicar, 'Under-nourishment of Native school children', Kingwilliamstown, 22nd July 1936. Macvicar Papers, Cory Library, Rhodes University, PR 3095–3153, (Two-i).

[19] Letter from Neil Macvicar, Vice-President, The South African Health Society, to the Minister of Finance, Cape Town, 26th February 1937. Macvicar Papers, Cory Library, Rhodes University, PR 3095–3153, (Two-I).

[20] Letter from Neil Macvicar, Vice-President, South African Health Society, to Dr Malan, Parliament, Cape Town, 27th February 1937. Macvicar Papers, Cory Library, Rhodes University, PR 3095–3153, (Two-i).

agricultural development. As a result, home life was disintegrating. The gold mines were the greatest problem. 'The mining industry has been built up upon cheap migrant labour and cannot afford to change the system. If wages adequate to support the miners' families at or near the mines were to be paid, many of the lower-grade mines would have to close down.' Macvicar also referred to 'a slave-holding mentality' which was still a factor in parts of the Northern Provinces, where white farmers looked to government to force men into their service. The incorporation of the territories would hand over captive populations to the Chamber and white farmers.[21]

During Macvicar's years at Lovedale, there were two important enquires into the problem of gold mining and tuberculosis which set out to answer the same question: Was tuberculosis being spread from the mines to rural areas and if so how could it be prevented? Macvicar was the leading authority on those questions. When the Tuberculosis Commission began in 1912, Macvicar had been at Lovedale for eight years and had treated thousands of men returning from the mines. He had an impressive list of publications in medical journals and was probably the only physician of his generation to have a higher degree specialising in tuberculosis. Surprisingly, the Commission did not visit Lovedale, Macvicar was not called as an expert witness and his published work was not cited in the Commission's report.[22] Instead of drawing on Macvicar's knowledge, the Commission relied on hearsay to accuse the Lovedale Missionary College of spreading tuberculosis among its pupils.[23] The SAIMR enquiry, whose report was released in 1932, similarly ignored Lovedale and did not call Macvicar as a witness. On his own initiative, Professor Lyle Cummins, who acted as an expert advisor to the 1932 SAIMR study, did visit the Lovedale mission and was greatly impressed. On Macvicar's retirement, Cummins wrote: 'I mentioned the work at Lovedale in my [recent]

[21] Letter from Dr Neil Macvicar, Johannesburg, to A. Creech Jones MP, Under Secretary for the Colonies, House of Commons, London, England, 4th October 1945, Macvicar Papers, Cory Library, Rhodes University PR 3095–3153, (One-ii), pp. 6; 8.

[22] See Annexure 'B', 'Return of Places Visited, Inspections Made, and Witnesses Examined by the Commission', *in Report Tuberculosis Commission, 1914.* U.G. 34-'14, Cape Town: Government Printer, 1914. Union of South Africa, p. 304.

[23] *Report Tuberculosis Commission, 1914.* U.G. 34-'14, Cape Town: Government Printer, 1914. Union of South Africa, p. 113.

Address at the Empire Conference because I think it is about the best and most valuable work of its kind I have come across, and the spirit of the work at Lovedale appealed to me as exactly what is needed'.[24]

Basil A. Dormer in Durban

Basil Dormer was born in Cathcart in the Eastern Cape in 1903. After graduating from Wits, he served as medical superintendent of Nelspoort Sanatorium for two years before spending a brief period of further study in England. In 1939, he became the first superintendent of King George V Jubilee Hospital in Durban, where he remained for thirty years. During his career at King George, Dormer was the Union government's Chief Tuberculosis Officer. He also served on UN Expert panels. Dormer promoted the introduction of Bacillus Calmette-Guerin (BCG), a vaccine to prevent tuberculosis. For almost ten years, with a Durban colleague F.J. Wiles, he conducted tuberculosis surveys using mobile X-ray units. He was a foundation member of the South African National Tuberculosis Association (SANTA) and worked closely with K.G.F. Collender in developing mass miniature radiography.[25] Under Dormer, the treatment provided at the King George Hospital was innovative. With so many families living close to poverty, the admission of a breadwinner to hospital was often a catastrophe. For that reason, it was difficult to persuade patients to enter treatment and those who did often absconded. With the introduction of isoniazid, Dormer realised the possibility of out-patient care. From 1953, he became the first South African physician to treat patients while they remained at work.

From the early 1940s, Dormer published a number of articles on what he viewed as a looming public health crisis. Like Macvicar, he identified that threat as arising initially from mining and migrant labour. The discovery of diamonds and then gold saw men from Europe and black

[24] Letter from Professor S.L. (Lyle) Cummins, Department of Tuberculosis, Welsh National School of Medicine, Cardiff, to Dr Neil Macvicar, 15th July 1937. PR 3095–3153, Macvicar Papers, Copy Library, Rhodes University, (Two-i).

[25] T. Collins. 'The History of Southern Africa's First Tuberculosis Epidemic'. *The South African Medical Journal*, Vol. 62, No. 21, 13th November 1982, p. 786.

workers from isolated rural areas flood into Kimberly and Johannesburg. Large numbers of black miners who contracted tuberculosis went home to spread infection and die.[26] What Dormer termed the South African Tuberculosis Epidemic was precipitated by labour migration, exposure to silica dust and overcrowded compounds.

Over time, the black population was dispersed between the Reserves, those living on white farms and those in urban and industrial centres. By the 1920s, tuberculosis was endemic in the Reserves, where the incidence was low and infection tended to run the same course as in Europeans.[27] South Africa's industrial revolution, which was accelerated by the Second World War, saw rural populations who were tuberculin positive pour into urban areas, where a lack of adequate housing, low wages and poor nutrition led to an epidemic. In the period 1947 to 1948, the tuberculosis allocation was £689,000, or 16.8 per cent of the total health budget of £4.1 million. Seven years later the budget had grown to £10,000,000 and the tuberculosis allocation to £3 million, or 30 per cent of the much larger budget.[28] While the full beneficial effect of the drugs was never fully realised, the catastrophic impact of TB on the population was blunted by streptomycin.

In the towns, most tuberculosis was rapid and fatal, an acuteness many physicians attributed to racial susceptibility. Dormer was sure that living and working conditions rather than race determined the outcome. To find a remedy, he collated the mortality data from Cape Town, Durban and Port Elizabeth. Dormer found that the incidence among blacks, while always high, had doubled between 1938 and 1946. By the latter date, it was more than twenty times higher than in Europeans.[29] For every recorded case of active tuberculosis, Dormer estimated there were probably nine others which missed detection. Dormer was notable among his contemporaries for his insistence that tuberculosis was a social disease which, although manifested in individuals, had its causes in the

[26] B.A. Dormer. 'Tuberculosis in South Africa'. *British Journal of Tuberculosis and Diseases of the Chest,* Vol. 50, No. 1, 1956, p. 53.

[27] B. A. Dormer. 'Tuberculosis in the Bantu'. *The South African Medical Journal,* 25th May 1946, p. 263.

[28] Dormer, 'Tuberculosis in South Africa', p. 60.

[29] Dormer, 'Tuberculosis in the Bantu', p. 264.

community. This made the family the most important unit for controlling infection.[30] Dormer was critical of the mines for repatriating highly infectious men into vulnerable communities. He pointed out that the mines' response to managing tuberculosis was in most respects the opposite of contemporary public health conventions. The mines were far better than any other industry in identifying tuberculosis cases but made no attempt to isolate or educate patients.

Dormer noted that black miners should have been the least likely members of their communities to develop tuberculosis. They were healthy young men who were medically screened before they entered the mines, and while under contract they were well fed. In 1942, among the 355,614 black miners, there were 986 cases of tuberculosis and 869 repatriations. Of the 117 cases who were treated in mine hospitals, 116 died. This gave an official tuberculosis annual death rate of 32 per 100,000. However, as Dormer pointed out, if 90 per cent of repatriates also perished, the death rate was a calamitous 252 per 100,000.[31] Dormer was sure that MMR would enhance case finding. However, he pointed out that the new technology was only the first step in a prevention programme. 'The mining industry leads all other industry in South Africa in its care of labour. It is hoped that in the near future all miners discovered to have tuberculosis will be treated at the mine hospitals until they … are non-infectious before rehabilitation.'[32] Dormer believed that patient education was the most important health control measure and that money would be better spent on feeding programmes for the poor than in treating advanced cases. He was confident that if the population was well fed and patients were educated about how to manage the disease, there would be a rapid decline in the number of new cases.[33]

Dormer spent his career within state medical institutions where he often treated men repatriated from the mines. Yet he offered no

[30] See B.A. Dormer, J. Friedlander and F.J. Wiles. 'Tuberculosis Control: The Importance of the Family Unit'. The *South African Medical Journal*, 22nd November 1941, pp. 453–454.

[31] Dormer, 'Tuberculosis in the Bantu', p. 264.

[32] Dormer, 'Tuberculosis in South Africa', p. 60.

[33] For a brief history of the public health response B.A. Dormer, J. Friedlander and F.J. Wiles. 'A South African Team Looks at Tuberculosis'. *Proceedings of the Transvaal Mine Medical Officers' Association*, Vol. XXIII, No. 257, November 1943, pp. 109–113.

explanation as to why they had such high rates of infection. In his numerous articles, he makes no mention of dust exposure or silicosis influencing the tuberculosis rate. Nor did he acknowledge that the mines' refusal to pay living wages impoverished whole communities in labour-sending regions. Despite those limitations, Dormer was a progressive who supported what Packard terms 'the environmentalist understanding of tuberculosis'.[34]

Eustace H. Cluver and the Stratford Commission

Eustace H. Cluver was born in the Cape in 1906 and studied medicine at Oxford. On his return to South Africa, he was appointed Professor of Physiology at the University of the Witwatersrand.[35] Cluver's primary interest was in preventive medicine. He joined the Union Health Department and eventually became Secretary for Health. Cluver was a progressive, and under his leadership the department developed primary health care centres for rural communities. In 1940, Cluver became Director of the SAIMR, and three years later he served on the Stratford Commission.

In 1933, while as an assistant officer with the Health Department, Cluver wrote a report on the lack of compensation for black miners with lung disease. 'The frequency with which natives coming to Johannesburg are found to be suffering from phthisis after having been employed on the mines indicates that a considerable number of natives must leave the mines at the completion of their contract period with unrecognised but compensable silicosis.' Cluver acknowledged that it was difficult to detect early silicosis at an ordinary exit examination. Such disease may later progress rapidly, and when the recruit sought to return to the mines he would be rejected at an initial examination in the territories. These men,

[34] R.M. Packard. *White Plague, Black Labour: Tuberculosis and the Political Economy of Health and Disease in South Africa.* Berkeley: University of California Press, 1989, pp. 234–35.

[35] In Memoriam Eustace Henry Cluver. *South African Medical Journal*, Vol. 62, 24th July 1982, p. 144. See also Ann Cluver Weinberg. *Looking for Goodness: Exploring Eustace Cluver and his Family 1657–1982.* Johannesburg: Sapler Press, 2008.

he argued, did not receive the compensation to which they were entitled. Having discussed the problem with the Director of Native Labour, Cluver suggested: 'A possible solution seemed to be [state] supervision of the initial examination of mining recruits in the Native Territories. At this examination natives found to be suffering from phthisis, should, if it appears that they had previously worked on the mines, undergo further examination with a view to compensation.'[36] Nothing came of his recommendations.

Of all the public inquiries, the Commission of 1941–1943, chaired by the former Chief Justice James Stratford, was arguably the most important. Cluver served on the Commission and his influence is obvious in those sections of the final report dealing with tuberculosis. Discussion of black miners occupied almost a third of the report, which described a 'disquieting state of affairs in regard to compensation, medical examinations, and after care'. The report was more sympathetic to the plight of migrant labour than its predecessors and supported the principle that occupational disease was a cost of production which should be paid by the industry. It rejected the assumption underlying much of the existing legislation, namely that because of their lower living standards, blacks with silicosis did not suffer the same deprivation as whites. On the contrary, black miners so depended upon their capacity for hard labour that once disabled by lung disease, they could not earn a living. The Commission's Report noted that because of landlessness and overstocking, the Reserves were increasingly dependent on mine wages. While the wages of white miners had risen, those of blacks had not. If wages were below subsistence, then compensation awards based on wages were bound to be inadequate. Stratford also found that the life expectancy of black miners with tuberculosis was so short that for the purposes of compensation they ought to be regarded as totally incapacitated.[37]

---

[36] Letter from Dr E.H. Cluver, Assistant Health Officer, Union Health Department, Johannesburg, to The Secretary for Public Health, Pretoria, 18th May 1933. Subject: Phthisis Return, Pass Office, Johannesburg. South African National Archives, GES 1106, 1930–1953, TB in Mine Natives & Native Labourers.

[37] *Report of the Miners' Phthisis Acts Commission, 1941–1943.* Pretoria: Government Printer 1943, pp. 25–26.

Dr Peter Allan, the Secretary for Public Health, told the Commission that the mines produced more tuberculosis than any other industry. If miners with tuberculosis returned home untreated, they were a danger to their families: 'These sufferers are entitled to free medical services'. Blacks with tuberculosis should receive a pension, free board, lodging and medical care in special sanatoria supported by the Compensation Fund. Compensation should cover lost wages, wages in kind, medical care, and vocational rehabilitation. To end the injustices of the compensation system, the Commission recommended that X-rays of black miners be made compulsory and that all examinations be carried out by the Bureau rather than by mine medical officers or the WNLA.[38]

The Commission's most radical recommendations concerned tuberculosis. Stafford agreed with Dr Allan that the gold mines produced more tuberculosis than any other industry and cited the high rate of tuberculosis among repatriated workers as evidence of the inadequacy of pre-employment medical examinations.[39] These criticisms were consistent with the dissenting voices of Macvicar, Cluver and Allan over the previous twenty-five years. What is notable about Stratford is that its recommendations for an overhaul of the compensation system were costed by the Government Actuary. Those estimates give an indication of the extent to which labour-sending communities subsidised the gold mines.

Stratford's calculations for white miners were built around two assumptions: that less than a third would contract silicosis and that the average total compensation for a miner and his dependants was £4100. In the period from 1938 to 1941, out of 20,000 white miners there were on average 275 new cases each year. On that basis, the total cost over twenty years would be £22.55 million, and the levy would rise to £2.2 million, an increase of around £800,000 per year. For black miners, the calculations of lost earnings were based on a monthly wage of £3 plus food and quarters set at 1s a day, a monthly total of £4 10s. Blacks worked on average ten months per year, so the annual loss was £45, which over a career

[38] *Report of the Miners' Phthisis Acts Commission*, pp. 14–17; 25; 29–29.

[39] *Report of the Miners' Phthisis Acts Commission*, pp. 14; 29.

lasting thirteen years totalled £486. Assuming 2000 new cases of tuberculosis and silicosis per year, the cost over twenty years would be £19.44 million, or £972,000 per year. In rejecting the Chamber's protests that the mines could not afford to pay, the Commission concluded that if that were true the mines must be so dangerous that they should be closed.[40]

The Stratford Commission calculations were based on the official data, which seriously under reported the incidence of occupational injury. While the Commission calculations for black miners were based on a silicosis *incidence* rate at less than 0.2 per cent per annum, a 1949 survey conducted by the Silicosis Medical Bureau on 1120 'non-working' black miners found twice as many cases of silicosis than were discovered in the whole population of 228,000 'working miners'.[41] Four decades later, several post-apartheid epidemiological studies (detailed in Chap. 14) estimated the *prevalence* of silicosis among black miners to be in the range of 3 to 36 per cent. But perhaps the best basis for comparison emerged from the more rigorous (but short-lived) examination regime instituted under Gerrit Schepers. As noted in the following section, between 1952 and 1956 the compensation rate for whites rose five-fold and that for blacks three-fold. If these rates were used to calculate the mines' combined annual levy, it would have risen to almost £14 million. For comparison, the net profit of Anglo American and its subsidiaries in 1963, the final year of Stratford's estimates, was just under £12 million.[42]

The Stratford Report was the high-water mark of the South African commissions. The rise to power of the National Party in 1948 saw the number of such inquiries decline. The inquiries also changed in character, and the kind of criticisms made by Stratford disappeared. Stratford was not, however, the last attempt to change mine medicine and the compensation system.

---

[40] *Report of the Miners' Phthisis Acts Commission*, pp. 9; 30–32.

[41] *Report of the Silicosis Medical Bureau for the Year Ending 31st March 1950*. Pretoria: Government Printer, 1950, pp. 5–6. The Bureau attributed the far higher rate of silicosis among non-working miners to the progressive nature of the disease. It also conceded that in many cases, the presence of silicosis was not being detected during the exit examination.

[42] Anglo American Corporation. *Annual Report for 1963*, p. 8.

Gerrit W.H. Schepers and the Miners' Phthisis Medical Bureau

Between 1946 and 1950, Anglo American spent more than £65 million on the development of thirteen new mines in the Orange Free State.[43] It was a massive investment designed to ensure the corporation's future. The development of those new fields coincided with a period of intense conflict over compensation. That conflict between the Bureau, the Chamber and the Mine Workers Union (MWU) was played out through a series of Commissions and the passing of four pieces of legislation. The crisis began with the *Silicosis Act No. 47 of 1946* and ended ten years later with the *Pneumoconiosis Act No 57 of 1956.* The underlying problem was the refusal of the Bureau to compensate white miners who suffered what the MWU believed was clearly evident disease. The barriers facing black miners were even greater, but under apartheid they lay on the margins of the MWU's struggle.

The one doctor whose early career spanned these struggles was G.W.H. Schepers. Schepers was born in 1914 and spent his childhood in the Orange Free State. He studied medicine at Wits University, where he was taught by the famous pathologist Sutherland Strachan.[44] Schepers obtained Bachelor of Medicine and Surgery degrees in 1938, a Doctor of Science in 1945 and a Doctor of Medicine in 1948. In 1944, he joined the Miners Phthisis Medical Bureau, where he remained until 1954. Schepers spent 1949 in the USA as a Harkness Fellow. He found the US scientists more knowledgeable about occupational lung disease than their Johannesburg counterparts. For the first time, Schepers learned about the importance of emphysema and chronic bronchitis as problems afflicting hard rock miners. On his return to South Africa, Schepers pressed for the inclusion of those diseases under the award.[45] Academically, Schepers was an outstanding scientist. However, by his own admission, he did not get

---

[43] Chairman's Report in the *Annual Report of Anglo American for the Year ended 31*st *December 1950,* p. 2.

[44] This account of Schepers' career is based in part on Dr Schepers Deposition at Alexandria, the US District Court for the Eastern District of Virginia 29th June 1981 in Re: Asbestos Litigation Locke v Johns-Manville C/P No. 77-1.

[45] Memo from Dr Gerrit Schepers and Emile Schepers to Jock McCulloch, 23rd August 2008, Great Falls, Virginia, USA.

on well with his new colleagues, most of whom were British and middle-aged. Having seen at first hand the cursory medical examinations of white miners at the Bureau and black miners at the WNLA depot, Schepers became an outspoken critic of the compensation system. That brought him into direct conflict with the Bureau Chairman, Dr S.W. Verster.[46] According to Dr Schepers, the Chamber of Mines wanted to minimise the cost of compensation, but it did not exert any direct influence on the Bureau's daily operations. Instead, it 'ran the Bureau's head doctors' through the office of the Minister of Mines. The Bureau, Schepers claimed, was in effect a tool of the mining houses, engaged in covering up the extent of occupational disease.[47]

White miners were processed at the Bureau in groups of around 300 a day. The workloads were crushing, and in 1951 the Bureau's ten interns each carried out 5472 examinations.[48] Allowing for other duties, including a daily team meeting, that allowed at most twelve minutes to examine a miner, read his X-rays and make a clinical judgement. Around 80 per cent of miners were Afrikaans speakers, yet Schepers was the only intern who knew Afrikaans. That made the taking of accurate histories more difficult. When he arrived at the Bureau, decisions about compensation were based entirely on X-rays. The Bureau's equipment was adequate for identifying large silicotic densities or advanced tuberculosis but could not detect the smaller lesions characteristic of early-stage disease. Consequently, it was common for a man to be too sick to work but to have his claim for compensation rejected.[49] Black miners suspected of having silicosis or tuberculosis were usually accessed at the WNLA depot by Bureau doctors. Schepers was the only intern who spoke Zulu and Tswana, and so he volunteered for those weekly rounds. The huge number of black miners meant the medicals conducted at the WNLA were even less adequate than those of whites.

[46] Dr Gerrit Schepers, transcripts of evidence before the Beyers Commission, 29th and 30th January 1952. TEBA Archives, University of Johannesburg, pp. 505–506.

[47] Memo from Dr Gerrit Schepers and Emile Schepers to Jock McCulloch, 23rd August 2008.

[48] *Report of the Commission of Enquiry into the Functioning of the Silicosis Medical Bureau and the Silicosis Medical Board of Appeal (Beyers Commission)*. Pretoria: Government Printer, 1952, p. 7.

[49] Memos from Dr Gerrit Schepers and Emile M. Schepers to Jock McCulloch, 16th and 17th August 2008.

After Schepers met the Secretary of the MWU, Daan Ellis, the two men joined forces in lobbying the South African Parliament for legislative change. On Schepers' advice, the Union bought a costly X-ray machine to carry out its own medical assessments. Schepers built up a series of a hundred cases of white miners whom the Bureau had wrongly denied compensation. He also obtained tissue specimens from black miners. Mine doctors were required by law to send the lungs of deceased miners to WNLA, but the latter had no pathology service, so the tissues were usually discarded.[50] In its final report in 1952, the Commission of Enquiry into the Functioning of the Silicosis Medical Bureau and the Silicosis Medical Board of Appeal (Beyers Commission) agreed with Schepers that the Bureau's medical reviews did not conform to the Silicosis Act. Beyers also questioned the merits of X-rays as a means for assessing compensation. Against objections from Commission members A.J. Orenstein and S.F. Oosthuizen, Beyers supported the introduction of pulmonary disability as a compensable disease. At the request of the deputy prime minister, Schepers wrote the original draft of the Act.[51] The Silicosis Amendment Act (Pulmonary Disability Act) No. 63 of 1952, which recognised emphysema and bronchitis as compensable, was passed while Beyer was still in session.[52] To improve the handling of compensation cases, a new arm of the Bureau, the Pulmonary Disability Committee (PDC) was created, and Schepers was appointed as its director. The PDC developed a more rounded approach to medicals based on the science Schepers had seen in the USA. The new approach saw a sudden rise in the number of successful claims. For whites, there were 306 initial awards in 1953. By 1955 that had grown five-fold. In 1953, there were 1748 initial awards to black miners; in 1954 that had risen to 3099 and in 1956 there were 4922,[53] an increase of over 280 per cent. Most of those new awards were for silicosis and tuberculosis. What changed between 1952 and 1956 were not the work conditions underground but Schepers' directorship of the PDC.

[50] Memo from Dr Gerrit Schepers and Emile Schepers to Jock McCulloch, 1st October 2008.

[51] Memo from Dr Gerrit Schepers and Emile Schepers to Jock McCulloch, 23rd August 2008.

[52] Memos from Dr Gerrit Schepers to Jock McCulloch, 16th and 17th August 2008.

[53] *Report of the Pneumoconiosis Board for the Year ended 31*st *March 1957*. Pretoria: Government Printer, 1957, p. 15.

To resolve what it perceived as an ongoing crisis on the mines, in 1954 the Department of Mines convened an inquiry under the chairmanship of Professor S.F. Oosthuizen from the University of Pretoria.[54] Oosthuizen spent much of his time investigating complaints about the compensation system. The most serious allegations concerned the poor quality of the Bureau's medicals and the role of the newly created Pulmonary Disability Committee (PDC) under Schepers' directorship. Oosthuizen found that the Bureau was failing to accept legitimate claims while the PDC was accepting claims that had little merit. Oosthuizen also submitted a confidential report to the Minister of Mines, Dr A.J.R. van Rhijn, accusing Dr Schepers of seeking to gain control of the Bureau.[55] Oosthuizen warned the minister that Schepers' continued employment at the PDC would lead to serious industrial unrest.[56] A month before the Report was tabled, Schepers left South Africa to become Director of the Saranac Laboratories in up-state New York, the most prestigious pulmonary research centre in North America. Three years later, he joined the Du Pont Corporation as a pathologist and later worked as Chief of Cardiovascular Disease with the Veterans' Administration in Washington. During more than four decades in the USA, Dr Schepers enjoyed success both as a researcher and as an administrator. He never again worked in South Africa.

The crisis facing the mining houses and the Bureau was finally resolved by the *Pneumoconiosis Act No 57 of 1956.* That legislation extended the compensation system to asbestos and coal mines and in that sense it was progressive. Its principal effect, however, was to remove pulmonary disability from the statue books. The PDC was shut down and the Bureau was restructured. After the 1956 Pneumoconiosis Act came into effect, the compensation rates returned to their previous levels as if by magic.[57]

---

[54] *Report of the Departmental Committee of Enquiry into the Relationship Between Silicosis and Pulmonary Disability and the Relationship Between Pneumoconiosis and Tuberculosis.* Pretoria: Government Printer, 1954, p. 3.

[55] *Supplementary Confidential Report of the Departmental Committee of Enquiry into the Relation Between Silicosis and Pulmonary Disability. Departmental Committee to Inquire into the Definition of Silicosis & Chest Diseases (Oosthuizen) Departmental Committee 1954.* South African National Archives, Pretoria, F 33/671, pp. 14, 30.

[56] *Supplementary Confidential Report*, p. 40.

[57] *Report of the Pneumoconiosis Board for the Year ended 31*st *March 1957*, p. 11.

During his years at the Bureau, Schepers saw daily the misery of white miners who were denied compensation and he wanted that suffering to end. He was outraged by the injustices heaped upon black miners who in going underground faced the destruction of their health and the ruin of their families, and all for wages set below subsistence. Schepers claimed to be apolitical; like Macvicar, he believed that his motivations came from the *New Testament*.[58] Unlike Macvicar and other previous dissenters, however, Schepers managed to bring about legislative change, albeit for only two years. He was an insider armed with a high level of technical knowledge about the awarding of compensation and the daily operations of the Bureau. He aligned himself with the MWU, which had powerful allies in Parliament, but his aims were far wider than those of the Union Secretary Daan Ellis. In January 1963, Schepers gave a paper to a general meeting of the MWU in Johannesburg. The mines could not operate without migrant labour. Almost half of those men would die of tuberculosis, while a large number would contact tuberculosis combined with silicosis. Schepers told his audience that the MWU had a moral responsibility to ensure that black miners were protected.[59]

Conclusion

The dissenters shared first-hand experience of disease in miners, and they were aware of the lack of biomedical care in labour-sending areas. They understood the impact of low wages in those communities, and they were aware of the failings of mine medicine and the lack of access to compensation. They all regarded the official disease rates as inaccurate. Macvicar turned the Victoria Hospital into something of a sanctuary for miners and their families, and a model of how to treat tuberculosis. However, Victoria had little influence on rural health care programmes. Macvicar was isolated from Johannesburg and the legislative process. His omission

[58] Author interviews with Dr Gerrit Schepers at Great Falls, Virginia, 23rd to 28th October 2010.

[59] Dr G.W.H. Schepers, transcript of speech to the Mine Workers Union, Johannesburg, January 1963, p. 25. In possession of the author.

as an expert witness from the 1914 and 1932 enquiries into tuberculosis suggests he was feared by the industry. Cluver played an important role at the Stratford Commission while Dormer was an innovative clinician. Given their numbers and status, it is curious that the dissenters had so little influence over mine medicine. Their voices were often drowned out by Bureau data which showed the mines were safe and by the opinions of the most influential expert witness of all, A.J. Orenstein.

References

Collins, T.F.B. 'The History of Southern Africa's First Tuberculosis Epidemic'. *The South African Medical Journal,* Vol. 62, No. 21, 13th November 1982, pp. 780-788.

Dormer, B.A. 'Tuberculosis in South Africa'. *British Journal of Tuberculosis and Diseases of the Chest,* Vol. 50, No. 1, 1956, pp. 52-60.

Dormer, B.A. and F.J. Wiles. 'Tuberculosis in the Bantu'. *The South African Medical Journal,* Vol. 20, No. 10, 25th May 1946, pp. 262-264.

Dormer, B.A., J. Friedlander and F.J. Wiles. 'Tuberculosis Control: The Importance of the Family Unit'. *The South African Medical Journal,* Vol. 15, No. 22, 22nd November 1941, pp. 453-454.

Dormer, B.A., J. Friedlander and F.J. Wiles. 'A South African Team Looks at Tuberculosis'. *Proceedings of the Transvaal Mine Medical Officers' Association,* Vol. 23, No. 257, November 1943, pp. 109-113.

Lunde, M.J. An Approach to Medical Missions: Dr Neil Macvicar and the Victoria Hospital, Lovedale, South Africa circa 1900–1950. PhD thesis, University of Edinburgh, 2009.

Macvicar, Neil. 'The Prevalence of Certain Diseases among the Natives of the Ciskei'. *The South African Medical Journal,* Vol. 6, No.22, 26th November 1932, pp. 721-724.

Macvicar, Neil. 'Tuberculosis Among the South African Natives'. *South African Medical Record,* Vol. 6, No. 11, 10th June 1908, pp. 161-176.

Macvicar, Neil. 'Tuberculosis Amongst the Coloured Population of South Africa'. *The South African Medical Record,* Vol. 8, No. 4, 26th February 1910, pp. 39-45.

Murray, Jill, Tony Davies and David Rees. 'Occupational Lung Disease in the South African Mining Industry'. In 'Research and Policy Implementation'. *Journal of Public Health Policy*, Vol. 32, Suppl. 1, June 2011, pp. S65-79.

Packard, R.M. *White Plague, Black Labour: Tuberculosis and the Political Economy of Health and Disease in South Africa.* Berkeley: University of California Press, 1989.

Weinberg, Ann Cluver. *Looking for Goodness: Exploring Eustace Cluver and his Family 1657–1982.* Johannesburg: Sapler Press, 2008.

12

The Career of A. J. Orenstein: 1914–1960

South Africa was the first state to compensate silicosis or miners' phthisis (1911) and pulmonary tuberculosis (1916) as occupational diseases. It was the first to establish a government body of medical specialists, the Miners' Phthisis Medical Bureau (1916), to adjudicate compensation claims. Despite their apparent objectivity, both the science and the legislative control of the mines were the outcome of an intensely political process. The complexities and controversial aspects of this process can be seen vividly in the career of Alexander J. Orenstein, who is widely regarded as the founder of South Africa's occupational medicine. His career was certainly unique in terms of its longevity and in the degree of his influence over mine medicine. As Superintendent of Sanitation for Rand Mines Ltd, Orenstein dominated mine medicine in the period from 1914 until the mid-1960s. According to conventional wisdom, he saved the lives of many thousands of black migrant workers by improving hygiene, nutrition, accommodation, medical care and access to

J. McCulloch, P. Miller, *Mining Gold and Manufacturing Ignorance*,
https://doi.org/10.1007/978-981-19-8327-6_12

compensation.[1] Dr E.H. Cluver, himself a major figure in public health, viewed Orenstein's influence as profound.[2]

The ban on Tropical labour, imposed in 1913, was a blow to an industry which had come to depend upon a constant flow of migrant workers. To have it lifted, in 1913 the Chamber invited Major William Gorgas, famous for his work on the Panama Canal, to find a remedy. In his report, Gorgas made a number of recommendations to improve medical care, hygiene, rations and living conditions. While the other mining houses ignored the Gorgas Report, the Chair of Rand Mines Ltd, Evelyn Wallers, was determined to reduce the death rates. On Gorgas' recommendation, he appointed A.J. Orenstein to oversee sanitation. Orenstein reported directly to Wallers, who encouraged him to reorganise the company's health services.[3]

Alexander J. Orenstein (1879–1972) was born in Odessa, in Russia. Soon after, his parents emigrated to the USA. In 1905 Orenstein graduated in medicine from Jefferson College in Philadelphia and became a naturalised US citizen. After a brief period with the American Armed Forces, he was posted to the Panama Canal. For seven years, he assisted William Gorgas in reducing the dreadful mortality rates from pneumonia, yellow fever and malaria. In 1914, he joined the Rand Mines Ltd. Little is known about Orenstein as neither his personal papers nor what would have been his voluminous professional correspondence have survived. What has survived is a glowing reputation. Following his death in July 1972 at the age of 92, the Chamber of Mines initiated the A.J. Orenstein Memorial Lecture at the University of the Witwatersrand in recognition of his achievements.

[1] See 'Obituary A.J. Orenstein'. *British Medical Journal*, 19th August 1972, p. 478 and 'Obituary A.J. Orenstein'. *Proceedings of the Mine Medical Officers' Association of South Africa*, Vol. LII, No. 413, May–August 1972, pp. 1–2. See also Phillip Tobias. 'Darwin, "Descent" and Disease'. *Transactions of the Royal Society of South Africa*, Vol. 40, Part 4, December 1972, p. 239.

[2] Dr E.H. Cluver. 'Development of Health Services on the Gold Mines'. First A.J. Orenstein Memorial Lecture, 18th September 1974, p. 6. Adler Medical Museum, University of the Witwatersrand. See also Bruce Fetter. 'Changing Determinants of African Mineworker Mortality: Witwatersrand and the Copperbelt, 1911–1940'. *Civilisations*, Vol. 41, No. 1/2, 1993, pp. 347–359.

[3] A.P. Cartwright, *Doctors of the Mines: A History of the Work of Mine Medical Officers*. Cape Town: Purnell and Sons, 1971, pp. 38–39.

Orenstein's obituary in *The South African Medical Journal* is full of praise for a man: '[w]ho was to a great extent responsible for bringing order in the routine examination of miners and for their proper compensation if they contracted silicosis'.[4] Through the role he played at national and international forums, Orenstein also exerted a major influence on research. The transcripts of the numerous commissions of enquiry on which he served, or before which he appeared as an expert witness, show that through the force of his personality Orenstein's views often prevailed. From 1917, he served on the Miners' Phthisis Prevention Committee, which produced important reports in 1919 and 1937. Orenstein was the first vice-president of the Mine Medical Officers Association, which he co-founded in 1921 and of which he remained an active member for fifty years. Between 1928 and 1932, Orenstein served on the Tuberculosis Research Committee and wrote substantial parts of its influential reports. In 1925 and again in 1938, he was a delegate to the ILO conferences in Geneva. In 1930, he was Conference secretary and edited the proceedings of the ILO's Silicosis Conference held in Johannesburg. In 1956, at the age of 77, Orenstein became the foundation Director of the Pneumoconiosis Research Unit within the Department of Mines, a position he held for three years. He continued to work as a consultant with Rand Mines Ltd until his death in 1972.

Orenstein enjoyed a number of significant advantages over contemporaries such as Anthony Lanza (USA), J.S. Haldane (UK) and Dr J.H.L. Cumpston (Australia), who worked on pneumoconiosis. The Johannesburg research community was arguably the best in the world, and prior to Orenstein's arrival in 1914 a series of commissions into miners' phthisis had confirmed once again the risks of silica exposure and the synergy between silica dust and tuberculosis.[5] In addition to having access to a world-class research community, Orenstein had the opportunity to conduct his own research. His employer, Rand Mines Ltd, was one of South Africa's most profitable mining companies. If Orenstein had deemed it necessary, it could well have afforded to fund research. Finally,

[4] 'Obituary A.J. Orenstein'. *The South African Medical Journal*, 19th August 1972, p. 1188.

[5] Jock McCulloch. *South Africa's Gold Mines and the Politics of Silicosis*. Oxford: James Currey, 2012, pp. 1–13.

South Africa's gold mines were the first to be subject to a comprehensive system of state regulation, which resulted in the collection of massive volumes of data on dust levels and disease rates. The system, which began with the Miners' Phthisis Act of 1911 and was extended by the subsequent Acts of 1912, 1916, 1919 and 1925, had no parallel anywhere in the world.

Despite the quality of the Johannesburg research community and the state regulation of the mines, there were a number of factors which worked against mine safety. The gold mines were South Africa's most important industry and dominated employment, foreign exchange and state revenue. The relatively stable price of gold meant that individual mining houses did not face one of the fundamental pressures of a competitive commodity market. Rather, through the Chamber of Mines, they were able to pursue an agreed set of policies regarding recruitment, risk, occupational disease and workers' compensation. While that gave the mining houses an unusual degree of political power, the industry was fragile. The pricing of gold made production costs even more important than in other industries, since increases in working costs could not be passed on to consumers. Apart from wages and expenditure on the compounds housing migrant workers, occupational health and compensation were areas where costs could be contained.

When Orenstein arrived in Johannesburg, the mine hospitals were overcrowded and lacked equipment and trained staff. Some mines did not even provide casualty rooms for men injured underground.[6] Mine doctors were part-time employees, and they were regarded by managers as members of their administrative staff and therefore subject to their authority. The mine hospitals were no better than barracks, and nursing was carried out by poorly trained male orderlies. The operating theatres were poorly lit and had neither steam sterilisers nor X-ray equipment. There were few surgical instruments to care for a workforce among whom traumatic injuries were common.[7] In addition to silicosis, tuberculosis and bacterial pneumonia, meningitis was a major cause of death. Within

[6] See Julie J. Baker. 'The Silent Crisis': Black Labour, Disease, and the Economics and Politics of Health on the South African Gold Mines, 1902–1930. PhD, Queen's University, Ontario 1989, pp. 76–138.

[7] Cartwright, *Doctors of the Mines*, p. 39.

a year, Orenstein had reduced overcrowding in Rand Mines' compounds, established a central sanitation and safety committee and installed X-ray equipment at the Crown Mines Hospital. He introduced first-aid courses and, in the face of strong opposition from mine managers, appointed black nurses to replace inefficient male orderlies. He also improved miners' rations and persuaded the Rand Mines board to install water-borne sewage. In line with Gorgas' recommendations, between 1917 and 1919 Orenstein re-organised four central hospitals from the existing services within the Rand Mines Ltd group. He also obtained funding for additional 96 beds at the City Deep hospital.[8] The improvements dramatically reduced the official mortality rates from pneumonia. Despite the clear benefits of these measures, the other mining houses for a long time chose not to follow Orenstein's initiatives. The issue of providing separate bunks in compounds to reduce the risk of infection, for example, continued to be fought out in committees well into the 1930s.[9] As always, the sticking point was cost.

In addition to his clinical work, Orenstein quickly developed a second set of responsibilities, which took up more and more of his time. The medical system was riven with tensions and between 1910 and 1954 there were more than fifteen commissions and committees of enquiry into occupational disease. Those commissions were driven by the white MWU and its two main concerns. The first was the struggle faced by its members in gaining compensation for silicosis. The second was the high tuberculosis rate among black miners, which the MWU believed put white miners and their families at risk. As forums where the issue of risk was open to public scrutiny, the Commissions were a point of vulnerability for the industry. The Chamber always made elaborate submissions backed up by expert testimony. Its favoured witness, A.J. Orenstein, believed he was far more knowledgeable about black miners than were the Bureau physicians who dealt principally with whites.[10]

---

[8] Baker, 'The Silent Crisis', pp. 108–111.

[9] Cartwright. *Doctors of the Mines*, p. 44. See also 'Sanitary Conditions on the Gold Mines. GES 934 708 Correspondence Gold Producers Committee'. SANA.

[10] See Letter from A.J. Orenstein, Rand Mines to A.B. Du Toit, Chairman of the Miners' Phthisis Medical Board 21st December 1925. SANA NTS 6720 33/315 Vol. 1 Miners Phthisis.

Limits to Improving Public Health Practices

At the end of the nineteenth century, the work of Thomas Legge, John S. Haldane and Thomas Oliver established the synergy between silica dust exposure and tuberculosis, a finding accepted by the 1907 Industrial Diseases Committee in Britain.[11] In South Africa, that body of knowledge was further endorsed by the Miners' Phthisis Commissions of 1902 and 1912, as well as the Tuberculosis Commission of 1914. Within four years of arriving in Johannesburg, Orenstein began attacking this orthodoxy. He denied the relationship between silica exposure and tuberculosis and refuted claims that the mines were spreading tuberculosis to labour-sending communities. His assaults on established medical knowledge continued after the Second World War, whenever the issues of tuberculosis or silicosis were raised. It is ironic that while Orenstein's early professional success was built on his relationship with William Gorgas, his South African career was dedicated to dismantling some of the key foundations of the report Gorgas prepared for the Chamber in 1914.

In July 1919, the Low-Grade Mines Commission was established as a result of lobbying by Evelyn Wallers, who at the time was President of the Chamber of Mines, and William Gemmill, the Chamber's Secretary. Many mines were struggling to make a profit and Wallers wanted the government to lift the ban on Tropicals.[12] The Commission was chaired by the Government Mining Engineer, Sir Robert Kotze, and its eight members included Wallers and Gemmill. While its final report was sympathetic to the Chamber, the Commission stopped short of recommending the reintroduction of Tropical labour.[13] A.J. Orenstein appeared as an expert witness.

To justify the return of Tropicals, Orenstein had to convince the Commission that the mines were safe. He acknowledged that acquired

[11] See Arthur McIvor. 'Germs at Work: Establishing Tuberculosis as an Occupational Disease in Britain, C.1900–1951'. *Social History of Medicine*, Vol. 24, No. 4, 2012, pp. 812–829.

[12] David Yudelman. *The Emergence of Modern South Africa: State, Capital, and the Incorporation of Organised Labor on the South African Gold Fields, 1902–1939*. Cape Town: David Philip, 1984, p. 138.

[13] See *Report of the Low-Grade Mines Commission*. Cape Town: Government Printer, 1919, pp. 25–37.

immunity to infection was a vital element in the disease rates for both tuberculosis and pneumonia.[14] The other factor was the physical condition of labour. Mining requires a high level of fitness and if a man was run down infection was more likely.[15] Orenstein was confident that fatalities could be prevented by improved hygiene and nutrition. He proposed that Tropicals be housed in cubicles within compounds (as had been done at the Rand Mines Ltd), be provided with blankets and be given government approved rations.[16] Orenstein was asked what he considered a reasonable mortality rate if the ban was lifted. He replied that the answer depended upon the death rate in the areas from which labour was recruited. Although he had no data, Orenstein suggested the natural annual death rate in Central Africa would be around 30 per 1000 per annum, excluding children. If that was so, then bringing men to the mines would save lives.[17] This was a spurious argument given that miners were a physical elite who were the least likely members of their home communities to die prematurely. Orenstein ended his evidence with an appeal to the Commission: 'You cannot undertake any large enterprise without certain deaths and the sum total of the deaths, even if it does exceed the death rate among these people, will not be sufficiently great to make it prohibitive, it is important to the [gold] industry to have the natives from the tropics'.[18] Since Tropicals would die if they stayed at home, they may as well die on the mines.

Much of Orenstein's testimony, including the data he cited on pneumonia deaths prior to the ban, was fraudulent. During the first months of 1913, the official returns put the annual mortality rate at well over 100 per 1000 per annum; Orenstein told the Commission the rate was 13.[19]

---

[14] Dr Orenstein, evidence before the Low-Grade Mines Commission, Minutes of Evidence 18th July 1919. SANA K160 Low Grade Mines Commission, Vol. 1, pp. 416–418.

[15] Orenstein, evidence, 1919, p. 392.

[16] Orenstein, evidence, 1919, p. 388.

[17] Orenstein, evidence, 1919, pp. 399–400.

[18] Orenstein, evidence, 1919, pp. 422; 437.

[19] See Memo from Prime Minister Botha, Cape Town, to the Governor General 12th May 1913. Draw Attention to the Misleading Nature of Statistics. SANA GG 1541 50/301 Mortality Among Tropical Natives.

Pneumonia was still a major problem in 1919; Orenstein's own data on tuberculosis at City Deep mine (discussed below) was compelling evidence that the ban should remain.

Deaths Underground

Beginning in 1911, Miners' Phthisis Acts made it illegal to employ a man with tuberculosis, and the system of medical surveillance was largely designed to exclude infected men from the mines. In theory, the system was so rigorous it was unthinkable that a seriously ill man could continue to work underground. For that reason, the sudden deaths of three black miners at Rand Mines in less than twelve months led to an enquiry.

On 27 May 1925, a miner named Fish became seriously ill while working underground at the Crown Mine. On that day Fish coughed up blood and died in a skip before he could be brought to the surface. A post-mortem established tuberculo-silicosis as the cause of death. Between February and May Fish had lost 7 lbs in weight but he had not been referred for further examination as required under the legislation. His workmates said Fish had been unwell for some time and that he coughed constantly.[20] The second death was that of John who had worked at Crown Mines for just over five years. A post-mortem revealed a major cavity in his right lung as the cause of death.[21] The final death was that of Tom who died suddenly at City Deep on 9 January 1926. Over the previous two years he had been admitted to hospital on four occasions following accidents. Tom died in a skip while being hauled to the surface. The attending medical officer, Dr Dodds, suggested that Tom may have suffocated as the men were packed into the skips 'like sardines'.[22] A post-mortem showed a 'marked degree of tuberculosis'. At the City Deep

[20] Minutes of Evidence and Proceedings Committee of Enquiry into the Circumstances Surrounding the Deaths of the Native Labourers Fish, Tom and John, 23rd February 1926, Appendix L, p. 4. SANA. GES 1005 9 17 A TB in Mines Natives.

[21] Report of the Medical Committee appointed to Enquire into the Deaths of Certain Mine Natives 8th February 1926 to 26th May 1926, pp. 1–2. SANA. GES 1005 9 17 A TB in Mines Natives.

[22] Report Dr G.M. Kapp, MPMB Native Tom, 21st January 1926. SANA. GES 1005 9 17 A TB in Mines Natives.

hospital tuberculosis cases were kept in general wards, and it seems likely that is where Tom became infected. The deaths of Fish, Tom and John suggested a serious breach of mine regulations, and the Department of Public Health established a Committee of Enquiry into the fatalities. Dr Watkins-Pitchford, head of the Bureau and the most distinguished pulmonary specialist in South Africa, was appointed chair.

The Committee soon identified a further six cases between January 1925 and February 1926. Most of the dead came from two mines, Crown and City Deep, and in one instance the deceased had been employed for just five weeks. No similar death had occurred in a white miner since the Bureau was established in 1916, an outcome Watkins-Pitchford attributed to the routine use of X-rays. Watkins-Pitchford also had no doubt about the importance of the cases. 'The Government very rightly think that if the fact of these deaths in such circumstances were made public, it might cause a serious scandal—that the boy should have had this disease and yet have continued to work until he died of it; that is the way the public press would put it.'[23]

After two months of deliberations, the Committee found that Rand Mines Ltd. accounted for most of the 'overlooked' tuberculosis cases.[24] It also found the company's claim that the presence of terminally ill men posed no threat to other miners as both 'dangerous and disingenuous'. The Committee concluded: 'Even a few tuberculotics on the mines constitutes a dangerous focus of infection and their presence involves a potential loss of life and money the magnitude of which one can scarcely overestimate'.[25] Watkins-Pitchford made a number of recommendations for the improvement of entry, periodic and exit medicals. In particular, he wanted X-rays to be used routinely with black miners.

---

[23] Minutes of Evidence and Proceedings Committee of Enquiry into the Circumstances Surrounding the Deaths of the Native Labourers Fish, Tom and John. 10th March 1926, p. 21. SANA. GES 1005 9 17 A TB in Mines Natives.

[24] Report of the Medical Committee appointed to Enquire into the Deaths of Certain Mine Natives, 8th February 1926 to 26th May 1926, p. 20. SANA. GES 1005 9 17 A TB in Mines Natives.

[25] Comments by the Committee on Dr Orenstein's Memorandum undated, April 1926? p. 8. SANA. GES 1005 9 17 Mines Natives.

As the senior medical officer at Rand Mines Ltd., Orenstein was outraged by Watkins-Pitchford's report, and he challenged every one of the Committee's findings. Orenstein did not agree that the nine deaths reflected badly on the medical system or on the officers involved. Neither did the presence of a few open cases in a compound justify a large expenditure of money. In particular, Orenstein objected to the suggestion that Tom may have contracted tuberculosis while being treated in hospital for an injury. Tuberculosis was not, according to Orenstein, a dangerous communicable disease. 'We cannot accept that tuberculosis, even among natives, is so highly infectious that a few weeks' residence in hospital might be the source of infections.'[26] Radiography was expensive and time consuming and Orenstein was sure that if X-rays were introduced for black miners, other methods of detection would be neglected, thereby leading to more early cases being overlooked. To X-ray every black miner at regular intervals would in itself be expensive. It would also require increasing the complement of medical officers four or five-fold.[27]

The Watkins-Pitchford report bought no change to mine medicals; black miners continued to die underground from tuberculosis. The enquiry did, however, have one major casualty and that was Dr Wilfred Watkins-Pitchford. On 17 May 1926, he resigned as Chair of the Bureau and as Director of the SAIMR.[28] The official reason for his resignation was ill health. Watkins-Pitchford returned to England, which he had left twenty-five years previously. That decision is puzzling: it is hard to imagine that England offered a better climate or diet for a sick man than did South Africa. Watkins-Pitchford lived on in apparent good health until his death in 1952.[29] During that period he was an active member of the British Medical Association but with one exception he made no further

[26] Letter from Dr Orenstein, Rand Mines Ltd to the Secretary, Committee of Enquiry, Miners' Phthisis Medical Bureau, 19th April 1926, pp. 3–4, 6. SANA. GES 1005 9 17 Mines Natives.

[27] Report of the Medical Committee appointed to Enquire into the Deaths of Certain Mine Natives, 8th February 1926 to 26th May 1926, pp. 5–6, 13. SANA GES 1005 9 17 A TB in Mines Natives.

[28] Marais Malan. *In Quest of Health: The South African Institute of Medical Research, 1912–1973.* Johannesburg: Lowry Publishers, 1988, p. 37.

[29] See 'Obituary Wilfred Watkins-Pitchford'. *British Medical Journal,* 11th October 1952, pp. 834–835.

contribution to the literature on silicosis or miners' phthisis on which he was a world authority. The enquiry into Rand Mines had effectively ended his career.[30]

As Infectious as Toothache

In May 1930 Orenstein appeared as a witness before the Commission into Miners' Phthisis. He had been with Rand Mines for almost fourteen years and with the resignation of Dr Watkins-Pitchford had become the Johannesburg expert on occupational disease. Like so many others, the Commission, chaired by Mr James Young, was the result of agitation by the white MWU, which claimed that the high tuberculosis rate among black miners put its members at risk.[31] During his testimony, Orenstein made numerous references to research that he and his colleagues at Rand Mines Ltd had completed on tuberculosis, but no data from those studies was tendered in evidence.[32] That research was not cited in SAIMR's 1932 Tuberculosis Report which Orenstein co-authored, nor was it presented at any of the monthly meetings of the MMOA Orenstein habitually attended.[33] It seems likely that it never existed.

Most of Orenstein's evidence was in defence of the medical examinations of black miners. Orenstein opposed the use of X-rays on the grounds that mine medical officers were expert in identifying early-stage disease. Three studies conducted by Rand Mines Ltd. had shown that: 'Any attempt to make the examination of blacks standardised on a basis of white experience would be a horrible mistake'.[34] Besides, even if cases

[30] Jock McCulloch. 'Medicine, Politics and Disease on South Africa's Gold Mines'. *The Journal of Southern African Studies*, Vol. 39, No. 3, 2013, pp. 543–556.

[31] *Report of the Miners' Phthisis Commission of Enquiry 1929–30*. Union of South Africa. Pretoria: The Government Printer, 1930. U.G. Number 38, 1930.

[32] See, for example, Dr Alexander Orenstein, evidence before the Miners' Phthisis Commission, Minutes of Proceedings 27th May 1930, SANA K105 Miners Phthisis Commission Correspondence Vol. 3, pp. 562; 572.

[33] See *Tuberculosis in South African Natives*. The South African Institute for Medical Research, Johannesburg, 1932.

[34] Dr Alexander Orenstein, evidence before the Miners' Phthisis Commission, Minutes of Proceedings, 27th May 1930, SANA K105 Miners Phthisis Commission Correspondence Vol. 3. 1930, pp. 563–565; 572.

were missed, there was little risk to white miners. 'Tuberculosis to the European is not an infectious disease in adult life in any sense of the word any more than toothache … but it is an infectious disease to such natives as have not been tubercularised.'[35] Orenstein made no mention of exposure to silica dust which, according to contemporary science, put all hard rock miners at a greatly enhanced risk of disease.

One commissioner suggested that the WNLA doctors were overworked, and he asked whether the examinations of black miners should be handed over to the Bureau. Orenstein strongly opposed putting the medicals of blacks on the same footing as for whites. The existing medicals were far superior to anything that could be achieved by a centralised organisation.[36] When asked about the number of tuberculosis cases amongst serving miners, Orenstein replied: 'I should say nil in the course of a year', meaning that all cases were quickly identified, segregated and returned to rural areas.[37] Orenstein also rejected the suggestion that black miners be examined every six months, as were whites.

There was, according to Orenstein, no evidence that tuberculosis was being spread by infected men in the compounds. Given that infected black miners were repatriated, Orenstein was asked if those men were likely to spread disease to their home communities. 'There is very little risk', he replied, 'as such men soon died'. When asked if repatriated miners should be educated about the risk of infecting their families Orenstein again replied no. It was 'extraordinarily difficult to teach the native anything by word of mouth', and 'giving the native a piece of paper, as you would a European, is absolutely hopeless'. Orenstein also rejected the idea of providing treatment in rural areas as it would be 'perfectly hopeless' to establish sanatoria for blacks. Paradoxically, he favoured information programmes to educate the white public about the risks of infection.[38]

The most telling aspect of Orenstein's evidence was his dismissal of the connection between silica dust and tuberculosis. Orenstein

[35] Orenstein, evidence 1930, p. 569.

[36] Orenstein, evidence 1930, pp. 571–573; 577.

[37] Orenstein, evidence 1930, pp. 593–594.

[38] Orenstein, evidence 1930, pp. 581–585.

acknowledged the medical orthodoxy that hard rock miners have a lower resistance to tuberculosis than do other workers, an orthodoxy which was well established by the late nineteenth century. He told the Commission that it did not apply in South Africa, where complex statistical analysis, too difficult for lay people to understand, showed that the conditions underground were no more conductive to tuberculosis than were those on the surface. Consequently, the infection rate among white miners was no higher than for the general population. There were, he admitted, certain other occupations in which tuberculosis was common, such as home tailoring or sweat shops. Those employed to put the tops on tam-o-shantas [a type of a hat] were at particular risk, as were type-setters working in the dark.[39] Orenstein made no mention of silica dust.

Orenstein's evidence was in sharp contrast to that of his colleague, Dr L.G. Irvine, who had replaced Watkins-Pitchford as Bureau Chair. According to Irvine, the data showed that of those white miners who received awards from the Bureau for simple tuberculosis, 25 per cent were subsequently diagnosed with silicosis and therefore re-classified for the purposes of compensation. Of those white miners who were compensated for tuberculosis and at death were given a post-mortem, around half were found to also have silicosis.[40] In South Africa as in the hard rock mines of the USA, Western Europe and Australia, tuberculosis and silicosis were inseparable.

Post-mortems

Four months before Orenstein appeared before the Young Commission, Dr W.O. Fischer, a medical officer with the Rand Mines, published an important study which puts Orenstein's evidence into context. The article was based on 1402 post-mortems conducted at the City Deep Native Hospital, one of the Rand Mines' institutions which Orenstein helped upgrade, in the period from 1922 to 1928. The majority of cases were

[39] Orenstein, evidence 1930, pp. 586; 590.

[40] Confidential Memo from Dr L.G. Irvine, Chair of the Miners' Phthisis Bureau, to the Miners' Phthisis Commission, 7th May 1930. SANA K162 Miners' Phthisis Commission 1929.

admitted to hospital as the result of accidents. The seriously injured were a random group, although one would expect accidents to be more common among novice miners, who may be over-represented in Fischer's cohort. The subjects were aged between twenty and forty years. In terms of geographical origins, they were divided equally between South Africans and East Coasters (Mozambicans). In addition to the autopsies performed by Fischer, 73 of the examinations were conducted by Dr Orenstein, Fischer's immediate superior.

Fischer's results were startling. Of all deaths on the mine during that period 302, or 21.5 per cent, were due to tuberculosis, and many of those men had died while working underground. Of that group as a whole, 42 had what he called 'general tuberculosis', 48 pulmonary and 136 acute miliary, with the remainder occurring in other organs. There were also 30 cases (or just over 2 per cent) of tuberculosis with silicosis. Presumably, in addition to those who died at work, there would have been many more who succumbed to tuberculosis on the way home and after being repatriated. Fischer also notes that a total of 468 (or 33 per cent) of deaths were due to lobar pneumonia and another 125 (or 9 per cent) to bronchial pneumonia, suggesting that the ban on Tropicals which was in place at the time was saving lives.[41]

Fischer's data was a fragment of a broader set of post-mortems, all of which produced similar results.[42] Under the Miners' Phthisis Acts Consolidation Act of 1925, any miner who died suddenly was subject to a post-mortem and the results were published in the Bureau's annual report. Each year over 500 miners perished on the mines and most of the autopsies were performed on that group. The Bureau's data shows far higher rates of silicosis and tuberculosis than were identified in living miners. That in itself is not surprising, as it had long been acknowledged that post-mortems may uncover lung disease missed in living subjects.

[41] W.O. Fischer. 'A Preliminary Report on 1,402 Consecutive Autopsies on Native Mine Workers'. *Journal of the Medical Association of South Africa*, 28th September 1929, pp. 511–512.

[42] See *Report of the Miners' Phthisis Medical Bureau for the twelve months ending 31st July 1924*. Pretoria: Government Printer, 1925, p. 4, *Report of the Miners' Phthisis Medical Bureau for the year ended the 31st of July 1928*. Government Printer, Pretoria, 1929, p. 7; and *Report of the Miner's Phthisis Medical Bureau for the three years ending 31st July 1941*. Pretoria: Government Printer, 1944, p. 7.

What is surprising is the dramatic difference between the two sets of figures, with the Bureau's post-mortem data from 1924 until 1950 revealing disease rates 100 times higher than the official morbidity figures. Fischer's article, like the Bureau's data, shows that contrary to the evidence Orenstein presented to the Young Commission, he was aware that tuberculosis and pneumonia remained a grave problem.

The International Setting

The problem of silicosis was one of the first to which the International Labour Office (ILO) turned its attention after it was formed; from 1921 it collected data and monitored the science. Following the Johannesburg Conference of 1930, silicosis was placed on the ILO's schedule of occupational diseases. Ten countries, including Britain and Brazil, ratified the convention; South Africa did not. The Second ILO Silicosis Conference was held in Geneva in September 1938 and attracted delegates from a dozen countries including the USA, Australia and Britain. Dr L.G. Irvine and Dr A.J. Orenstein represented South Africa. Orenstein was elected Conference Chairman.

The Conference provides some insight into Orenstein's behaviour at such forums. When speaking about medical examinations, Orenstein would without acknowledgment use the data from one small and unrepresentative cohort, namely white miners, when referring to mine workers in general. As a result, the Conference was greatly impressed by his depiction of the South African system. One delegate, Dr Langelez, commented that if criteria as strict as those applied to entry medicals in South Africa were adopted in Belgium, it would immediately lead to a labour shortage.[43] Like the other delegates, Dr Langelez had no way of knowing that the medicals to which Orenstein referred to applied only to white miners, who represented at most ten per cent of the workforce.

One of the most important questions the delegates considered was that of diagnosis. After much discussion, the Conference endorsed the

[43] *Silicosis: Proceedings of the International Conference held in Geneva from 29th August to 9th September 1938*. London: International Labour Office, 1938, p. 77.

findings of the 1912 Miners' Phthisis Commission that the diagnosis of silicosis requires an employment history, a complete clinical examination and an X-ray. The examinations should be carried out by specially trained medical staff and preferably, as in South Africa, those examinations should be conducted by a government Bureau.[44] Neither Orenstein nor Irvine told the Conference that the medicals of black miners met none of those requirements. There were no X-rays, there was no clinical examination and no work or medical histories were taken.

Orenstein's most mischievous claim concerned silicosis. He explained to the Conference that African miners did not work continuously as did Europeans, and as a result very few developed silicosis. The total number of cases among migrant workers with fifteen- or sixteen-years' underground experience was too low to provide reliable data.[45] The delegates left Geneva believing there was virtually no lung disease among a workforce which numbered over 300,000. For the first time in modern history, it seemed, a major industry had eradicated silicosis.

After serving with distinction during the Second World War, Orenstein returned to his post with Rand Mines Ltd. The most important post-war change to mine medicine was the introduction of treatment for tuberculosis, a medical advance which posed a new challenge for the industry. Chemotherapy was effective but it was also expensive: the regime usually took two years to complete and required careful monitoring. As always, the mining houses wanted to minimise their costs and they were determined that the state should pay.

Following a damning Report into the industry's management of tuberculosis by the Oosthuizen Committee,[46] the Secretary for Health wrote to the Chamber in July 1954. He noted that since recruits were given a pre-employment medical, those who fell ill with tuberculosis must have contracted the disease on the mines. If those men were treated successfully, it would benefit the communities from which the industry recruited

[44] *Silicosis: Proceedings 1938*, pp. 96–97.

[45] *Silicosis: Proceedings 1938*, p. 63.

[46] See *Report of the Departmental Committee of Enquiry into the Relationship Between Silicosis and Pulmonary Disability and the Relationship Between Pneumoconiosis and Tuberculosis*. August 1954. F 33/671 Treasury SANA.

its labour. The Chamber's legal advisor, B.T. Tindall, was unsympathetic. Many migrant workers in other industries returned to the Native Territories with infective tuberculosis, and therefore treatment by the mines alone would not halt the spread of disease. Tindall added that since gold mining was the only industry liable to compensate tuberculosis, it would be unfair of government to single out the mines for further expenditure. In Tindall's view, the industry should not pay.[47]

In October, the Group Medical Officers Committee met to discuss how best to respond to the Secretary's request. The meeting was chaired by Dr A.J. Orenstein and included WNLA's senior medical officer Dr Frank Retief and the Chamber's Legal Adviser. Contradicting a century of medical orthodoxy, the Group concluded that even in cases where miners had heavily dusted lungs there was no conclusive evidence that pulmonary tuberculosis was associated with silicosis. They recommended that the Chamber go no further than acknowledge that the presence of silica in the lungs favoured the development of infection. The Committee also agreed that the exit examinations were adequate, citing as proof that over the previous year only three former miners in the Territories had been compensated. A better explanation is that former miners faced such obstacles in lodging a claim very few bothered and even fewer succeeded. Orenstein suggested that as a compromise, the industry might agree to an X-ray at the final examination for that small group of black miners with two years' continuous service. Two weeks later the Group Medical Officers met again to prepare a Statement of Evidence. They agreed to submit that the incidence of tuberculosis uncomplicated by pneumoconiosis was lower in black miners than among the general population.[48] The flow of disease was from the rural areas to the mines. Dust exposure was not the cause of tuberculosis.

---

[47] Memorandum from B.T. Tindall, Legal Adviser to the Manager Transvaal and Orange Free State Chamber of Mines, Johannesburg, 17th July 1954. TEBA Archives, University of Johannesburg. WNLA 20L March Diseases & Epidemics Tuberculosis Witwatersrand Native Labour Association 1953–Nov. 1954, p. 1.

[48] Minutes of the Meeting of the Sub-Committee of Group Medical Officers held on Monday, 25th October 1954, in the Chamber of Mines Building. TEBA Archives, University of Johannesburg. WNLA 20L March 1953–Nov. 1954 Diseases & Epidemics Tuberculosis.

Conclusion

A.J. Orenstein's career spanned a period which saw major safety improvements in the hard rock mines of Australia, the USA and Western Europe, most often as a result of trade union pressure. Unlike mine officials in those countries, Orenstein did not have to adjust to a gradual liberalisation of work regimes or industrial practices. On the contrary, black migrant workers in South Africa had even fewer rights under apartheid than they did prior to 1948. Orenstein's career was also unusual in that he spent virtually all of his working life with a mining company. The other prominent figures in mine medicine worked initially as mine medical officers before taking senior appointments with state agencies. L.G. Irvine became Chair of the Bureau, Spencer Lister became Director of the SAIMR while Andrew Watt spent the second half of his career as senior medical officer with Rand Mutual Life Assurance Company.

During his first ten years in Johannesburg, Orenstein saved thousands of lives by improving hygiene, nursing and nutrition. The remainder of his career left a darker legacy. Orenstein was the key architect of a medical orthodoxy about miners' phthisis which may have led to hundreds of thousands of deaths. The Orenstein orthodoxy can be easily summarised. The system of mine medicine was so effective that by 1920 South Africa's gold mines had all but eradicated silicosis, the oldest of the occupational diseases. Exposure to silica dust played no role in tuberculosis, which was not spread from the mines to rural areas but was brought to the mines by infected migrant workers. Finally, the mines benefitted blacks who returned to their homes far healthier than when they left. Orenstein repeated that orthodoxy at the monthly meetings of the Mine Medical Officers Association; to committees such as the Gold Producers Committee, a key organisational unit within the Chamber on which he often served as chairman; and at the numerous commissions of enquiry before which he appeared as an expert witness. It also featured in his presentations to international audiences such as the ILO Silicosis Conferences in Johannesburg, Geneva and Sydney. In those forums, no specialist had more influence than Orenstein.

The evidence Orenstein presented to his peers was formulaic. When discussing tuberculosis, he usually referred to the rates among white miners, which were far lower than those among blacks, to argue that the mines were free of infection. When silicosis was the issue, he used the confirmed rates among black miners, who were rarely compensated for the disease, to claim that there was no silicosis. There were also numerous occasions on which Orenstein simply lied. He told the Young Commission in 1930, for example, that the exit examinations of black miners at the WNLA were conducted by Bureau specialists.[49] Under the Miners' Phthisis Acts from 1916, the Bureau was in theory responsible for 'the overall supervision' of medicals at individual mines and at the WNLA compound, but because of the workloads that never happened. The only black miners examined by Bureau interns were that tiny proportion set aside for compensation. Even with those men the medicals were at best cursory.[50]

During a fifty-year career, Orenstein produced little research. His authority came from his association with Gorgas, his position at Rand Mines Ltd. and his longevity. Over time, he became the moderator of knowledge of miners' phthisis to a lay audience. While common sense suggested that conditions on the mines produced occupational disease, he often repeated, highly complex statistical analyses proved that was not the case. Two questions dominated Orenstein's performances as an expert witness. They were the origins of tuberculosis on the mines and the best way to manage infection. In answering those questions, Orenstein racialised the science to a far greater degree than did any of his contemporaries. He did so by using notions of racial weakness to override the play of well-established environmental factors such as dust exposure, malnutrition and unhygienic compounds to explain the incidence of tuberculosis among migrant workers. Orenstein was highly successful as an expert witness and while often challenged, his preferred approach to repatriations, the education of black miners and their families and the provision of sanatoria in labour-sending regions remained largely unchanged until after the Leon Commission in 1994.

[49] Orenstein, evidence 1930, p. 570.

[50] See Jock McCulloch. 'Hiding a Pandemic: Dr G.W.H. Schepers and the Politics of Silicosis in South Africa'. *The Journal of Southern African Studies*, Vol. 35, No. 4, 2009, pp. 835–848.

Occupational diseases are negotiated political categories which usually arise from the complex play of social, political, cultural and economic circumstances.[51] Orenstein was a master at articulating those forms of knowledge to a lay audience. He worked tirelessly over a long career to erase from public debate the knowledge accumulated by South African scientists in the period before his appointment in 1914. The result was a medical fraud which allowed the gold mines to masquerade as safe when in fact they were highly dangerous, especially for migrant workers. The contrast between Orenstein's success in improving sanitation and his baleful influence on the management of tuberculosis identifies the limits to mine medicine. His testimony as an expert witness tells us what the industry would and would not tolerate in terms of medical discourse. The distance between the image of occupational health Orenstein helped to manufacture and the one which emerged during the recent miners' class action is a testament to the changed constellation of political forces.

References

Baker, Julie J. 'The Silent Crisis': Black Labour, Disease, and the Economics and Politics of Health on the South African Gold Mines, 1902–1930. PhD thesis, Queen's University, Ontario, 1989.

Cartwright, A. P. *Doctors of the Mines: A History of the Work of Mine Medical Officers*. Cape Town: Purnell and Sons, 1971.

Dembe, Allard E. *Occupation and Disease: How Social Factors Affect the Conception of Work-Related Disorders*. New Haven: Yale University Press, 1996.

Fetter, B. 'Changing Determinants of African Mineworker Mortality: Witwatersrand and the Copperbelt, 1911–1940'. *Civilisations*, Vol. 41, No. 1/2, 1993, pp. 347–359.

Malan, Marais. *In Quest of Health: The South African Institute of Medical Research, 1912–1973*. Johannesburg: Lowry Publishers, 1988.

[51] See Allard E. Dembe. *Occupation and Disease: How Social Factors Affect the Conception of Work-Related Disorders*. New Haven: Yale University Press, 1996. See also Paul-André Rosental. 'Silicosis and Global Public Health'. In Rosental, ed. *Silicosis: A World History*. Baltimore: Johns Hopkins University Press, 2017, pp. 1–13.

McCulloch, Jock. 'Hiding a Pandemic: Dr G.W.H. Schepers and the Politics of Silicosis in South Africa'. *The Journal of Southern African Studies*, Vol. 35, No. 4, December 2009, pp. 835–848.

McCulloch, Jock. 'Medicine, Politics and Disease on South Africa's Gold Mines'. *The Journal of Southern African Studies*, Vol. 39, No. 3, 2013, pp. 543–556.

McCulloch, Jock. *South Africa's Gold Mines and the Politics of Silicosis*. Oxford: James Currey, 2012.

McIvor, Arthur. 'Germs at Work: Establishing Tuberculosis as an Occupational Disease in Britain, C.1900–1951'. *Social History of Medicine*, Vol. 24, No. 4, 2012, pp. 812–829.

Rosental, Paul-André, ed. *Silicosis: A World History*. Baltimore: Johns Hopkins University Press, 2017.

Tobias, P. 'Darwin, "Descent" and Disease: An Essay on the Centenary of the Publication of The Descent of Man (1871)', *Transactions of the Royal Society of South Africa*, Vol. 40, Part 4, December 1972, pp. 239–260.

Yudelman, D. *The Emergence of Modern South Africa: State, Capital, and the Incorporation of Organised Labor on the South African Gold Fields, 1902–1939*. Cape Town: David Philip, 1984.

Part V

The End Game

13

Technologies, Care and Repatriations: 1926–1966

The work environments on the Rand were constantly changing. Over time, the mines became deeper, and their sources of labour shifted between internal and transnational migrants. The technologies used by medical officers in dealing with workers also changed. Prior to 1950, there was no effective chemotherapy for tuberculosis. The recommended treatment consisted of an enhanced diet, isolation and skilled nursing, something out of reach of labour-sending communities on the verge of starvation. The centrepiece of the mines' management of tuberculosis was repatriation of sick workers, without notification of local authorities, isolation of those with infective disease, or education of families on how to avoid infection. In the 1950s, the development of mass miniature radiography (MMR), together with new drugs for the treatment of tuberculosis, radically changed the available technologies of diagnosis and care.[1]

[1] Chemotherapy for the treatment of tuberculosis was discussed in Chap. 7. The Bacillus Calmette-Guerin (BCG) vaccine provided another tool for tuberculosis prevention. See I.W.F. Spencer, M. Coster, A. McPhail and M. Richter. 'A Field Experience of Mass Percutaneous BCG Inoculation as an Immunizing and Diagnostic Procedure'. *The South African Medical Journal*, Vol. 38, No. 12, 11th April 1964, pp. 223–233.

J. McCulloch, P. Miller, *Mining Gold and Manufacturing Ignorance*,
https://doi.org/10.1007/978-981-19-8327-6_13

Mass miniature radiography was an advanced technology which the Chamber developed at great expense as part of its efforts to control occupational lung disease. In May 1955, it became compulsory at entry medicals.[2] MMR was designed for case finding in large populations and was not intended to replace normal radiographic or clinical examination. In theory, a full-sized radiograph was always taken to confirm abnormalities identified in a miniature film. After the Second World War, the technology was quickly adopted in many industrial states: by the mid-1950s, it was used extensively by the Ministry of Health in England and by the armed services in the USA, Canada and Australia to screen for tuberculosis. In defending one of the cases which was later amalgamated into the miners' class action, Anglo American claimed that mass miniature radiography was but one element in a health surveillance and treatment programme conducted by employers.[3] It was, however, the most important element, as it was the primary tool for case finding and therefore for compensation.

The original research into MMR was conducted at the WNLA X-ray Department by Major K.G. Collender. Trained in England as an electrical engineer, Collender joined the WNLA in 1926 and began working on mass radiography.[4] He made slow progress and the construction of a twin pilot plant producing the world's first miniature chest X-rays was not completed at the WNLA Depot until 1939. The plant was used initially for new entrants, who accounted for less than 20 per cent of mine labour. The small negatives proved far more reliable than the stethoscope. Of those cases which received compensation from the Silicosis Bureau in 1946, almost 90 per cent were discovered on Collender's 24 × 18 mm films.[5] In the meantime, Collender completed plans for a multiple plant with the capacity to examine all of the 330,000 recruits who passed

[2] Dr F. Retief. 'Miniature Radiography of Mine Native Labourers at the WNLA Depot'. *Proceedings of the Transvaal Mine Medical Officers' Association*, Vol. XXII, No. 250, April 1943, p. 181.

[3] The South Gauteng High Court, Johannesburg, Case No. 18267/04, Alpheos Zonisele Blom, Plaintiff, and Anglo American South Africa Limited, Defendant. Defendant's Plea to Plaintiff's Amended Particulars of Claim. Dated at Johannesburg 4th October 2011, pp. 32–33.

[4] B.A. Dormer and K.G. Collender. 'Miniature Radiography'. *The Lancet*, 10th June 1939, pp. 1310–11.

[5] P. Marais and Z. Botha. 'Dr K.G.F. Collender: Inventor of Mass Miniature Radiography'. *Adler Museum Bulletin*, Vol. 31, No. 2, 2005, pp. 27–30.

through the WNLA compound each year. After a developmental phase lasting twenty years, MMR for all black miners began in 1955, just after effective chemotherapy for tuberculosis came on stream. By 1959, the 36 hospitals for black miners on the Witwatersrand housed 43 miniature and 26 large X-ray installations.[6]

There were several factors at play in the Chamber's decision to develop MMR. The most important was the persistent criticism from the white MWU that its members were at risk of tuberculosis infection from black miners. There was also criticism from medical and lay authorities, including those represented at the 1942 Lansdown and Stratford Commissions, of the mine's refusal to use X-rays with black miners, a failure which they blamed for the spread of tuberculosis. The third factor in the introduction of MMR was the international research community of which the Chamber was an important member. The ILO Conference on Pneumoconiosis, held in Sydney in April 1950, attracted delegates from the United Kingdom, the United States and South Africa.[7] Dr A.J. Orenstein was elected Vice-Chairman and served as one of five nominated experts. William Gemmill, who had attended ILO conferences since the 1920s, represented the Chamber and reported on the proceedings. It was an important moment for the gold industry. Anglo-American, which had invested heavily in the Free State mines, had recently opened the state-of-the-art Oppenheimer Hospital at Welkom. The UN and the ILO were championing labour rights in the fading colonial empires. On the biomedical front, effective chemotherapy for tuberculosis had recently become available. Perhaps most significant of all, apartheid was less than two years old. The ILO Conference agenda was to review the state of knowledge and if possible, define a minimum international standard for compensation. The delegates agreed that in making such assessments, three factors had to be considered: the worker's history of dust exposure, a clinical examination, and radiographic findings. The importance of each factor varied from case to case, but the delegates did agree on the

[6] *Report of the Pneumoconiosis Bureau for the Period 1st April 1958 to 31st March 1959*. Pretoria: Government Printer, 1960, p. 7.

[7] For a coverage of the proceedings see *Collected Papers from the Third International Conference of Experts on Pneumoconiosis 1950*. Geneva: International Labour Organization, 1953.

importance of an X-ray.[8] Six months later, the Chamber finally decided to introduce MMR at initial examinations.

The impact of the technology on improving the quality of examinations was compromised by staff workloads. On average, a mine medical officer at the WNLA would read between 800 and 1000 films in an hour. In 1965, Dr J.A. Louw proudly told a Mine Medical Officers Association meeting that he could review up to 2000 plates in 60 minutes.[9] The size of the plates presented another problem. In a study from 1963, Sluis-Cremer from the NIOH found that only 24 per cent of slight or moderate cases of pneumoconiosis were picked up with miniature X-rays, in contrast to 48 per cent with full-scaled plates.[10] Thirty years later, the Leon Commission into occupational health found that the use of MMR was one of the major reasons for the under-reporting of silicosis.[11]

Whatever its effectiveness, the adoption of MMR did not determine the uses to which it was put. The most persistent factor cited as spreading tuberculosis from the mines was the repatriation of infected men. In 1954, the pathologist Professor Sarel Oosthuizen completed a highly critical report of the industry.[12] He found the medicals of black miners inadequate and recommended that a radiological examination be carried out at entry and exit medicals and at three monthly intervals during service. Oosthuizen was particularly scathing about the repatriation of tubercular miners and recommended that suspected cases be kept under observation in mine hospitals until they were no longer infectious.[13] He

---

[8] Report by W. Gemmill, 'International Labour Organization, Third International Conference of Experts on Pneumoconiosis', 20th April 1950, pp. 3; 5. TEBA Archives, W.N.L.A. 20D Diseases & Epidemics, Pneumonia, 1950.

[9] Dr J.A. Louw. 'The Miniature Chest Radiograph: Its Uses, Value and Problems.' *Proceedings of the Mine Medical Officers' Association*, Vol. XLV, No. 392, October-December 1965, p. 81.

[10] G.K. Sluis-Cremer and J.C. Wagner. 'A pathological-radiological correlation in 108 cases of asbestosis proved at post-mortem'. *Proceedings of the International Congress of Occupational Health,* Madrid, Vol. 14, No. 2, 1964, pp. 608–610.

[11] See *Report of the Commission of Inquiry into Safety and Health in the Mining Industry (Leon Commission).* Pretoria, Department of Minerals and Energy Affairs, 1995, pp. 16–17.

[12] Report of the Departmental Committee of Enquiry into the Relationship Between Silicosis and Pulmonary Disability and the Relationship Between Pneumoconiosis and Tuberculosis August 1954. SANA F 33/671 Treasury. See also Chap. 4, pp. 8–9.

[13] Memorandum from the Legal Adviser, B.T. Tindall, dated 2nd May 1955, p. 16. TEBA Archive, WNLA 14/3, Mass Miniature Radiography, General File, March 1954 to January 1956.

also proposed that it should be an offence for a mine owner to discharge an employee suffering from active tuberculosis. It was the most severe criticism of mine medicine since the Stratford Commission ten years earlier. In response to the report, the Group Medical Officers Committee (GMO), chaired by Orenstein, noted that since all new entrants were shortly to be given a miniature radiograph at the WNLA Hospital, that recommendation would be met. However, the Committee opposed periodic X-ray examinations as both unjustified and expensive. It also opposed Oosthuizen's recommendation that it be an offence for a mine owner to discharge an employee suffering from active tuberculosis.[14] A subsequent meeting agreed in principle that miners certified by the Bureau with tuberculosis should be admitted to hospital for observation or treatment but cited several factors which made such policy impractical.[15]

In all, the evidence suggests that the Chamber's use of MMR at *entry* medicals identified those with easily diagnosable lung disease who were then repatriated without compensation. The technology was not used to enhance the awarding of compensation by more effective *exit* medicals. Perhaps the most important reason for the MMR initiative was the large number of returning miners, who were valued because of their skill and experience. Those workers had cumulative dust exposure over several contracts and were therefore at a high risk of developing tuberculosis and silicosis. The WNLA hoped that the detection and exclusion of men with early-stage occupational lung disease would improve mine safety and save on compensation.[16] The technologies of case finding were greatly enhanced; the technologies of care were strained even further.

[14] Minutes of a Meeting of the Sub-Committee of Group Medical Officers held on 21st June 1955, in the Chamber of Mines Building, p. 3. TEBA Archive, WNLA 14/3, Mass Miniature Radiography, General File, March 1954 to January 1956.

[15] Minutes of a Meeting of the Sub-Committee of Group Medical Officers 4th July 1955, in the Chamber of Mines Building, p. 1. TEBA Archive, WNLA 14/3, Mass Miniature Radiography, General File, March 1954 to January 1956.

[16] Dr F. Retief. 'Miniature Radiography of Mine Native Labourers at the WNLA Depot'. *Proceedings of the Transvaal Mine Medical Officers' Association*, Vol. XXII, No. 250, April 1943, p. 186.

Overcrowding and X-rays at the WNLA Hospital in the 1950s

As MMR came on stream, there was an urgent need to increase the number of beds at the recently expanded WNLA hospital. During 1953, a total of 4504 black miners were referred to the Bureau for assessment. In the following year that number rose sharply to 5765, but of those only half were compensated.[17] The industry was left with the cost of housing an increasing number of men who clogged up the WNLA hospital while awaiting their claims to be processed. The Hospital had just over 1000 beds but in May 1955 it held 1700 patients, approximately half of whom had claims before the Bureau. The average period of detention for miners with claims was 53 days. The use of MMR for all recruits at the WNLA was due to begin, and the number of compensation cases was expected to increase. The WNLA's Chief Medical Officer urged that only those with a reasonable chance of success be referred to the Bureau.[18] The residue should be repatriated.

In January 1956, over 37,000 recruits were radiographed at the WNLA compound. The WNLA's Chief Medical Officer reported with pride: 'This we consider is a unique record, not equalled or even approached, anywhere else in the radiological world'.[19] His enthusiasm was not shared by James Gemmill, who replaced his father as general manager of the WNLA. Gemmill was concerned that the MMR would lead to a sudden increase in the number of recruits rejected at Johannesburg. The initial data indicated that around 5000 men would be eliminated each year from the work force. Gemmill also feared that the white Mine Workers'

---

[17] See Jock McCulloch. 'Hiding a Pandemic: Dr G.W.H. Schepers and the Politics of Silicosis in South Africa'. *The Journal of Southern African Studies*, Vol. 35, No. 4, 2009, pp. 835–848.

[18] Memorandum from the Chief Medical Officer, WNLA 20th May 1955, on Overcrowding of WNLA Hospital. TEBA Archive, WNLA 14/3, Mass Miniature Radiography, General File, March 1954 to January 1956.

[19] Report from the Association's Chief Medical Adviser WNLA, June 1956. Subject: Miniature Radiography on 70 mm Film; With 12" × 15" Film of Abnormal Findings in the Initial Examination of all Mine Entrants at the WNLA X-ray Department. TEBA Archive, WNLA 14/3, Mass Miniature Radiography General, February 1956 to November 1956.

Union would once again take up the issue of the high tuberculosis rate.[20] Between November 1954 and February 1955, when only novices and men with long service were X-rayed, and from November 1955 to February 1956, when *all* entrants were X-rayed, the percentage of rejects increased fivefold.[21] There was, however, no sudden increase in the number of compensation awards. Instead, a larger number of recruits were repatriated. There was also a rise in the numbers certified by the Bureau with tuberculosis but not deemed eligible for compensation.[22]

Where James Gemmill saw a threat to the industry, the Department of Health saw an opportunity to address a public health problem. In August 1956, a conference was held in Pretoria to discuss MMR. Those present included the Secretary for Health, Dr le Roux and James Gemmill from the WNLA. Mr Worroll, a Legal Adviser, represented the Chamber. Dr le Roux believed that the Pneumoconiosis Act of 1956, which made initial, periodic and exit radiological examinations compulsory, gave the mines and his department an opportunity to identify and treat early-stage tuberculosis.[23] Le Roux wanted the system expanded so that every black miner was radiologically examined every six months, and those infected isolated in sanatoria. He reminded the meeting that the detention and treatment of infected miners was in the national interest, and that the cost involved should therefore be borne by Government.[24] Gemmill and

[20] Memorandum from J.A. Gemmill, General Manager, WNLA to the General Manager, Chamber of Mines, Johannesburg, 5th June 1956. Subject: Statistics—Miniature Radiological Examination. TEBA Archive, WNLA. 14/3, Mass Miniature Radiography General, February 1956 to November 1956.

[21] Letter from Frank Retief, Chief Medical Officer, Hospital, Johannesburg, to The General Manager, WNLA Limited, Johannesburg, 1st June 1956. Subject: Statistics-Miniature Radiological Examination. TEBA Archive, WNLA. 14/3, Mass Miniature Radiography General, February 1956 to November 1956.

[22] Memo A.T. Milne, General Manager to G.P.C. Members, 27th September 1957, Memorandum to the Round Table Conference on Vulnerable Mines: Recruiting and Medical Examination of Mine Native Labourers, pp. 6–7. TEBA Archives, WNLA 14/3, Mass Miniature Radiography General, Dec. 1956 to Jan. 1958.

[23] *Report of the Pneumoconiosis Bureau for the Period 1st August 1956 to 31st March 1958*. Pretoria: Government Printer, 1958, pp. 8–9.

[24] Note from James Gemmill, General Manager, WNLA Limited, Johannesburg, 30–31st August 1956. Subject: Tuberculosis- Native Mine Labourers (Act 57 of 1956: Conference in Pretoria, 29th August 1956.) TEBA Archive, WNLA 14/3, Mass Miniature Radiography General, February 1956 to November 1956.

Worroll both rejected the idea of X-rays every six months on the grounds that such examinations were unnecessary. They also warned that Le Roux's proposal to send black miners to sanatoria, in either Johannesburg or elsewhere, would have an adverse effect on recruitment.[25]

Limitations of Mass Miniature Radiography

The MMR technology did have a number of critics. The Northern Rhodesia Commission on Silicosis of 1949 was unimpressed by Collender's invention. 'We are satisfied that full-scale radiographic examination, if financial and other practical circumstances permit its arrangement, is in every way to be preferred to the use of a miniature plant which cannot, for example, be relied upon to show the presence of silicosis in the very early stage'. The Commission recommended that periodical examinations should be carried out annually, and that they should include both radiographic and clinical reviews.[26] On leaving the mines, every recruit should be fully examined by the Bureau rather than by a mine medical officer. The WHO's guidelines for the Control of Tuberculosis in Undeveloped countries suggest it was also unimpressed by MMR. The WHO endorsed the conventional wisdom that a reduction in the tuberculosis rate can best be achieved by improvements in nutrition, housing, education and occupational health. In a comment which was probably directed at the Chamber, the WHO noted, 'It is of little value simply to take thousands of X-ray films of the lungs of people and then do nothing more about the matter. Even for the assessment of morbidity, dependence upon the X-ray pictures alone is inadequate'.[27]

Under the Pneumoconiosis Act of 1956, asbestos mines, which had formerly been designated 'registered mines' under the 1946 Silicosis Act,

[25] Memorandum, the Chamber of Mines, 21st August 1956. Subject: Pneumoconiosis Act—Medical Examination of Native Labourers. TEBA Archive, WNLA 14/3, Mass Miniature Radiography General, February 1956 to November 1956.

[26] *Northern Rhodesia Report of the Commission on Silicosis Legislation*. Lusaka: The Government Printer, 1949, pp. 11–12.

[27] World Health Organization 1950. 'Plan for Control Programmes: Suggestions for the Control of Tuberculosis in Countries with Undeveloped and Underderveloped Programmes', pp. 19–20. SWA RCS 2251, Secretariat, Native Labour Tuberculosis Control.

became Controlled Mines, thereby making it illegal to employ men with tuberculosis. The Act required mine owners to conduct initial, periodic and final medicals. However, in contrast to their approach to the gold mines, the Department of Health insisted on the use of full-sized X-ray plates for black and Coloured labour on asbestos mines. The Department cited the error rate in MMR diagnosis of between 25 to 35 per cent to argue against the use of miniature plates.[28] The Asbestos Producers' Association complained that as the result of the Native Affairs Department requiring wet-drilling, the reduction of dust in the mills and the building of compounds, many mines were already under threat. However, the Association's appeals for permission to use miniature films were rejected. It is not clear why the Department enforced a higher medical standard on the asbestos mines. It may have viewed MMR as a necessary compromise on the Rand because of the sheer volume of labour. Alternatively, the policy may have reflected the relative political influence of the two industries. The asbestos mines were in isolated rural areas; they employed few white miners, and at their peak, they accounted for just 3 per cent of the value of mineral exports.[29]

Repatriations and the Spread of Tuberculosis

Repatriations of miners with infective tuberculosis were first questioned soon after the passing of the original Miners' Phthisis legislation in early twentieth century but the issue had never been resolved. On the eve of Second World War, Dr Peter Allan, who had become Deputy Chief Health Officer, again raised concerns about the spread of tuberculosis from the mines. He believed that to contain infection it was essential to keep track of cases. 'Once a case has been notified he should never be lost

[28] Memorandum from Mr Douglas, Asbestos Producers Association, 1st September 1956. Subject: Pneumoconiosis Act No. 57 of 1956. SANA, F 5-353 Treasury Mines, Silicosis, Pneumoconiosis Act, 1956, Asbestos Mines.

[29] For a history of the mines see Jock McCulloch. *Asbestos Blues: Labour, Capital, Physicians and the State in South Africa*. London: James Currey/ Indiana University Press/ Juta Press, 2002.

sight of'.[30] Sanatorium treatment was expensive, but Allan wanted all tuberculosis patients, black and white, to have institutional care as was done in England. The Poor Law hospitals admitted many advanced cases; in Birmingham in 1937, over half of tuberculosis deaths occurred in institutions. The removal of those patients from the general population was, in Allan's opinion, perhaps the most important public health measure taken in Britain, a view supported by subsequent historical research.[31]

In 1942, the Deputy Chief Health Officer in Durban, Dr F.W.P. Cluver, reiterated concerns that repatriated miners were a primary source of infection. The return of those men to the Reserves without his Department's knowledge made it impossible to control the spread of disease. 'No organisation has the right to repatriate cases of infectious diseases back to their kraals without the knowledge of the Health Department. Surely this would not be done in the case of leprosy or plague, and yet, tuberculosis is highly infectious and causes more deaths per annum than all the other formidable epidemic diseases put together'.[32] Cluver wanted notifications issued so that local health inspectors could monitor family members and instruct patients on how to avoid infecting others. The Miners' Phthisis Bureau notified all cases of tuberculosis in whites to the Secretary for Public Health, while black miners were notified to the Director of Native Labour who dealt with compensation. As the result of sustained lobbying by Dormer and others, in 1941, the Director Native Labour agreed to notify the Deputy Chief Health Officer in Durban of all black miners repatriated to Natal and Zululand.[33] For reasons which are unclear, the Director then failed to do so, making it all but impossible

[30] Dr Peter Allan. 'The Treatment of Pulmonary Tuberculous from the Public Health Aspect'. *The South African Medical Journal*, Vol. 13, No. 9, 13th May 1939, p. 317.

[31] Leonard G. Wilson. 'The Historical Decline of Tuberculosis in Europe and America: Its Causes and Significance'. *Journal of the History of Medicine and Allied Sciences*, Vol. 45, No. 3, 1990, pp. 366–396.

[32] Memorandum from Dr F.W.P. Cluver, Deputy Chief Health Officer, Durban, to unnamed recipients, Undated, Nov 1941? Subject: Repatriation of Native Tuberculotics from the Witwatersrand. SANA, GES 1106, 409/178, 1930–1953, TB in Mine Natives & Native Labourers.

[33] Letter from Senior Assistant Health Officer, Dept. of Public Health, Johannesburg, to The Secretary for Public Health, Pretoria, 28th July 1943. Subject: Miners' Phthisis. SANA, GES 1106, 409/178, 1930–1953, TB in Mine Natives & Native Labourers.

for Cluver's office to trace cases.[34] In July 1943, the Durban Town Clerk asked the Bureau for notifications as no 'repatriation advice' had been received by his office for over a year.[35]

In February 1944, Dr Peter Allan again wrote to the Chamber asking for assistance in tracing repatriated miners with tuberculosis. The Deputy Chief Health Officer at Durban had an agreement that the Director of Native Labour would issue notifications of men repatriated to Natal, so they could be followed up. 'The problem is obviously very serious and the numbers of tuberculous Natives returning to the territories may seriously influence the position there'.[36] Allan acknowledged that any form of isolation would have to be enforced and may be detrimental to recruiting. For that reason, the Chamber' co-operation was essential. There is no surviving reply, but a month later Dr Allan wrote a brief note to the Minister on how best to address the tuberculosis problem. The first step was to determine the number of men who returned infected to the Territories. Allan wanted those cases followed up by a series of dedicated clinics in the Transkei.[37] He envisaged the Umtata Tuberculosis Hospital becoming a centre from which itinerant medical officers could work. There should also be huts near clinics where cases requiring nursing could be housed.

Allan's proposal was not taken up, but his letter did have some effect. Within weeks, the Chamber appointed a Sub-Committee consisting of representatives from the major mining houses, the WNLA and the NRC to review repatriations. The Sub-Committee found there was no evidence

[34] Letter from Dr F.W.P. Cluver, Deputy Chief Health Officer, Health Department, Durban, to the Acting Under Secretary for Public Health, Union Health Department, Pretoria, 20th March 1942. Subject: Repatriation of Native Tuberculotics from Witwatersrand. SANA, GES 1106, 409/178, 1930–1953, TB in Mine Natives & Native Labourers.

[35] Letter from Deputy Chief Health Officer, Durban, to the Secretary for Public Health, Pretoria, 17th July 1943. Subject: Miners' Phthisis. SANA, GES 1106, 409/178, 1930–1953, TB in Mine Natives & Native Labourers.

[36] Letter from Dr Peter Allan, Secretary for Public Health, Pretoria, to The Secretary, Chamber of Mines, Johannesburg, 4th February 1944. Subject: Repatriation of Native Tuberculotics from Mines to Native Territories. SANA, GES 1106, 409/178, 1930–1953, TB in Mine Natives & Native Labourers.

[37] Memorandum from Dr Peter Allan, Secretary for Public Health, Cape Town, to The Minister of Health, 11th April 1944. SANA, GES 1106, 409/178, 1930–1953, TB in Mine Natives & Native Labourers.

to support criticism of the mines. Blacks, it noted, were heavily tuberculised and it was well known that any change from the normal leisure of village life may cause a physical breakdown, lighting up a pre-existing infection. At worst, repatriations contributed to 'the endemicity of tuberculosis in the Native Territories'.[38] The solution lay in education, which reformers like Dr Macvicar strongly advocated, and in the creation of health centres in the rural areas where the more serious cases could be hospitalised. Such clinics were the responsibility of government, not the mines. Government should also address the ignorance, malnutrition and lack of medical services associated with tuberculosis. The Sub-Committee concluded that the introduction of MMR would probably see an increase in those rejected at entry medicals, and in consequence a fall in the number of repatriations.

The one constant regarding repatriations is how little changed over time. During 1945, the daily papers in Lourenço Marques featured a number of articles on returning miners spreading tuberculosis to rural areas of Portuguese East Africa (Mozambique). The General Manager of WNLA, Mr G. Lovett, was unsympathetic. Tuberculosis was endemic among natives and infection on the mines had been constantly falling; the current rate being a mere 0.21 per cent. The claim that the mines were responsible for the spread of disease was unfounded.[39] Breaks in underground service protected migrant labour against miners' phthisis and tuberculosis. Tuberculosis was a disease of squalid, overcrowded communities. It was not a disease of the mines.

The WNLA also had to counter criticism of its treatment of men repatriated with traumatic injuries. In June 1951, the District Manager at Lourenço Marques wrote to head office about two paralysed miners who had been repatriated to Ressano Garcia. Both had worked six or seven contracts on the mines and their families were upset that they had received

---

[38] Report of the Special Sub-Committee appointed by the Gold Producers' Committee on 21st February 1944, upon the subject of the repatriation of Native tuberculotics from the mines, 7th & 14th March 1944, p. 1. SANA, GES 1106, 409/178, 1930–1953, TB in Mine Natives & Native Labourers.

[39] Letter from G.O. Lovett, General Manager WNLA, to the District Manager, WNLA Lourenço Marques, 5th November 1945. TEBA Archives. WNLA 20L Tuberculosis February 1931 to February 1957 Diseases and Epidemics.

no compensation.[40] Two Portuguese farmers who were present when the men arrived at the depot made disparaging comments about the mines. The District Manager suggested that some form of *ex-gratia* payment would help to allay further criticism. Less than a year later, the District Manager of WNLA in Lourenço Marques wrote again to head office in Johannesburg about the need to review the management of disabled miners. Seriously injured men arriving at Lourenço Marques by train were transported through the town on stretchers in the back of a van, and then dispatched by ship to their villages. That spectacle had aroused unfavourable publicity and local medical authorities were demanding that such cases be treated humanely. The District Manager suggested that WNLA purchase a second-hand ambulance to transport the men to the boat.[41] The head office declined.

In July 1952, the newly appointed Chair of the Silicosis Medical Bureau, Dr S.W. Verster, visited hospitals in the Transkei and Ciskei to assess the health of former miners. In his report to the Department of Mines, he observed, 'Old silicotic natives more often than not develop a chronic open tuberculosis, and they help not only to infect their own relatives by indiscriminate spitting and coughing, but also infect the countryside at their beer parties. Such cases must be isolated in tuberculous colonies, hospitalisation is useless.'[42] Young males from 12 to 14 years of age, he noted, were often infected in this way, and were also often cured with appropriate treatment. No records or means of identification were kept of such cases. This should be done, he advised, as these young males are the future recruits for the mines, and their earlier infection explained why novice miners sometimes develop tuberculosis so soon after they go underground. Silica dust invariably activates any old

[40] Letter from the District Manager, WNLA Lourenço Marques 28th June 1951 to the General Manager WNLA, Johannesburg, 22nd April 1952. TEBA Archives. WNLA 35 Repatriations and Rejects: East Coast Natives March 1925 to August 1957.

[41] Letter from the District Manager, WNLA Lourenço Marques, 22 April 1952, to the General Manager WNLA, Johannesburg, 22nd April 1952. WNLA 35 Repatriations and Rejects: East Coast Natives March 1925 to August 1957.

[42] Verster's report is quoted in Letter from Acting Secretary for Mines, Department of Mines, Pretoria, to The Secretary for Health, Pretoria, 21st August 1952. SANA, GES 1106, 409/178, 1930–1953, TB in Mine Natives & Native Labourers.

quiescent tubercular lesions and these cases should never be recruited'.[43] Verster was confident that such cases could be traced through the aid of headmen, district surgeons and magistrates. Those patients should be allotted to a local service, which would be responsible for periodical examinations. That would also facilitate post-mortem examinations, extremely few of which were being carried out. The Secretary for Mines, who reviewed Verster's report, wanted all men repatriated with silicosis or tuberculosis to be periodically X-rayed. There was no change in response to these recommendations.

In 1953, the Deputy Chief Health Officer in East London wrote yet another letter to head office about the spread of disease from the mines. He noted that large numbers of men from the region were recruited and that a certain percentage of them were repatriated with pulmonary tuberculosis. The Tembuland Hospital in Umtata and the Macvicar Hospital at Alice were the only tuberculosis hospitals for non-Europeans, but they had few isolation beds. As an average of five men were repatriated each week, it was impossible to accommodate those patients in hospital. The Deputy Chief Health Officer estimated that 70 per cent of repatriates with tuberculosis died shortly after their arrival home. To prevent the medical system being overwhelmed with 'moribund and incurable patients' he suggested that the last X-rays taken by the NRC in Johannesburg be forwarded to the regional tuberculosis officer. The local authority could then arrange for incurable patients to be isolated at their respective villages.[44] Again, none of that happened.

The efforts of medical officers to control the spread of tuberculosis continued. In 1964, Dr I.W.F. Spencer and a group of colleagues from the Johannesburg City Health Department issued a set out guidelines for prevention. Based upon the public health conventions used in the UK for almost half a century, the guidelines provide a way of judging the mines' management of tuberculosis, and their repatriation policies in particular.

[43] Letter from Acting Secretary for Mines, Department of Mines, Pretoria, to The Secretary for Health, Pretoria, 21st August 1952. SANA, GES 1106, 409/178, 1930–1953, TB in Mine Natives & Native Labourers.

[44] Letter from Deputy Chief Health Officer, Union Health Department, East London, to The Secretary for Health, Pretoria, 2nd June 1953. Subject: Tuberculosis Sufferers Repatriated by the WNLA. SANA, GES 1106, 409/178, 1930–1953, TB in Mine Natives & Native Labourers.

Spencer explained that the first aim of any control programme is to inhibit the entry of new cases into the infective pool, and to identify, remove, isolate, and treat cases from the pool. The size of the pool will determine the best epidemiological instrument, be it chemotherapy, BCG vaccination, or mass radiography. The poorer the social setting, the larger the infective pool. For that reason, in South Africa the infective pool was much greater in blacks.[45]

Spencer's concerns were shared by the South African National Tuberculosis Association (SANTA) which wrote to the Chamber's General Manager, James Gemmill, in October 1964. SANTA wanted enhanced monitoring to ensure that repatriated miners received regular and adequate treatment to prevent the spread of disease and to guard against drug resistance. There was a lack of facilities in the Territories, and SANTA wanted to establish a system to follow up cases. It offered to create a mobile treatment team to work in collaboration with District Surgeons and to second a Field Officer to Umtata to follow up patients in need of continued treatment.[46] As a first step, the SANTA wanted notifications from the NRC of infected miners returning to the Territories. While James Gemmill welcomed the SANTA's efforts to improve treatment, he deemed its proposals unnecessary. According to him, the NRC had, for almost twenty years, already notified the Department of Health of all repatriated cases.[47] James Gemmill's response was at odds with the Chamber's own internal correspondence. In July 1964, for example, the Chamber's legal advisor wrote a memo on repatriations. He was concerned that the failure to report to the local authority a case of tuberculosis discovered at the WNLA Depot would subject the individual medical officer involved, or the Association itself, to the penalties prescribed

[45] Spencer et al., 'A Field Experience of Mass Percutaneous BCG'.

[46] Letter from National Secretary, South African National Tuberculosis Association, Johannesburg, to J.A. Gemmill, General Manager, Chamber of Mines Building, Johannesburg, 26th August 1964. Subject: Follow Up of Mine Recruits Repatriated to the Bantu Territories. TEBA Archive, WNLA 20 D Diseases and Epidemics, August 1959 to February 1966.

[47] Letter from J.A. Gemmill, General Manager, Chamber of Mines, Johannesburg, to The National Secretary, S.A. National Tuberculosis Association, Johannesburg, 9th September 1964. Subject: Follow Up of Mine Recruits Repatriated to the Bantu Territories. TEBA Archive, WNLA 20 D Diseases and Epidemics, August 1959 to February 1966.

under Section 38 of the Pneumoconiosis Compensation Act of 1962.[48] Under that Act, medical officers were required to notify the local authority of any tuberculosis cases. Since the WNLA was the authority at whose insistence the medical examination was conducted, failure to notify the local authority would render the Association itself, and not the examining medical officer, guilty of an offence and liable to a fine not exceeding R200.

The Impact of Repatriations

There are few surviving records of the effects of the Chamber's repatriation policies, not least since the industry did not carry out post-employment studies of miners. Previous chapters dealt with the spread of tuberculosis through the southern African region. Those accounts are supplemented by anecdotal evidence from District Surgeons in the Reserves inside South Africa. Dr E. Pullois, a medical officer in the Glen Gray District, made a submission to the Stratford Commission in February 1942, in which he described the effect of mine recruiting on the region. The state of public health was extremely poor: there were 70,000 people in the district but only two doctors and one mission hospital. There were frequent epidemics of typhus and typhoid, and occasional outbreaks of plague and smallpox. Although there were no records of the tuberculosis rate, Pullois was in no doubt the disease was increasing alarmingly. Poor housing, under-nourishment and actual starvation were largely responsible. Pullois also noted 'the highly unsatisfactory condition' of men returning from the mines. The mines recruited heavily from Glen Grey, but those who were repatriated with silicosis or tuberculosis were left to their own devices. 'In the vast majority of cases, there is absolutely no "after-care"; no periodical examination, or anything of the sort. The patient is left to deteriorate slowly or rapidly as the case may be, but he is nobody's responsibility and nobody's care… no protective measures

[48] Memorandum from Legal Adviser, Transvaal and Orange Free State Chamber of Mines, Johannesburg, to The Secretaries, Witwatersrand Native Labour Association, 10th July 1964. Subject: Pneumoconiosis Compensation Act, 1962. TEBA Archive, WNLA 20 D Diseases and Epidemics, August 1959 to February 1966.

whatever are provided for the rest of the population. If statistics were worked out, I feel sure that one would be appalled by the number of open tubercular cases that are left to spread the disease'. The impact on children was devastating. 'I am certain that this very unsatisfactory position is partly responsible for the extremely high infantile mortality amongst natives, as the children are apt to be infected by massive doses with no possible chance of a slow immunisation process.'[49] The regulations under which the government was responsible only for tubercular cases in a 'communicable form' was strangling efforts to provide care. The vast majority of families could not pay for hospital treatment, while cases in a 'communicable form' were mostly hopeless. More than once, Pullois had refused to admit patients who offered every hope of cure because the Government's provision of 2s 6d a day did not apply. Such patients returned home to infect others.

Dr A.R.R. Mears, who was a District Surgeon at Tsolo in the Transkei in the 1940s, has left an account of a surgeon's life which explains why, after leaving the mines, so few men were referred to the Bureau for assessment. A part-time District Surgeon's patients would come from all over the district, with acute patients travelling distances of 30 or 50 kilometres on sledges or scotch carts. To make a gross annual income of £1500 to £2000, a District Surgeon needed to treat 60 to 80 fee-paying patients a day.[50] Such a medical practice was challenging. Few District Surgeons spoke indigenous languages and the overheads were high. The upkeep of a car, for example, was expensive, given the poor roads and the distances travelled. Sometimes a Surgeon had to ride on horseback for miles to reach a patient. Obtaining basic supplies of drugs was often difficult.

The part-time District Surgeon had several sources of income. He provided essential services to the state at an annual retaining fee of around £300, without which few doctors could survive. Those services included prison visits, assisting with police enquiries, conducting post mortems for courts and giving evidence, the supervision of District Nurse clinics, and

---

[49] Letter from Dr E. Pullois, Medical Practitioner, Cape Town, to Senator R. Jones, The Senate, Cape Town, 25th February 1942, pp. 1–2. BNA S332/5/1 Phthisis: Miners' Compensation for B.P. Native Mine Workers, S332/5/1.

[50] Dr A.R.R. Mears. 'The Part-Time District Surgeon in the Transkeian Territories'. *The South African Medical Journal*, Vol. 19, No. 21, 10th November 1945, p. 403.

vaccination tours. In addition, conducting examinations for the NRC provided on average 10 per cent of an annual income. Processing miners' phthisis claims for the Bureau was an elaborate process and required making medical examinations. 'The average D.S. resents this work because he feels that it is insurance work, and as such is deserving of a fee'. District Surgeons were also not paid a fee for post-mortems, which made them servants of the Miners' Phthisis Bureau. They were required to remove the lungs, but no containers were supplied: the Bureau suggested Surgeons use oatmeal tins. Post-mortems were an inconvenience and often meant a financial loss for the Surgeon. They might involve a journey of hundred kilometres or more on the worst roads, while the Surgeon's fee-paying patients waited for his return. Often, by the time he arrived, the deceased was decomposing, making it difficult if not impossible to identify the cause of death. He was merely being used to collect lungs. Mears writes, 'This policy of neglecting the living for the dead is neither in his nor this patient's interest.' Consequently, in some districts, very few claims were processed. Mears was also highly critical of the compensation system. Any native repatriated for suspect silicosis, he wrote, 'should receive full compensation without any post-mortem as he can never be accepted for work on the Mines again'. District Surgeons should be paid 10s 6d for each miners' phthisis examination and sputum collection, and £3 3s for each post-mortem. 'He is bound hand and foot to [the Public Health Department] who can at will increase his duties without consultation. In fact, it is a whole-time appointment on part-time pay'.[51]

The Bureau Director, Dr J. Smith, was aware that very few claims for compensation were lodged by former miners. The X-ray plants at most rural hospitals were inadequate, and he doubted whether local hospitals had qualified radiographers.[52] In December 1946, Mr J.F. Muller from the Department of Native Affairs shared his concerns about 'the deplorable lack of medical facilities' available for carrying out the requirements of the Act. District Surgeons received little assistance and Muller demanded that at the very least, containers for the collection of viscera be

[51] Mears, 'The Part-Time District Surgeon', pp. 404–406.

[52] Dr J.M. Smith, Chairman, Miners' Phthisis Medical Bureau, Comments upon Medical Aspects of the Miners' Phthisis Acts Commission Reports, Johannesburg, February 1945, p. 3. SANA, K105 Miners' Phthisis 1941–48, Vol.1 Correspondence, The Stratford Commission.

provided. Surgeons often used biscuit tins, and as a result, the lungs of deceased miners arrived at the Bureau in a state of decomposition.[53] After a decade of complaints, the Secretary for Health directed District Surgeons to use a 10 per cent solution of formalin and water to preserve the organs for examination.[54] The problem of jars remained.

Two WNLA Doctors

District Surgeons like Drs Pullois and Mears treated returning miners, but their view was limited by the small stage on which they worked. Conditions in the labour reserves such as the Transkei varied from district to district. The two institutions which oversaw mine medicals, disease management and compensation were the Bureau and the WNLA Compound in Johannesburg. Fortunately, we have first-hand accounts of how both institutions worked during the period 1944 until 1980.

Dr G.W.H. Schepers joined the Miners Phthisis Medical Bureau as an intern in 1944. He remained there until 1954, when he migrated to the US to take up the chair at the prestigious Saranac Research Laboratories. Black miners did not attend the Bureau. The decisions on their X-ray results and hence their right to compensation were in theory made by mine medical officers under the supervision of visiting doctors from the Bureau. According to Schepers, that did not happen. The Bureau interns felt overworked, and they viewed the weekly visits to the WNLA compound as an onerous task which they did their best to avoid. Schepers was interested in the work. He spoke three indigenous languages and he volunteered in the hope of improving the certification system. As he later recalled: 'I cannot remember a single diagnosis for silicosis at the WNLA until I did the work'.[55]

[53] Letter from J.F. Muller, Secretary for Mines to G. Mears, Secretary for Native Affairs, Cape Town, 13th February 1947. Subject: Silicosis Act 1946. SANA, K105, Miners' Phthisis 1941–1948, Silicosis Vol. 7, Act 1946, Natives Files.

[54] Circular No. 4 of 1953, from Secretary for Health to All Magistrates and Native Commissioners in the Union, 20th February 1953. SANA, GES 412/17A, Medical Bureau.

[55] This account of mine medicine is based on a series of interviews Jock McCulloch conducted with Dr Gerrit Schepers at his home in at Great Falls, Virginia, USA, 23rd–28th October 2010. Gerrit Schepers died in September 2011 at the age of 97. See also McCulloch, 'Hiding a Pandemic'.

There were no medicals at the WNLA, just the processing of large numbers of men to meet a legislative requirement. The miners were lined up naked and a doctor would run a stethoscope over each chest, doing ten men a minute. The other test was weight loss. The X-ray equipment was adequate, but there was a massive volume of work. The X-rays were read by the WNLA doctors, and a small number of cases, who in Schepers' words 'were in the process of dying', were set aside for review. Medical officers had no indigenous languages and were unable to ask black miners the simplest questions about their symptoms and work experience. Schepers recalls that there was a lot of disease and that in many cases the black miners he saw 'had simply been worked to death'. After two years underground, a man's health was usually ruined: 'Whites survived on average three years after retirement. Blacks who had worked three consecutive contracts were usually dead a year after they left the mines'. The rate of tuberculosis was very high, but in most cases, infected miners were repatriated without compensation. Schepers recalls there was a 'horrendous spread of tuberculosis from the mines', a fact that was common knowledge at the Bureau.

The view of mine medicine from the WNLA side is strikingly similar. Dr Oluf Martiny was born in a small town in the Free State where his father was a doctor. Having graduated from Wits University in 1949, he worked briefly in Denmark and then at Missouri State Tuberculosis Sanatorium in the USA, where many of his patients were stone masons with silicosis. On returning to South Africa in 1954, Martiny joined the WNLA, where he remained until his retirement thirty years later. Working for the WNLA gave medical officers plenty of free time to pursue other interests, something Martiny could not have done in private practice. The major drawback was the poor pay, which meant that like his colleagues Martiny augmented his income by doing assurance examinations.[56]

There was a massive workforce on the mines, with more than 300,000 men in the system at any one time. They included new recruits coming

[56] This account Oluf Martiny's career is based on an interview Jock McCulloch conducted with Dr Martiny at Forest Town, Johannesburg on 27th April 2011. The interview is supplemented by Oluf Martiny's unpublished manuscript, My Medical Career, Johannesburg, November 1995 to September 1999.

in from rural areas, experienced miners who had completed their contracts, and sick men awaiting repatriation. Most of those men passed through the WNLA compound. In Martiny's words, it was like a bucket constantly filling up. Every morning, five to seven doctors would start their day by examining recruits who were lined up naked. For most of the year, the examinations began at 7 am and finished by 10 or 11 am. However, during January and February, with thousands of novices and ex-employees returning from leave, they started at 6 am. On the busiest day, Martiny and his colleagues examined 12,000 recruits in a five-hour period.[57] Management considered those examinations the most important part of the day's work. Martiny thought them a waste of time.

When Martiny started at the WNLA, mass miniature X-rays were being introduced. The recruits were X-rayed using three cameras, working in tandem. They were then ushered into the examining rooms, ten at a time, to be checked for defects such as damaged limbs or weak lungs. The orderlies would mark the chests of those requiring further review with a wax crayon. After an X-ray, the recruits were photographed and fingerprinted by the Department of Bantu Affairs, and their identity documents processed. They then travelled by train from the Booysens station, next to the WNLA, or they were bussed to their assigned mines.

On Tuesdays, there was an influx of medical repatriates from the mines, but insufficient accommodation in the hospital to house them. They included men who had been seriously injured in accidents, old miners who had returned to work but were ill, and cases of extreme silicosis in drillers with less than a year underground. The patients would stay until they were fit to travel. During the day, most of the men spent their time outdoors under the trees and it was only at night that the overcrowding became obvious. There would be patients in the beds, on the floor, under the beds or wherever they could find a place to sleep. The men were issued with one blanket to wear during the day as well as for sleeping at night, summer and winter. They were not given clean night shirts or fresh blankets even when they had stayed for weeks. Patients slept on the cement with a thin felt pad for a mattress. In moving through

[57] Martiny, My Medical Career, p. 7.

the wards at night, orderlies had to step over sleeping bodies and hear the constant coughing and spitting.

The dormitories which housed the migrant workers on their way to or from the mines were dilapidated sheds. The rooms measured five by six metres, with four or more men to a room sleeping in double bunks, with a coal stove for cooking and heating. The men made lockers out of discarded boxes, in which they kept their personal possessions and food, including raw meat.[58] They slept on cement platforms where they lay side by side with strangers. There were no mattresses, and they had to provide their own blankets. Most of the men preferred to sleep outdoors on the dirty tarmac, where there were no bedbugs, and the air was clean. Even in the severe Johannesburg winters, many refused to sleep in the dormitories. Pneumonia and influenza were common. Martiny complained to the WNLA's assistant general manager, but nothing was done.[59]

The WNLA doctors were expected to keep recruits with tuberculosis out of the mines. They were also expected to diagnose tuberculosis in serving miners. Most had early-stage disease and were ideal candidates for treatment. The Bureau's reviews of compensation claims could take several weeks, and the WNLA doctors used the opportunity to begin chemotherapy, which had only recently become available. On discharge, each patient was given a letter of referral to his nearest rural clinic on the assumption that treatment would continue. The state was responsible for the men once they left the mines, but there was no treatment in the rural areas. In Martiny's words, the repatriated miners 'were doomed'.[60]

At Libode in the Transkei, the local doctor provided treatment for only a month, as there were insufficient beds to cope with the large number of patients. Those who improved while in hospital were often prematurely discharged to make way for new admissions. They in turn would relapse, be re-admitted and given the same drugs to which they had become resistant. In Lesotho, which was a major labour-sending region, the local hospitals often ran out of drugs, and had to wait for weeks and even months before they received fresh supplies. Those breaks in treatment again led to

[58] Martiny, My Medical Career, pp. 20–21.

[59] Interview with Dr Oluf Martiny, 2011.

[60] Martiny, My Medical Career, p. 39.

drug resistance. Martiny spoke with the Chamber's president about that problem, but his appeals were ignored.[61]

Soon after his appointment in 1954, Martiny began a survey to trace men with tuberculosis who had been repatriated over the previous ten years. The Chamber's recruiting arms, WNLA and the Native Recruiting Corporation, had around 150 field officers and close links with rural hospitals so in theory it should have been easy. That was not so, and it was only with the help of the Department of Health that Martiny eventually discovered that 90 per cent of repatriated miners had not continued treatment.[62] Presumably, many of those men relapsed and died. A number of those who survived would have developed drug resistant tuberculosis, a serious and widespread problem in contemporary southern Africa.[63]

Conclusion

The fragments of medical histories make harrowing reading. Combined, they represent a compelling picture of the relationship between policy and practice, population health and available statistics. South African legislation conformed to contemporary standards for dealing with occupational lung disease, and tuberculosis in particular. In many respects, it led the world in conducting systematic examinations of the mine workforce, collecting dust samples and establishing scientific institutes to study pneumoconiosis. South Africa pioneered mass miniature radiography, and quickly began using the new drugs for treating tuberculosis. Many of the medical officers employed in the region were highly qualified and conscientious professionals who left a record of acute observation of local health issues, and diligent correspondence with relevant authorities about crucial improvements.

In practice, little of this translated into stemming the epidemic of tuberculosis which was engulfing southern Africa. While the mining

[61] Interview with Dr Oluf Martiny, 2011.

[62] Martiny, My Medical Career, p. 40.

[63] See *The Global Tuberculosis Report 2018*. Geneva: The World Health Organization, 2018.

houses repeatedly claimed they conformed to the law, there was little evidence they ever notified local authorities of the thousands of miners with infectious tuberculosis they repatriated. Although many administrators and doctors stressed the importance of nutrition, isolation of infectious cases and community education, the racialised political economy of mining provided them with few resources to implement such measures. And while many doctors adopted the new chemotherapy, the actual use of the drugs by their patients resulted in the development of drug-resistant TB.

Combined, these factors had repercussions far beyond southern Africa. The myriad of failures to perform exit medical examinations of miners or to notify local authorities of tuberculosis cases, systemic repatriation without compensation, the onerous burden of proof on miners who sustained occupational injury, the lack of medical capacity in the labour-sending regions and the absence of follow-up studies of repatriated miners combined to create the statistical invisibility of lung disease. In a world increasingly dependent on statistics to guide state policy, such ignorance helped justify inaction. On a world stage, it helped hide an epidemic of occupational lung disease associated with South Africa's gold mines, and manufacture ignorance about the epidemiology of pneumoconiosis.

References

Allan, P. 'The Treatment of Pulmonary Tuberculosis from the Public Health Aspect'. *The South African Medical Journal*, Vol. 13, No. 9, 13th May 1939, pp. 317–319.

Dormer, B.A. and K.G. Collender. 'Miniature Radiography'. *The Lancet*, Vol. 233, No. 6041, 10th June 1939, pp. 1309–1311.

Marais, P. and Z. Botha. 'Dr K.G.F. Collender: Inventor of Mass Miniature Radiography'. *Adler Museum Bulletin*, Vol. 31, No. 2, 2005, pp. 27–30.

McCulloch, Jock. 'Hiding a Pandemic: Dr G.W.H. Schepers and the Politics of Silicosis in South Africa'. *The Journal of Southern African Studies*, Vol. 35, No. 4, December 2009, pp. 835–848.

McCulloch, Jock. *Asbestos Blues: Labour, Capital, Physicians and the State in South Africa*. London: James Currey/ Indiana University Press/ Juta Press, 2002.

Mears, A.R.R. 'The Part-Time District Surgeon in the Transkeian Territories'. *The South African Medical Journal,* Vol. 19, No. 21, 10th November 1945, pp. 402–406.

Spencer, I.W.F., M. Coster, A. McPhail and M. Richter. 'A Field Experience of Mass Percutaneous BCG Inoculation as an Immunizing and Diagnostic Procedure'. *The South African Medical Journal,* Vol. 38, No. 12, 11th April 1964, pp. 223–233.

Wilson, Leonard G. 'The Historical Decline of Tuberculosis in Europe and America: Its Causes and Significance'. *Journal of the History of Medicine and Allied Sciences,* Vol. 45, No. 3, July 1990, pp. 366–396.

14

Things Fall Apart—Independent Research, Asbestos Litigation and the Gold Miners' Class Action: 1983–2019

Under apartheid, the focus of mine legislation was to improve the lot of white miners and their families. When the white MWU became concerned that its members had a heightened risk of bronchitis and heart disease, in 1949 Dr Peter Allan was appointed to chair a commission to investigate. The Gold Producers Committee, which submitted written and verbal evidence, was represented by several senior officers including Dr A.J. Orenstein. According to the GPC, the mines had never been safer and the improvements in air quality over the previous ten years were well in advance of those required under the Mines Acts. The Government Mining Engineer agreed. The routine inspections by his Department were, he claimed, probably the most rigorous in the world.[1]

The Commissions of Enquiry appointed during the 1960s and 1970s were notable for their support of the industry and its policies. The testimony of the chief group medical officer of Goldfields before a Commission in 1964 is representative of the tone. Dr Pieter Smit complained that while the government was legally responsible for tuberculosis, it was the

[1] *Report of the Commission of Enquiry Regarding the Occurrence of Certain Diseases, other than Silicosis and Tuberculosis, Attributable to the Nature of Employment in and about Mines.* Pretoria: Government Printer, 1951, pp. 6–7. U.G. 22/1951.

J. McCulloch, P. Miller, *Mining Gold and Manufacturing Ignorance*,
https://doi.org/10.1007/978-981-19-8327-6_14

mining industry which was treating miners. The industry provided far better care and facilities than were available elsewhere in South Africa: 'In this country, Mr Chairman, I must say the liberality of diagnosis and compensation far exceed anything I have seen anywhere else on the continent and in the United Kingdom'. The Chairman agreed.[2]

Until the early 1990s, those who were critical of the mines found little support. The Commission of Enquiry on Occupational Health of 1976 reported: 'Although much could probably still be done about industrial health in the mining industry, there is little in the gold mining industry about which the Republic need be ashamed'. The history of industrial health of the gold mines 'speaks of a sensitivity and a willingness on the part of the authorities as well as of the industry itself to accord humanitarian considerations their rightful place at all times'. The exemplary conditions on the mines were 'due to the fact that the worker's safety and health were their constant concern, and created an undeniable climate of industrial peace'. As a result, the Commission noted, the incidence of silicosis and tuberculosis had fallen sharply.[3]

Official approval went hand in hand with hardening of the relevant science. By 1920, the Chamber had fashioned a medical orthodoxy which held that tuberculosis was brought to the mines by recruits, a position from which the industry never wavered. A commentary from 1944 by the Gold Producers Committee, for example, explained that the mines were not the source of tuberculosis. On the contrary, tuberculosis was endemic in the Native Territories. With a change of environment and occupation, latent tuberculosis became active. At the same time, however, owing to the good rations and living conditions, the mines had probably saved many thousands of men who would otherwise have contracted tuberculosis had they remained in their rural homes.[4]

The Chamber's orthodoxy was still in vogue in 1989. In that year R.L. Cowie, a senior medical officer at the Anglo American hospital at

[2]Transcript of evidence, Dr Peter Smit, Chief Group Medical Officer of Goldfields, before the Commission of Enquiry Regarding Pneumoconiosis Compensation 1964, pp. 689; 694. South African National Archives K269.

[3]*Report of the Commission of Enquiry on Occupational Health.* Pretoria: Government Printer, 1976, pp. 6–7; 11.

[4]Gold Producers' Committee Comments upon Report of Miners' Phthisis Commission, February 1944, p. 15. South African National Archives, K105 Miners' Phthisis 1941–48, Vol. 1, Correspondence, The Stratford Commission.

Welkom, published an essay explaining why, despite the new generation chemotherapy, the rate of tuberculosis in black miners remained high. From 1912, he noted, the mines reduced dust levels and silicosis was transformed into a chronic disease. The tuberculosis rate in white miners was falling while the disease remained prevalent in blacks. Cowie admitted that black miners had far higher exposures to dust than did whites and that they remained at risk of accelerated silicosis and tuberculosis. That period ended in the 1950s, with the development of isoniazid. However, treatment was tedious and prolonged. A miner with silicosis could survive tuberculosis, but the outcome was poorer for black miners because treatment took eighteen months and was rarely completed. The introduction of short-course therapy on the Free State mines in 1977 was an important advance: 'In many respects the treatment of black gold miners with pulmonary tuberculosis is more efficient and more structured than that for white miners'.[5]

In what Cowie termed an anomaly, however, 'the apparently adequate management' of black miners had no influence on the prevalence of disease, which remained at around 500 cases per 100,000 person years. According to Cowie, the anomaly was due to the large infective pool in the labour-sending population. When visiting home, some miners were exposed to infection and when they returned to the mines, they became ill. That was why surveillance and treatment on the mines could not eradicate the disease. Cowie concluded that the only solution was to abolish the migrant labour system, and thereby shield miners from infection.[6] Like those who preceded him, Cowie made little mention of dust exposures, relying instead on the idea that racial susceptibility played a role. He also provided no epidemiological or other evidence to support his claims about the path of transmission. In contrast, a recent study using DNA finger printing shows that most infection occurs on the mines and is spread by persistently infectious individuals who have previously failed

---

[5] R.L Cowie. 'The Five Ages of Pulmonary Tuberculosis and the South African Goldminer'. *The South African Medical Journal,* Vol. 76, 18th November 1989, p. 567.

[6] Cowie, 'The Five Ages of Pulmonary Tuberculosis', p. 567.

treatment. The longer a worker remains in the mines, the more likely he is to be infected.[7]

Silicosis and the Orthodoxy About Intermittent Employment

Major discrepancies in the silicosis rates between white and black miners were the other key element of the companies' narrative about mine safety. As a senior SAIMR researcher, Dr Mavrogordato, put it at the 1930 Silicosis Conference: 'Observation had shown that the smaller incidence of silicosis among natives, as compared with Europeans, was due to the intermittent employment; natives who were employed continuously developed silicosis more rapidly than Europeans.'[8] That orthodoxy was repeated by the Chamber whenever the costs of the migrant labour system were questioned. In 1943 the Witwatersrand Mine Native Wages (Lansdown) Commission, which was in other respects so critical of the industry, found that the return of men to the Reserves after a period on the mines tended to lower the incidence of miners' phthisis.[9] In 1951 the Allan Commission found that the migrant labour system probably accounted for the rarity of silicosis in black miners.[10]

The Chamber's claims that the low silicosis rates in black miners provided evidence of the benefits of oscillating migration remained persuasive into the twenty-first century. Writing in 2007, May Hermanus remarked that for most of the twentieth century, miners worked for limited periods before returning to the rural areas. From the mid-1980s,

[7] P. Godfrey-Faussett, P. Sonnenberg, S.C. Shearer, M.C. Bruce et al. 'Tuberculosis Control and Molecular Epidemiology in a South African Gold-mining Community'. *The Lancet*, Vol. 356, No. 9235, 23rd September 2000, pp. 1066–1071.

[8] *Silicosis: Records of the International Conference held at Johannesburg 13–27th August 1930.* London: ILO, 1930, p. 45.

[9] *Report of the Witwatersrand Mine Native Wages (Lansdown) Commission on the Remuneration and Conditions of Employment of Natives on the Witwatersrand Gold Mines.* Pretoria: Government Printer 1943, p. 24.

[10] *Report of the Commission of Enquiry Regarding the Occurrence of Certain Diseases, other than Silicosis and Tuberculosis, Attributable to the Nature of Employment in and about Mines (Allan Commission).* Pretoria: Government Printer, 1951, p. 14. U.G. 22/1951.

however, many remained in employment for more than two decades, thereby enduring prolonged exposure to dust and developing occupational diseases in greater numbers.[11] Rodney Ehrlich agreed: 'It is highly plausible, as argued first by Jean Leger and later by others, that there has indeed been a significant rise in the incidence of silicosis in the latter part of the century owing to what has been called labour stabilization'.[12] Those comments suggest that the Chamber's version of the industry's history is accurate, and that both the current high disease rates and labour stabilisation are novel. What actually happened is more complex.

The industry restructuring in the 1980s did indeed result in longer contracts and greater frequency of uninterrupted service. In 1976, the proportion of black miners who had worked for more than ten years in total was around 14 per cent; by 1990 it had risen to 37 per cent.[13] However, this did not constitute a radical break with past employment practices. As noted in previous chapters, most men kept returning to the mines until their health broke down. They now tended to survive longer after they began mining. New drugs for the treatment of tuberculosis helped, as did delays in the onset—and diagnosis—of silicosis. At the 1930 ILO conference in Johannesburg, Dr Mavrogordato from the South African Institute of Medical Research explained that the onset of silicosis in white miners now took thirteen years instead of eight to nine years. 'If the time taken to produce a clinical silicosis could be pushed up to twenty years', he mused, 'silicosis could be considered as eliminated on the Rand from the social point of view'.[14]

At the same time, the mines have always preferred to employ experienced men. From the beginning of the twentieth century, South Africans were recruited for a term of six months to perform 180 shifts, which in

---

[11] M.A. Hermanus. 'Occupational Health and Safety in Mining—Status, New Developments, and Concerns'. *The Journal of the Southern African Institute of Mining and Metallurgy*, Vol. 107, August 2007, p. 534.

[12] Rodney Ehrlich, Inaugural Professorial Lecture: The body as history: on looking at the lungs of miners, 26th September 2007, University of Cape Town, p. 6.

[13] Francis Wilson, Testimony, Volume 8. Leon Commission of Inquiry into Safety and Health in the Mining Industry, at Braamfontein, Johannesburg, 1st August 1994, pp. 734–36.

[14] International Labour Office (ILO). *Records of the International Conference Held at Johannesburg 13–27 August 1930*. Studies and reports, Series F (Industrial Hygiene) No. 13 Silicosis. Geneva 1930, p. 79.

practice usually took seven and a half months. East Coasters and Tropicals were recruited for twelve months, but their average stay was about seventeen. As early as 1906, Drs L.G. Irvine and D. Macaulay wrote that it was becoming common for East Coasters to remain for eighteen months continuously on the mines.[15] Of the 36,401 East Coasters recruited in that year, almost 60 per cent were returning men. By 1912 that had risen to over 75 per cent. In his Tuberculosis Commission Report from 1914, Dr Gregory noted that the risk of both silicosis and tuberculosis increased with the duration of employment and hence was most common among East Coasters. He recommended that steps be taken to limit the total duration of employment.[16]

The Miners' Phthisis Medical Bureau's Annual Report for 1924 noted that retaining the services of experienced workers beyond the usual contract of six to nine months meant a considerable number were permanent.[17] By 1930, of the 400,000 men who went through the system each year, around 80 per cent were re-engagements.[18] In 1937, William Gemmill told the Native Labour Committee in Johannesburg that the average age of black miners was 30 to 32 years, as the mines preferred experienced men.[19] In 1940, more than 80 per cent of men arriving in the WNLA compound had been on the mines before.[20] Mr Lovett, one of the Chamber's senior officials, told a high-level conference on pensions in December 1946 that only around 12 per cent of the workforce

[15] L.G. Irvine and D. Macaulay. 'The Life-history of the Native Mine Labourer in the Transvaal'. *Journal of Hygiene*, Vol. 6, No. 2, April 1906, p. 155.

[16] 'Minority Report by Dr Gregory' in *Report Tuberculosis Commission, 1914.* U.G. 34-'14, Cape Town: Government Printer, 1914. Union of South Africa, p. 270. See also Irvine and Macaulay. 'The Life-history of the Native Mine Labourer in the Transvaal', p. 155.

[17] *Report of the Miners' Phthisis Medical Bureau for the Twelve Months ending July 31, 1924.* Pretoria: Government Printer, 1925, p. 28.

[18] Spencer Lister ed. *Tuberculosis in South African natives with special reference to the disease amongst the mine labourers on the Witwatersrand (being the report of the Tuberculosis research committee, originally established by the Transvaal chamber of mines and later expanded into a joint committee by incorporation of representatives of the Union government).* Johannesburg: South African Institute for Medical Research, no. 30, 1932, p. 74.

[19] William Gemmill and Mr Wellbeloved, Chamber of Mines. Evidence presented before the Native Labour Committee, Johannesburg 13th December 1937, p. 5. SANA K 356 Native Labour Commission 1937.

[20] *Report of the Witwatersrand Mine Native Wages Commission*, p. 15.

entering the mines each year were novices.[21] In its 1947 monograph on the health benefits of migrant labour, the Chamber noted that on average a worker would serve seven or eight contracts, with almost 80 per cent of men returning each year.[22] Writing in 1972, Francis Wilson suggested that up to the age of 40, most black miners served five to seven contracts.[23] As Jack Simons noted in 1960, claims that oscillating migration reduced the risk of silicosis was a 'dangerous illusion'. He was sure that the ignorance of the actual incidence of pneumoconiosis and tuberculosis among African miners was due to the combined failure of industry and the state to carry out basic research.[24]

The Leon Commission in 1994 was the first post-apartheid inquiry into occupational health, and the first to take detailed evidence from black miners. The Chamber's initial submission restates the companies' virtues. It runs to twenty-one chapters and covers almost every aspect of mine safety. There are at most three references to dust, and there is no mention of silicosis.[25] The Commission's final report rejected at once the Chamber's submission, its version of the industry's history and the science on which that account was founded. Leon found that silica dust levels on the gold mines were hazardous, and that they had probably been so for more than fifty years. As more men remained in almost continuous employment, it suggested, silicosis and tuberculosis were more likely to be *detected*.[26] That proved correct: over the period from 1975 to 2009, the

[21] Minutes, Johannesburg, December 1946 of Conference on Pensions for Natives -Silicosis Act held in Pretoria on Thursday, 5th December 1946, p. 11. SANA, K105, Miners' Phthisis 1941–1948, Silicosis Vol. 7, Act 1946, Natives Files.

[22] *The Native Workers on the Witwatersrand Gold Mines.* Transvaal Chamber of Mines Publication P.R.D. No. 7, 1947, p .5.

[23] Francis Wilson. *Migrant Labour in South Africa: Report to the South African Council of Churches.* Johannesburg: The South African Council of Churches and SPRO-CAS, 1972, pp. 116–117. See also V.L. Allen. *The History of the Black Mineworkers in South Africa: Vol.1, The techniques of resistance 1871–1948.* Keighley: The Moor Press, 1992, p. 343 on Shangaans from Mozambique, who in 1936 comprised almost 30 per cent of the workforce, were the longest serving of all mineworkers and came close to what he called 'career mineworkers'.

[24] H.J. Simons. Migratory Labour, Migratory Microbes. Occupational Health in the South African Mining Industry: The Formative Years 1870–1956. Unpublished manuscript, 1960, p. 34.

[25] 'Submission of the Chamber of Mines to *The Commission of Inquiry into Health and Safety*', June 1994.

[26] *Report of the Commission of Inquiry into Safety and Health in the Mining Industry.* Pretoria, Department of Minerals and Energy Affairs, 1995, pp. 47; 51–53.

reported proportion of white miners with silicosis increased from 18 per cent to 22 per cent, while that of black miners rose tenfold from 3 per cent to 32 per cent. Part of that rise was due to an ageing workforce and increasing periods of service. Part was due to the improved medical surveillance and access to compensation, which came with majority rule.[27]

The Mines Health and Safety Act No. 29 of 1996, passed in response to Leon's recommendations, ushered in a new era of non-racial legislation. In a major departure from the ODMWA, it promoted employer, employee and government participation in the workplace, based on principles of co-operation and shared responsibilities, initiatives that were unthinkable under apartheid. In addition, new codes of conduct for medical surveillance and guidelines for dust measurement were developed. Unfortunately, the complex political and economic situation at the time blunted the impact of the legislation.

J.C.A. Davies and Independent Research

Alongside legislative change, majority rule saw the formation of progressive NGOs in Cape Town and Johannesburg, a reinvigoration of union activity and the emergence of a loose collective of researchers who worked on occupational disease but had no affiliation with the mining industry. That collective, based at the NCOH and Wits University in Johannesburg and the University of Cape Town, included J.C.A. Davies, Jonny Myers, Marianne Felix, Anna Trapido, Jill Murray, Neil White, Danuta Kielkowski and Rodney Ehrlich. Their work represented a break with the past in two important respects: it involved follow-up studies of occupational and environmental risk in mining communities; and the research was conducted with independent institutional backing. In the process, both the agenda and often the models of data collection changed. The previous generation of dissidents, such as Macvicar, Cluver and Allan, were critical of the compensation system but had no collective voice and

[27] Gillian Nelson. Living in the Shadow of a Dust Cloud: Occupational Respiratory Diseases in the South African Mining Industry, 1975 to 2009. PhD thesis, Faculty of Health Sciences, University of the Witwatersrand, Johannesburg, 2012, p. vii.

limited political influence. The predecessor of the NCOH, established in 1956 for the study of occupational disease and occupational hygiene and called the Pneumoconiosis Research Unit, was the institution where new research could be best carried out. There was indeed an aborted start with J.C. Wagner's work on asbestos miners in the mid-1950s. However, his discovery of the link between asbestos and mesothelioma provoked such opposition from the mining industry that research on asbestos at the PRU was brought to an abrupt halt, and Wagner himself was forced to leave South Africa.[28] Asbestos mining continued, as did the export and use of the fibre. The appointment of Professor J.C.A. Davies as Director of the NCOH in 1983 was at once a result and a catalyst of a change in direction.

J.C.A. Davies was born in Scotland in 1931. His father, a mining engineer, migrated to South Africa in the following year. The family lived in Johannesburg and Davies attended St. Johns College. Having matriculated in 1949, he trained at Guy's Hospital in London.[29] It was a period of social reconstruction and the National Health Service had just been created. After qualifying, Davies worked at the New Cross Hospital in West Midlands with lung cancer patients and later with ex-servicemen who had tuberculosis. The new chemotherapy had revolutionised treatment, and Davies was seeing tuberculosis patients at a time when the balance in fatalities was shifting to lung cancer. In 1959, Davies with his wife Deidre and his daughter returned to South Africa. By this time, apartheid had been established and Davies found the political climate intolerable. The family soon moved to Southern Rhodesia (Zimbabwe). Davies initially worked as a rural doctor, and then three years at the small farming settlement of Shangani, where he established a successful tuberculosis control programme. In February 1963, he was recruited to the national tuberculosis service. Davies was responsible for the Midlands and Victoria provinces, where the tuberculosis rates were relatively low in a population of around half a million. Davies started a tuberculosis regis-

---

[28] Jock McCulloch. *Asbestos Blues: Labour, Capital, Physicians and the State in South Africa*. London: James Currey/Indiana University Press/Juta Press, 2002, pp. 173–175.

[29] This narrative is based on a series of interviews with Professor J.C.A. Davies at the NIOH, Constitution Hill, Johannesburg on 21st November 1998, 5th November 2014 and 7th December 2015.

ter, an initiative pioneered in Denmark. With the aid of a large sanatorium at Driefontein, the tuberculosis programme he initiated reduced the death rate by over half. The systematic approach to data collection had another important outcome: it identified an excess of men who had a history of mining. In male wards, that excess ranged from 22 per cent to 75 per cent.[30] In contrast, the rates among the general population were low. The results lent support to D.H. Shennan's work on the Manicaland province in eastern Zimbabwe. The key to controlling tuberculosis in the general population, Shennan argued, was to control the disease on the mines.[31]

In 1965, Davies completed a Diploma of Public Health at the London School of Hygiene. It was a difficult time in Zimbabwe: in protest against the illegal Smith regime, the WHO had pulled out of the country. In 1974, the Chief Medical Officer at Harare, T.J. Stamps, later a cabinet minister in the Mugabe government, resigned and Davies replaced him. When Zimbabwe gained independence in 1983, there were no jobs in the public health sector for whites. Davies was approached by the retiring Director of the Medical Bureau for Occupational Diseases (MBOD) in Johannesburg, Frank Wiles, to apply, and was the successful applicant for the position. The NCOH was a multi-disciplinary organisation with good staff, but it was not producing much needed research.

The South African Medical Research Council took over the Pneumoconiosis Research Unit in 1970 and renamed it the National Research Institute for Occupational Diseases (NRIOD). In 1979, the Institute was expanded, renamed National Centre for Occupational Health (NCOH) and transferred to the Department of Health and Population Development. Researchers in the organisation faced severe restrictions on their work. All papers had to pass through a system of censorship; when a paper was to be given overseas, permission had to be obtained from the Department of Foreign Affairs. The most public incident involved the withdrawal of a paper on asbestos disease caused by

[30] M.L. Westwater and J.C.A. Davies. 'Follow up at Eighteen Months of 742 Tuberculosis Patients Notified in Midlands and South-Eastern Province in Six Months in 1964'. *The Central African Journal of Medicine*, Vol. 13, No. 8, 1967, pp. 175–176.

[31] D.H. Shennan. 'Towards the Rout of Tuberculosis in Rhodesia: The Work of the Tuberculosis Service'. *The Central African Journal of Medicine*, Vol. 11, No. 8, 1965, pp. 220–228.

environmental exposure, which Les Irwig and Hannes Botha were scheduled to present to an international conference in New York in June 1978.[32] In addition, the organisation was hamstrung by a simmering political dispute about its role. The director, Ian Webster, was a worker advocate who wanted to focus on researching workplace issues. In contrast, the president of the Medical Research Council (MRC), A.J. Brink, wanted the Institute to concentrate on high-technology medicine.

When, in 1983, Davies took up his appointment as Director of NCOH, he had been out of South Africa for thirty-five years and, as he put it, 'apartheid made him appear more radical than he had been in Zimbabwe'. Soon, he began playing an important role in shielding junior researchers from departmental interference. In his first annual report as Director, he announced that the Institute would address the major industrial hazards of silicosis, asbestos disease and hearing loss. Davies noted that hospital out-patient departments and private medical practitioners were not familiar with the compensation legislation and with lodging claims. Partly as a result, there was gross under reporting of pulmonary tuberculosis among miners in labour-sending communities, with most cases being shipped off to the Bantustans. Even under those circumstances, the existing data was disturbing. Among the lungs of 1663 black gold miners examined during 1982, 6.9 per cent were shown to have active pulmonary tuberculosis. The incidence of tuberculosis among in-service gold miners in the same year was 800 per 100,000. This was 80 times that of the USA and Scandinavian rates.[33]

Davies arrived at the NCOH with an important research question: What happened to miners after they left the mines? The prevailing orthodoxy at the NCOH was that black miners got tuberculosis while whites developed silicosis because of their continuous exposure. From his Southern Rhodesia experience, Davies knew that both groups contracted

---

[32] See Laurie Flynn, *New Scientist*, 22nd April 1982, pp. 237–239, and response by A.J. Brink, *New Scientist*, 26th May 1983, p. 575.

[33] Churchyard, G.J. and E.L. Corbett. 'Tuberculosis and Associated Diseases'. In R. Guild, R.I. Ehrlich, J.R. Johnson and M.H. Ross (eds). *Handbook on Occupational Health Practice in the South African Mining Industry.* Safety in Mines Research Advisory Committee (SIMRAC), 2001, p. 156; Department of Health and Welfare, *Annual Report on the Work of the National Centre for Occupational Health*, 1983, p. 13.

both diseases, although at different rates.[34] One of the first projects Davies sponsored was Shelley Arkles' work on spinal injuries among migrant workers from Lesotho. It was a study of the hidden costs of mining of a kind which had not been attempted since the work of Peter Allan in 1922.[35] While not dealing with it directly, the results were relevant to miners suffering from occupational lung disease. Ninety per cent of Arkles' sample of 64 repatriated miners had never been employed outside the mines. The situation of a disabled worker, Arkles noted, should not be compared to that of unemployed members of impoverished rural communities, but rather to the position such men occupied before their accident.[36] The majority of Arkles' respondents were married and had the benefit of family support. Almost 70 per cent had five or more dependents and despite their injuries they felt responsible for the maintenance of their families.

Arkles found that former miners were severely disadvantaged in an over-supplied labour market. A small disability, such as a reduction in lung function, could result in a total loss of employment. Almost 70 per cent of Arkles' study group were unemployed, a rate which appeared typical for disabled miners. 'Patients who are disabled and unemployed become disabled and unemployable purely from the passage of time.' Low educational levels had a negative impact on the miners' understanding of their rights to compensation. Within three months of repatriation, most men had spent their lump sum payouts on food, household goods and school fees. A quarter of the respondents spent between 11 and 22 per cent of their pensions on transport each month to collect the payments. In the absence of unemployment benefits, redundancy payments and adequate compensation, rural households were responsible for the care of injured men.[37] Arkles concluded that the mining industry should provide vocational training and assistance programmes for workers with serious disabilities. Her recommendations were ignored.

[34] Interview with J.C.A. Davies, 5th November 2014.

[35] R.S. Arkles, A. J. Weston, L. L. Malekela and M. H. Steinberg. *The Social Consequences of Industrial Accidents: Disabled Mine Workers in Lesotho.* National Centre for Occupational Health. NCOH Report Number 13/1990.

[36] Arkles et al., *The Social Consequences of Industrial Accidents,* pp. 26; 75.

[37] Arkles et al., *The Social Consequences of Industrial Accidents,* pp. 62; 56; 42; 67.

Arkles' project was followed by the work of Marianne Felix on the environmental impact of asbestos mining at Mafefe in the Limpopo province. Such a study had never been conducted, and Felix identified a serious environmental hazard in mining communities.[38] At about the same time, J-P. Leger published an important article which discussed, in considerable detail, the neglected epidemic of occupational disease among black miners, and the difficulties he encountered in reconciling the disparate and patchy available data sets.[39] In addition to sponsoring the work of Arkles and Felix, Professor Davies began his own research on asbestos and gold miners. In 1991 he visited several hospitals in the Eastern Transvaal, a region which supplied labour to the local asbestos mines and migrant workers to the gold mines. Large numbers of women worked as cobblers, hand processing the ore before it was fed into the mills. They had extreme exposure and high rates of asbestos disease, including mesothelioma. The hospital at Groothoek had good X-ray equipment, but Davies found the hospital staff were unaware of their obligations under the Act to report chest disease in miners. Nobody could recall the last case reported under the ODMWA. Neither were they submitting the organs of deceased miners for the purposes of compensation. Apparently, neither the state authorities nor the mining companies had noticed that over the preceding decades no organs had been submitted.[40] After leaving the industry, white miners and their families used the free benefit examinations diligently to access compensation. Black miners too had the *legal right* to free benefit medical examinations every two years, but *in practice* very few had access to the doctors at the MBOD (and autopsies on death) in Johannesburg.

Davies returned to the Eastern Transvaal in 1993 and began holding clinics for former miners. Initially, he examined 72 cases of lung disease, of which two-thirds were compensated. Soon, men and women began

---

[38] Maria Anne Felix. Environmental Asbestos and Respiratory Disease in South Africa. PhD, University of the Witwatersrand, 1997.

[39] J-P. Leger. 'Occupational Disease in South African Mines—a Neglected Epidemic'. *South African Medical Journal*, Vol. 81, 15th February 1992, pp. 197–201. See also J.C.A. Davies. 'A String of Studies: The Pathway to the Courts'. *Occupational Health Southern Africa*, Vol. 21, No. 1, 2015, p. 18.

[40] J.C.A. Davies. 'Sound an Alarm'. *The South African Medical Journal*, Vol. 84, No. 3, 3rd March 1994, pp. 133–134.

pouring into the clinic to register. Between November 1991 and June 1993, Davies submitted a total of 399 cases of former asbestos and gold miners for compensation. Of that group, 272 received awards for pneumoconiosis and/or tuberculosis. It was compelling evidence that the state-regulated system of medical surveillance and compensation was not working. Davies warned: 'There is almost certainly a large unfunded liability in the rural areas of the subcontinent from which migrants have been recruited'.[41] Davies reported on the results of his work in 1994, just prior to the final report from the Leon Commission on which he served, and almost ten years before litigation against the gold mining companies began.[42]

The Leon Commission put criticism of the compensation system firmly in the public domain and inspired further pathbreaking research. One such project emerged from a chance conversation at a health conference in Maputo in 1994. Two doctors from Botswana were keen to conduct a follow-up study of migrant workers. They approached Neil White and his colleagues Jonny Myers and Rodney Ehrlich from the UCT. The resulting study at Thamaga in Botswana showed massive rates of uncompensated silicosis and tuberculosis among former miners. The study of just over 300 former miners was based on individual questionnaires, chest radiographs, spirometry and medical examinations. The mean age of men in the sample was 56 and their mean duration of mine service 15 years.[43] On average, the men had worked on four different mines. The vast majority laboured in gold mines; the remainder in platinum, asbestos, chrome and diamond mines. Almost half now classified themselves as unemployed rather than retired.

As Steen notes, South Africa's mines have always been dangerous. An eighteen-year-old starting a mining career had a one-in-two to

[41] Davies, 'Sound an Alarm', p. 134.

[42] A. Trapido, N. Mqoqi, C. Macheke and B. Williams. 'Occupational Lung Disease in Ex-Mineworkers—sound a further alarm!'. *The South African Medical Journal,* Vol. 86, No. 5, May 1996, p. 559 and the reply by M.A.C. Le Grange. 'Occupational Lung Disease in Ex-Mineworkers'. *The South African Medical Journal,* Vol. 86, No. 7, July 1996, p. 841.

[43] T.W. Steen, K.M. Gyi, N.W. White, T. Gabosianelwe, S. Ludick, G. N. Mazonde, N. Mabongo, M. Ncube, N. Monare, R. Ehrlich and G. Schierhout. 'Prevalence of Occupational Lung Disease Among Botswana Men Formerly Employed in The South African Mining Industry'. *Occupational & Environmental Medicine,* Vol. 54, No. 1, 1997, pp. 19–26.

one-in-three chance of being permanently disabled by accident or disease. The group Steen and his colleagues studied conformed to this pattern: 190 of the 300 reported a period of hospitalisation or sick leave while on the mines. In all, a quarter of the Thamaga group had a history of tuberculosis, while 23 per cent had experienced a disabling occupational injury. The prevalence of pneumoconiosis was 26 to 31 per cent, and 7 per cent had progressive massive fibrosis, a crippling disease associated with reduced life expectancy. Many of the participants were entitled to compensation under South African law but had not received an award. If the 40 miners in the group with second-grade pneumoconiosis had received compensation, more than R1.2 million would have been remitted to Thamaga. Without awards, the social costs of disability were borne by families and the Botswana administration.[44]

The survey pointed to a failure by employers and the South African authorities to prevent and identify pneumoconiosis. The findings also confirmed the existence of a latent period between exposure and radiological abnormalities, which is a well-documented feature of silica dust exposure. Although pneumoconiosis occurred in some miners with less than five years' service, most commonly it was not detected until a decade after initial exposure. The study confirmed that, in the absence of follow-up examinations, pneumoconiosis was systematically under-reported, as the disease progressed after exposure has ended. Of the 80 Thamaga men who had previously received treatment for tuberculosis, 56 had been treated in Botswana and just 24 in South Africa. Only ten had received an award: tuberculosis could only be compensated under the Occupational Diseases in Mines and Works Act (ODMWA) if it occurred during service or within a year of leaving the mines. However, over half of the men with a history of the disease reported developing tuberculosis in their final year of mining, suggesting that they were repatriated after being diagnosed but were not compensated. Four new cases of pulmonary tuberculosis were also identified during the survey. Steen found that former miners in Botswana had a high prevalence of previously unrecognised pneumoconiosis due to inadequate radiographic surveillance or a failure by employers to act on X-ray results.

[44] Steen et al., 'Prevalence of occupational lung disease'.

Steen's study was soon followed by the work of Anna Trapido at Libode in the Eastern Cape, which identified a similar burden of disease.[45] Trapido's work originated in conversations between Tony Davies and a Wits epidemiologist Brian Williams. Both wanted to know what happened to black miners after they left the industry. Williams recruited Anna Trapido as a doctoral student and helped her find a sample which would provide a solid basis for investigating whether ex-gold miners had an excess rate of tuberculosis. Trapido began her research in Lesotho. She then visited another recruiting area for the mines, Libode in the Eastern Cape. The local TEBA officer was sympathetic and gave her access to miners' records. There were 11,000 recruits in the files and Trapido took a random sample of 500 men. Davies was staggered by the rates of uncompensated lung disease she uncovered.[46] After some delay, the work by Steen and Trapido was confirmed by a study of silicosis, tuberculosis and chronic airways disease in a group of 624 South African gold miners led by Gavin Churchyard and funded by AngloGold Corporation. The mean age of the subjects was 49 years, and mean period of employment duration 26 years. The study found silicosis in 24.6 per cent, past tuberculosis in 26 per cent, current tuberculosis in 6 per cent and airflow obstruction in 13 per cent of the men. In total, almost half of the miners had at least one of these respiratory conditions.[47]

In response to new research showing a high incidence of uncompensated silicosis and tuberculosis in former miners, the Chamber commissioned what it termed 'a proper study' of the prevalence of silicosis among miners retrenched from the President Steyn mine at Welkom. The project used a sample of 520 miners aged over 37 years. Less than half had a normal chest X-ray, and almost one in five had evidence of silicosis. The prevalence of silicosis increased significantly with increased length of

[45] Anna Trapido. An Analysis of the Burden of Occupational Lung Disease in a Random Sample of Former Gold Mineworkers in the Libode District of the Eastern Cape. PhD thesis, University of the Witwatersrand, 2000. See also Jock McCulloch. 'Counting the Cost: Gold Mining and Occupational Disease'. *African Affairs*, Vol. 108, No. 431, 2009, pp. 221–240.

[46] Interview with J.A.C. Davies, 5th November 2014.

[47] Brendan V. Girdler-Brown, N.W. White, R.I. Ehrlich and G.J. Churchyard. 'The Burden of Silicosis, Pulmonary Tuberculosis and COPD Among Former Basotho Goldminers'. *American Journal of Industrial Medicine*, Vol. 51, No. 9, 2008, pp. 640–647.

service.[48] In a parallel study of ex-miners from the same mine, Churchyard and his colleagues also found that HIV infection increased the incidence of tuberculosis fivefold, and silicosis increased the incidence of tuberculosis threefold. The presence of both HIV and silicosis increased the incidence of tuberculosis by fifteen times, in a so-called multiplicative interaction. The President Steyn sample suggested that even a negligible degree of silicosis was associated with an increased risk of tuberculosis.[49] The results, published several years after the initial data collection was completed, confirmed the reports from Thamaga and Libode, and came to feature prominently in the miners' subsequent litigation.

Similar findings emerged in Lesotho. In 2005, Lugemba Budiaki completed the first cross-sectional study of current and former Basuto miners attending tuberculosis clinics at Maseru's hospitals. The aim of the study of 421 adult male patients was to determine the proportion diagnosed with tuberculosis who were mineworkers, and to identify those who had been compensated. Six out of ten of the men completed their primary education, more than two-thirds were literate and just under 10 per cent had attended secondary school. Most had spent prolonged periods on the mines.[50] Of the participants, only 4 per cent were active miners. The unemployment rate among the cohort was very high, and this made access to compensation imperative for family survival. Close to 40 per cent of the 421 participants diagnosed with tuberculosis at Maseru clinics were former or current mineworkers. Of the group studied, 42 men had received compensation for tuberculosis, but 33 miners diagnosed in Maseru with compensable disease had not been compensated. Only 23 of the total had been diagnosed while on the mines before being repatriated

[48] Gavin Churchyard, Lindiwe Pemba, Bulewa Magadla, Kobus Dekker, Mark Vermeijs, Rodney Ehrlich, Jim te Water Naude and Jonny Myers. 'Silicosis Prevalence and Exposure Response Relationships in Older Black Mineworkers on a South African Goldmine'. Safety in Mines Research Advisory Committee, Final Report. *Health 606*, May 2003, pp. 20; 25.

[49] H.H. Park, B.V. Girdler-Brown, G.J. Churchyard, N.W. White and R.I. Ehrlich. 'Incidence of Tuberculosis and HIV and Progression of Silicosis and Lung Function Impairment Among Former Basotho Gold Miners'. *American Journal of Industrial Medicine*, Vol. 52, No. 12, Dec. 2009, pp. 901–908.

[50] L. Budiaki, Tuberculosis and Compensation: A Study of a Selection of Basotho Mineworkers from Maseru District, Master of Public Health thesis, University of the Witwatersrand, Johannesburg, 2005, pp. 36–37.

to Lesotho for treatment. Almost 80 per cent of the cohort had been diagnosed in Lesotho. Just over half had been given an exit medical examination, and almost 70 per cent were diagnosed with tuberculosis more than a year after leaving the mines, making them ineligible for compensation.[51] The men cited various reasons why they had not applied for compensation. Some said that they were not aware they had such a right, but the majority reported that as they were diagnosed long after leaving the mines, they could not make a claim.[52]

Research by Thuso Tlhaole reiterated that most disabled mineworkers lived with the threat of poverty.[53] The amounts successful claimants received were not adequate to support a family. Retired miners used their compensation on essential household items such as food, electricity and fuel. They also relied heavily on relatives and friends for support. Budiaki recommended that the occupational health unit in the Ministry of Health in Lesotho be strengthened, and that miners be made aware of their rights to compensation prior to taking up employment. Improved data collection and research were needed to identify occupational groups at risk. Any mineworker suspected or diagnosed with tuberculosis should have a sputum test and an X-ray examination. Those diagnosed with tuberculosis should be educated about how to manage the disease. The results also suggested the need for urgent reform to the mines' system of medical examinations, and for TEBA to be actively involved in the compensation process.

Budiaki's study complemented the work of Anna Trapido and Jaine Roberts on the Eastern Cape. All three researchers documented the extensive hardship suffered by miners and their families, and the high rates of uncompensated disease. After more than two decades of majority rule in South Africa, Budiaki concluded, there was still a lack of information among migrant workers about the risks of mining and their rights to

[51] Budiaki, Tuberculosis and Compensation, pp. 36; 25; 29.

[52] The miners could have lodged a claim if silicosis was also present, but they were not aware of this possibility and so did not seek later examination. Rodney Ehrlich, personal communication, 2 October 2019.

[53] Thuso Tlhaole, Social Protection Arrangements for Retired Mineworkers with Physical Disabilities in Maseru Urban, Lesotho. MA Thesis, University of Witwatersrand, Humanities Faculty, 2011.

compensation. Where compensation claims were made, the processing time varied between three months and three years.[54] In addition, some mineworkers had injuries such as loss of limbs or fingers for which they had not received compensation.

The Weak Link: Litigation Against the Asbestos Industry

The work by researchers such as Davies, Felix, Steen and Trapido shattered the orthodoxy that miners were at little risk of contracting lung disease. However, their research had no immediate impact on work practices or state regulation. What *did* bring change was the election of the ANC government in 1994. The new political conditions saw a dramatic shift in the legal possibilities for the victims of occupational injury. That shift was first evident in the successful litigation by asbestos miners.

The British companies Cape Plc and Turner & Newall mined asbestos in South Africa for almost a century. When the mines began closing in the mid-1980s, they left behind large numbers of men and women in the Northern Cape and the Northern Province with occupational disease.[55] Extensive lobbying by community groups eventually led to a claim against Cape Plc. The case opened in a London court in February 1997. The plaintiffs, represented by the British lawyer Richard Meeran, sought to hold Cape liable for its failure to provide a duty of care for its employees. Over the next three years, the legal process was devoted to the question of jurisdiction. Cape wanted the case heard in South Africa where the injuries had taken place and where the plaintiffs lived. The defence argued for the UK, where Cape's assets were held. On appeal, the Law Lords finally ruled in favour of the plaintiffs. The issue of a duty of care was then scheduled to be heard in London. While the Cape case was running, a second set of claims was lodged in South Africa against the corpo-

[54] Budiaki, Tuberculosis and Compensation, p. 38.

[55] J. E. Roberts. What is the Price of 80 KG's: The Failure of the Detection of, and Compensation for, Asbestos-related Disease: Social Exclusion in Sekhukhuneland. MA thesis University of Natal, December 2000.

rate giant Gencor by the civil rights lawyer Richard Spoor on behalf of more than 4000 men and women who had worked for Gefco, a fully owned subsidiary of Gencor. It was the first time that such a claim had been brought in South Africa for injuries sustained in the mining industry.

The asbestos and gold mining stories share much common ground. They are connected by the careers of the leading medical researchers in the UK and the US, who from the 1930s worked on both silicosis and asbestosis. They are connected by US corporations such as Johns Manville and Union Carbide, which over decades faced litigation from employees suffering from those diseases. Within South Africa, the connections are more intimate. The Anglo American Corporation and its associate De Beers are vast enterprises. By 1957, they controlled 40 per cent of South Africa's gold production, 80 per cent of the word's diamonds and a sixth of its copper. They also produced most of South Africa's coal. From the 1930s, Anglo American developed a maze of interlocking directorships, mutual agreements and restrictive trade practices. That complex structure, which is characteristic of the South African mining houses, makes identifying the ownership and control of subsidiaries difficult.[56]

Cape Asbestos Plc was a British firm with its head office in London. Yet, from its foundation, it was linked with De Beers. It was also related to Anglo American. During the Second World War the Oppenheimer company, Central Mining & Investment Corp., became the major shareholder in Cape. Central Mining was one of four holding companies within the Oppenheimer group, the others being Anglo American, De Beers and Rand Selection. By 1949, Central held the majority of seats on the Cape board. In 1969, the Central holding in Cape was superseded by Charter Consolidated, a British mining company also controlled by Oppenheimer interests, which held 63 per cent of Cape's shares.[57] In 1979, Cape's mines were sold for £15.5 million to Barlow Rand, an Anglo American company. They were then sold on again to General Mining, yet another Oppenheimer enterprise. Charter's controlling share

[56] Duncan Innes. *Anglo-American and the Rise of Modern South Africa.* London: Monthly Review Press, 1984.

[57] Geoffrey Tweedale and Laurie Flynn. 'Piercing the Corporate Veil: Cape Industries and Multinational Corporate Liability for a Toxic Hazard, 1950–2005'. *Enterprise and Society,* 8th June 2007, p. 7.

in Cape gave Anglo American a commercial interest in the asbestos industry. It also gave Anglo American's board reason to monitor the Cape litigation.

The links between the asbestos and silicosis crises may in part explain why, despite the public health costs, the Chamber, the Departments of Mines and Health, as well as successive national governments fought so hard during the 1970s and 1980s to save the asbestos mines from their own occupational health crisis.[58] Asbestos was an insignificant industry which at its peak in 1977 employed less than 40,000 men and women and accounted for only 3 per cent of the value of South Africa's minerals production. It is possible that the Chamber feared that recognition of asbestos disease would lead directly to recognition of uncompensated disease in gold miners. In the event, this is exactly what happened.

In the first two decades of the twentieth century, health crises on the gold mines saw the state, industry and white labour achieve an accommodation which was maintained for almost a century. That fragile agreement was constantly renegotiated through a series of Acts and Commissions which often led to concessions for white miners and their families. The successful litigation against Cape Plc and Gencor destroyed that precarious balance. Importantly, it forged a pathway to class action, an unprecedented form of redress allowable under the post-apartheid South African constitution.

From the 1930s, access to information was the key to the politics of asbestos. The industry in the US, the UK and elsewhere captured and re-framed the knowledge of risk through a variety of techniques. It suppressed or hid evidence of hazards, initially among miners and factory workers and later among those with bystander exposure.[59] It manufactured doubt by falsely claiming there was insufficient data to justify a ban. As that defence fell apart during the 1960s, the asbestos corporations secretly paid leading scientists to corrupt the legitimate research of their colleagues, thereby frustrating more stringent regulation of asbestos

[58] Jock McCulloch. 'Asbestos, Lies and the State: Occupational Disease and South African Science'. *African Studies*, Vol. 64, No. 2, 2005, pp. 201–216.

[59] Jock McCulloch and Geoffrey Tweedale. *Defending the Indefensible: The Global Asbestos Industry and its Fight for Survival.* Oxford: Oxford University Press, 2008.

mining and manufacture.[60] Widespread opposition to apartheid and the deplorable work and living conditions in the Northern Cape made South Africa's asbestos mines a point of fragility for the global industry. Mills were often located in the centre of small towns like Prieska, and residents were contracting asbestosis, a disease usually confined to filthy workplaces, from environmental exposure. Those towns were isolated and there were very few white miners and no trade union presence. There was minimal state regulation and therefore no data until 1954, when the asbestos mines were declared Registered Mines under Section 29 of the Silicosis Act.[61]

Asbestos had been mined in South Africa for almost forty years before any research was carried out into asbestos disease.[62] In 1930, Dr George Slade, who worked as a medical officer at a British owned mine in the Transvaal, completed what was probably the first study of respiratory disease in asbestos miners.[63] Slade's research showed that the mills were unsafe and there was widespread disease. His report brought no legislative response, but it did bring his career as a mine medical officer to an abrupt end.[64]

In 1954, after he was appointed to a fellowship at the Pneumoconiosis Research Unit (PRU), J.C. Wagner set out to establish whether asbestos mining in South Africa caused asbestosis and lung cancer. The project was novel, but Wagner received little institutional encouragement. Wagner soon discovered a large number of cases of mesothelioma in the Northern Cape and identified an association between the disease and asbestos exposure. In contrast to previous studies, he found that even slight exposure could produce tumours many years later, and that the risk

---

[60] Jock McCulloch. 'Saving the Asbestos Industry, 1960 to 2003'. *Public Health Reports*, Vol. 121, No. 5, 2006, pp. 609–614.

[61] P.H.R. Snyman. 'Safety and Health in the Northern Blue Asbestos Belt'. *Historia,* Vol. 33, No. 1, May 1988, pp. 31–52.

[62] F.W. Simson. 'Pulmonary Asbestosis in South Africa'. *The British Medical Journal*, Vol. 1, No. 3516, 26th May 1928, pp. 885–887.

[63] G.F. Slade. The Incidence of Respiratory Disability in Workers Employed in Asbestos Mining with Special Reference to the Type of Disability Caused by the Inhalation of Asbestos Dust, MD thesis, University of the Witwatersrand, 1930.

[64] Jock McCulloch. *Asbestos Blues: Labour, Capital, Physicians and the State in South Africa*. London: James Currey/Indiana University Press/Juta Press, 2002, pp. 121–122.

extended beyond the workplace. Part of the significance of Wagner's discovery was the fact that it broke down the barrier usually separating occupational from environmental injury.

When Wagner visited Europe on sabbatical in 1956, he was ordered by the Director of the PRU, A.J. Orenstein, not to mention the discovery of mesothelioma and its possible association with asbestos. Wagner ignored that instruction and discussed the matter with directors of the two major asbestos companies in Britain: T&N and Cape Plc. The directors dismissed Wagner's research as 'worthless' and suggested he follow a different line of inquiry. After he presented a paper on his findings at the international conference on Pneumoconiosis in Johannesburg in 1959, the Department of Mines commissioned the PRU to coordinate a survey of the Northern Cape.[65] The project was funded by the asbestos industry and supported by a small grant from the South African Cancer Association.[66] There was a great deal at stake for the British owned companies. If a comprehensive survey of the Northern Cape had been completed in the mid-1960s, the mining of crocidolite could well have ended, sending the global asbestos industry into a decline from which it may never have recovered.

The PRU survey of just over 2000 residents from the towns of Prieska, Koegas, Kuruman and Penge was carried out between November 1960 and February 1962. Because of the rarity of mesothelioma and the small sample size, the PRU did not expect to find any new cases.[67] By February 1961, the initial returns showed catastrophic living and work conditions in Prieska. The X-rays of adult residents, only a minority of whom had worked in the mines, identified a hazard for every person who lived in those towns. The study also revealed several new cases of mesothelioma, a remarkable result in such a small population. The authors of the report reached two conclusions: firstly, that people who lived or had lived in

[65] Letter from J.C. Wagner, Weymouth, Dorset, 26th November 1994 to Andrew Saxton, Ebsworth & Ebsworth, Solicitors, King Street, Sydney (in the possession of the author).

[66] *Report on the Progress of Mesothelioma Survey as at 30th April 1962.* Pneumoconiosis Research Unit, South African Council for Scientific and Industrial Research, unpublished, p. 2. NCOH Papers.

[67] Pneumoconiosis Research Unit. *Fifth Annual Report for the Year ending 31st March 1961.* South African Council for Scientific and Industrial Research, Johannesburg.

those towns were 'in danger of contracting asbestosis even though they have had no industrial exposure to asbestos dust inhalation'. Secondly 'there is an alarmingly high number of cases of mesothelioma with evidence that the condition is associated with an exposure to asbestos dust which need not be industrial'.[68]

The Northern Cape Asbestos Producers Advisory Committee attacked the survey and all funding by industry and the South African Cancer Association ended immediately.[69] The new PRU Director, Dr L.G. Walters, noted that three factors had brought the survey to an end: the reaction by certain members of parliament and a subsequent scare campaign in the press, the asbestos companies' concern about being able to recruit labour and their refusal to support further research.[70] The final report, which was completed in 1964, runs to sixty pages and contains just twelve lines on mesothelioma.[71] The early data and allied results were suppressed and subsequent researchers at the PRU were unaware of their existence.[72] It was as if those reports had never existed. Over the next fifteen years, asbestos production continued to rise, but no further research was done into the impact of asbestos on human health. There was no disclosure regarding the risk to mining communities, to consumers or to the general public. South Africa continued to mine and export crocidolite until 1996. That suppression bought the asbestos mines thirty years' grace. It may well have done the same for the gold mines.

In March 2003, Cape Plc. and Gencor reached out of court settlements in London and Johannesburg with plaintiffs seeking compensation for cancers caused by asbestos mining.[73] The Gencor case was the

---

[68] *Report on the Progress of Mesothelioma Survey*, 1962, pp. 2; 4.

[69] Pneumoconiosis Research Unit, *Fifth Annual Report for The Year Ending 31st March 1963*. South African Council for Scientific and Industrial Research, Johannesburg.

[70] Pneumoconiosis Research Unit, *Fifth Annual Report for the Year ending 31st March 1962.* South African Council for Scientific and Industrial Research, Johannesburg.

[71] 'Field Survey in the North-West Cape and at Penge.' *Transvaal Asbestosis and Mesothelioma Pneumoconiosis Research Unit Report No 1/64*. South African Council for Scientific and Industrial Research, Johannesburg, 1964, p. 3.

[72] Interviews with Dr Bill Harrison at Killara, Sydney, 15th October 1999, and with Dr Jennifer Talent at Fish Hoek, South Africa, 16th December 1999.

[73] Jock McCulloch. 'Beating the Odds: The Quest for Justice by South African Asbestos Mining Communities'. *The Review of African Political Economy*, Vol. 32, No. 103, 2005, pp. 63–77.

first time that South African miners had won a class action for workplace injuries. The London plaintiffs had breached the 'corporate veil'—in other words, pierced that aspect of limited liability law which gives parent companies legal protection against the liabilities of subsidiaries. Cape's lawyers had assumed that this veil would protect the company from its South African litigants. Documents which spilled into the public domain revealed that the South African Departments of Mines and Health had been aware for decades of the hazardous conditions on the mines but did nothing.[74] In effect, the degree of collusion between state agencies and the industry was so extreme that it threw into question the integrity of the regulatory system itself. The day after the R450 million Gencor Agreement was signed, the civil rights lawyer Richard Spoor announced he would turn his attention to the plight of gold miners suffering from silicosis.

The Gold Miners' Class Action

In 2013, Abrahams Kiewitz Attorneys, Richard Spoor Attorneys and the Legal Resources Centre separately commenced class actions against the gold mining companies on behalf of miners suffering from silicosis and/or tuberculosis. Three years later, and against strenuous opposition from the mining companies, the South African High Court approved the request to amalgamate the three separate actions into one. Importantly, the court also amended the common law regarding what is called 'the transmissibility of damages' so it was in line with South Africa's post-apartheid constitution, and in particular its clauses regarding human rights. In brief, the changes allow for the wives and other dependents of miners taking part in the class action to inherit damages awards in case the men die before the court process is completed.[75] This change is particularly important in terms of gender justice. The burden of care for men suffering from occupational lung disease is usually provided at home by

---

[74] McCulloch, 'Asbestos, Lies and the State'.

[75] Republic of South Africa, The High Court of South Africa, Gauteng Local Division, Consolidated Case Number 48226/12, 13 May 2016.

wives and daughters. It is they who lift, carry and bathe the disabled mineworkers, monitor their medications, accompany them to appointments and stay up at night to attend to their needs. By the same token, the intensity of care work means that they are not able to take on income-generating activities or participate in education. Any damages awarded post-mortem to the men they helped care for indirectly compensates them for the unpaid work they performed.[76] In May 2018, the parties to the dispute reached an historic agreement to settle the Silicosis and Tuberculosis Class Actions. The settlement was confirmed by the South African High Court in July 2019.[77]

By the time it was settled, the scope and magnitude of the amalgamated class action and the range of legal representatives involved had been unprecedented in South Africa. Indeed, the case is rated as one of the most complex multi-party class action settlements in the world.[78] The defendant mining companies represent almost the entire gold mining industry in South Africa.[79] The settlement negotiations involved companies with different approaches, needs, interests, financial capacities, risk profiles and cultures, who deal with unique geological circumstances. The companies are also competitors and potential adversaries in future litigation. The potential number of those meeting the disease eligibility requirement of the claim may range from seventeen thousand to half a million. The number of dependant applicants is expected to be in the tens of thousands.[80]

Besides its scope and complexity, the settlement agreement is highly innovative in terms of the remedies it offers to miners and their dependents. The use of class action, rather than the pursuit of individual claims to compensation, the mineworkers' legal team forcefully argued, was the only realistic avenue available to the workers in their search for justice. All

---

[76] Tracy-Lynn Field. 'Exacting Silicosis Justice Through the Class Action Mechanism'. *Mineral Economics,* Vol. 32, 2019, p. 218.

[77] For another detailed consideration of the class action and its antecedents, see Jason Brickhill. 'A River of Disease: Silicosis and the Future of Class Actions in South Africa'. *South African Journal on Human Rights,* Vol. 37, No. 1, 2021, pp. 31–58.

[78] Republic of South Africa, The High Court of South Africa, 2019, 1/p. 36.

[79] Republic of South Africa, The High Court of South Africa, 2016, p. 11 #7.

[80] Republic of South Africa, The High Court of South Africa, 2019, 1/pp. 35–36 #57, #58

the mining companies, they noted, were accused of having committed the same wrongs; it was neither economical, nor in the case of any individual mineworker affordable, to bring such evidence to trial action were he to sue in his individual capacity. Neither was it a good use of the country's court system. As Richard Spoor put it:

> [T]he majority of the class members are impoverished rural people, many of whom are in poor health, who are spread across the sub-continent and who have very limited access to the civil justice system. The very large proportion of class members who were migrant workers from Mozambique, Malawi, Lesotho and Swaziland [eSwatini], probably have no access to the South African justice system at all. ... Litigating on behalf of claimants located in remote rural areas and in neighbouring countries is particularly difficult. ... In many instances letters and notices must be delivered by hand, travel to and from these remote areas is slow, expensive and often unreliable. There are few if any local correspondent attorneys to rely upon and either the attorney must travel to see the client or vice versa. A simple matter such as arranging for a medical examination can take days to organize and involve claimants travelling hundreds of kilometres.[81]

The mining companies did not dispute the miners' difficult circumstances, but strenuously opposed amalgamating their claims for compensation into class action. The mineworkers through their counsel complained about being stonewalled without relent by the mining companies from the beginning and all the way through the litigation. The companies, they alleged, placed every possible obstacle to having the matter adjudicated and fought the application as vigorously and as aggressively as they possibly could. They spared neither effort nor resources in doing so. They have done this, even though they could not dispute the fact that, should the court refuse the class action to go ahead, and the mineworkers be forced to bring individual actions, the result without doubt would sterilise most of the individual claims.[82] In approving the 2018 settlement, Justices of the South African High Court singled out the potential costs and time of adjudicating individual claims as one of

[81] Republic of South Africa, The High Court of South Africa, 2016, p. 67.

[82] Republic of South Africa, The High Court of South Africa, 2016, pp. 127–128.

the key reasons for their decision. The alternative, they noted, involved hundreds or thousands of cases being brought, potentially in parallel to one another, each involving complex issues of prescription, negligence, causation and apportionment of damages. The costs and time associated with defending such cases would be astronomical. During the decade that the mammoth litigation would take, moreover, many of the claimants would die, and some of the mining companies become insolvent.[83]

The mechanisms for proving injury and receiving compensation constitute the other major innovation of the historic settlement. There is a gulf between the 'relaxed proof requirements for eligibility for compensation' it established and those which would be required under existing law if the cases went to trial.[84] For instance, the dependants of silicosis sufferers are not required to prove that silicosis is the cause of death of the breadwinner. It suffices if the person was certified to be suffering from Silicosis Class 2 or Class 3 before their death. Another innovation concerns tuberculosis claimants, who need not establish that they contracted tuberculosis as a result of exposure to dust, but only that they have contracted tuberculosis and have the requisite employment history on the gold mines. A fair proportion of these persons who will be eligible to receive a benefit under the settlement, the court noted, would not have been able to prove their claims and would not have been able to recover anything at the conclusion of trials under the existing provisions.[85] In addition, the settling companies and the class representatives agreed to adopt a wider definition of risk work so that a greater number of claimants may be eligible than was catered for during the certification proceedings. Previously risk work was restricted to underground mineworkers and their dependants; the new definition also includes certain surface work where there is potential exposure to excessive silica dust.[86] Finally, rather than placing obstacles in the way of those seeking compensation, the settlement fund provides for the tracing and tracking of class members, free advice, registration, medical examinations and where necessary

[83] Republic of South Africa, The High Court of South Africa, 2019, 1/p. 39 #61; pp. 51–52 #85.
[84] Republic of South Africa, The High Court of South Africa, 2019, 1/p. 52 #88.
[85] Republic of South Africa, The High Court of South Africa, 2019, 1/p. 32 #52.
[86] Republic of South Africa, The High Court of South Africa, 2019, 2/p. 14 #21.

travel. In all, the settlement has toppled some of the key pillars upon which employment in the South African gold mining industry was built. Its provisions not only represent a dramatic change in the legal rights of mine workers and their families throughout the region but have considerable symbolic importance as part of South Africa's transformation from minority rule. Securing the rights of migrant workers has been a highly significant part of that transformation.[87]

References

Allen, V.L. *The History of the Black Mineworkers in South Africa: Vol.1, The techniques of resistance 1871–1948*. Keighley: The Moor Press, 1992.

Arkles, R.S., A.J. Weston, L.L. Malekela and M.H. Steinberg. *The Social Consequences of Industrial Accidents: Disabled Mine Workers in Lesotho.* National Centre for Occupational Health. NC0H Report Number 13/1990.

Brickhill, Jason. 'A River of Disease: Silicosis and the Future of Class Actions in South Africa'. *South African Journal on Human Rights*, Vol. 37, No.1, 2021, pp. 31–58.

Budiaki, L. Tuberculosis and Compensation: A Study of a Selection of Basotho Mineworkers from Maseru District, Master of Public Health thesis, University of the Witwatersrand, Johannesburg, 2005.

Churchyard, G.J. and E.L. Corbett. 'Tuberculosis and Associated Diseases'. In R. Guild, R.I. Ehrlich, J.R. Johnston and M.H. Ross (eds). *Handbook on Occupational Health Practice in the South African Mining Industry.* Safety in Mines Research Advisory Committee (SIMRAC), 2001.

Churchyard, Gavin, Lindiwe Pemba, Bulewa Magadla, Kobus Dekker, Mark Vermeijs, Rodney Ehrlich, Jim Tewaternaude and Jonny Myers. 'Silicosis Prevalence and Exposure Response Relationships in Older Black Mineworkers on a South African Goldmine'. Safety in Mines Research Advisory Committee, Final Report, Health 606, May 2003.

Cowie, R.L. 'The Five Ages of Pulmonary Tuberculosis and the South African Goldminer'. *The South African Medical Journal,* Vol. 76, 18th November 1989, pp. 566–567.

Davies, J.C.A. 'Sound an Alarm'. *The South African Medical Journal*, Vol. 84, No. 3, 3rd March 1994, pp. 133–134.

---

[87] Interview with Richard Spoor, Convention Centre, Cape Town, 23rd October 2014.

Davies, J.C.A. 'A String of Studies: The Pathway to The Courts'. *Occupational Health Southern Africa*, Vol. 21, No. 1, 2015, pp. 18–21.

Felix, Maria Anne. Environmental Asbestos and Respiratory Disease in South Africa. PhD thesis, University of the Witwatersrand, 1997.

Field, Tracy-Lynn. 'Exacting silicosis justice through the class action mechanism'. *Mineral Economics,* Vol. 32, 2019, pp. 213–221.

Girdler-Brown, Brendan, V.N.W. White, R.I. Ehrlich, and G.J. Churchyard. 'The Burden of Silicosis, Pulmonary Tuberculosis and COPD Among Former Basotho Goldminers'. *American Journal of Industrial Medicine*, Vol. 51, No. 9, 2008, pp. 640–647.

Godfrey-Faussett, P., P. Sonnenberg, S.C. Shearer, M.C. Bruce, C. Mee, L. Morris and J. Murray. 'Tuberculosis Control and Molecular Epidemiology in a South African Gold-Mining Community'. *The Lancet*, Vol. 356, No. 9235, 23rd September 2000, pp. 1066–1071.

Hermanus, M.A. 'Occupational Health and Safety in Mining – Status, New Developments, and Concerns'. *The Journal of the Southern African Institute of Mining and Metallurgy,* Vol. 107, No. 8, August 2007, pp. 531–538.

Innes, Duncan. *Anglo-American and the Rise of Modern South Africa.* London: Monthly Review Press, 1984.

Irvine, L.G. and D. Macaulay. 'The Life-history of the Native Mine Labourer in the Transvaal'. *Journal of Hygiene*, Vol. 6, No. 2, April 1906, pp. 149–174.

Le Grange, M.A.C. 'Occupational Lung Disease in Ex-Mineworkers'. *The South African Medical Journal,* Vol. 86, No. 7, July 1996, p. 841.

Leger, J.-P. 'Occupational disease in South African mines – a neglected epidemic'. *South African Medical Journal.* Vol. 81, 15th February 1992, pp. 197–201.

McCulloch, Jock. *Asbestos Blues: Labour, Capital, Physicians and the State in South Africa.* London: James Currey/ Indiana University Press/ Juta Press, 2002.

McCulloch, Jock. 'Asbestos, Lies and the State: Occupational Disease and South African Science'. *African Studies,* Vol. *64,* No. 2, 2005a, pp. 201–216.

McCulloch, Jock. 'Beating the Odds: The Quest for Justice by South African Asbestos Mining Communities'. *The Review of African Political Economy*, Vol. 32, No. 103, 2005b, pp. 63–77.

McCulloch, Jock. 'Saving the Asbestos Industry, 1960 to 2003'. *Public Health Reports,* Vol. 121, No. 5, 2006, pp. 609–614.

McCulloch, Jock. 'Counting the Cost: Gold Mining and Occupational Disease'. *African Affairs*, Vol. 108, No. 431, 2009, pp. 221–240.

McCulloch, Jock and Geoffrey Tweedale. *Defending the Indefensible: The Global Asbestos Industry and its Fight for Survival.* Oxford: Oxford University Press, 2008.

Nelson, Gillian. Living in the Shadow of a Dust Cloud: Occupational Respiratory Diseases in the South African Mining Industry, 1975 to 2009. PhD thesis, Faculty of Health Sciences, University of the Witwatersrand, Johannesburg, 2012.

Park, H.H., B.V. Girdler-Brown, G.J. Churchyard, N.W. White and R.I. Ehrlich. 'Incidence of Tuberculosis and HIV And Progression of Silicosis and Lung Function Impairment Among Former Basotho Gold Miners'. *American Journal of Industrial Medicine*, Vol. 52, No. 12, Dec. 2009, pp. 901–908.

Roberts, J.E. What is the Price of 80 KG's: The Failure of the Detection of, and Compensation for, Asbestos-related Disease: Social Exclusion in Sekhukhuneland. MA thesis University of Natal, December 2000.

Shennan, D.H. 'Towards the Rout of Tuberculosis in Rhodesia: The Work of the Tuberculosis Service'. *The Central African Journal of Medicine*, Vol. 11, No. 8, 1965, pp. 220–228.

Simons, H.J. Migratory Labour, Migratory Microbes. Occupational Health in the South African Mining Industry: The Formative Years 1870–1956. unpublished manuscript, 1960.

Simson, F.W. 'Pulmonary Asbestosis in South Africa'. *The British Medical Journal*, Vol. 1, No. 3516, 26th May 1928, pp. 885–887.

Slade, G.F. The Incidence of Respiratory Disability in Workers Employed in Asbestos Mining with Special Reference to the Type of Disability Caused by the Inhalation of Asbestos Dust, MD thesis, University of the Witwatersrand, 1930.

Snyman, P.H.R. 'Safety and Health in the Northern Blue Asbestos Belt'. *Historia,* Vol. 33, No. 1, May 1988, pp. 31–52.

Steen, T.W., K.M. Gyi, N.W. White, T. Gabosianelwe, S. Ludick, G.N. Mazonde, N. Mabongo, M. Ncube, N. Monare, R. Ehrlich and G. Schierhout. 'Prevalence of Occupational Lung Disease Among Botswana Men Formerly Employed in The South African Mining Industry'. *Occupational & Environmental Medicine*, Vol. 54, No. 1, 1997, pp. 19–26.

Tlhaole, Thuso. Social Protection Arrangements for Retired Mineworkers with Physical Disabilities in Maseru Urban, Lesotho. MA thesis, University of Witwatersrand, Humanities Faculty, 2011.

Trapido, Anna. An Analysis of the Burden of Occupational Lung Disease in a Random Sample of Former Gold Mineworkers in the Libode District of the Eastern Cape. PhD thesis, University of the Witwatersrand, 2000.

Trapido, A., N. Mqoqi, C. Macheke and B. Williams. 'Occupational Lung Disease in Ex-Mineworkers – sound a further alarm!'. *The South African Medical Journal,* Vol. 86, No. 5, May 1996, p. 559.

Tweedale, Geoffrey and Flynn, L. 'Piercing the Corporate Veil: Cape Industries and Multinational Corporate Liability for a Toxic Hazard, 1950–2005'. *Enterprise and Society,* Vol. 8, No. 2, June 2007, pp. 268–296.

Westwater, M.L. and J.C.A. Davies. 'Follow up at Eighteen Months of 742 Tuberculosis Patients Notified in Midlands and South-Eastern Province in Six Months in 1964'. *The Central African Journal of Medicine,* Vol. 13, No. 8, 1967, pp. 175–176.

Wilson, Francis. *Migrant Labour in South Africa: Report to the South African Council of Churches.* Johannesburg: The South African Council of Churches and SPRO-CAS, 1972.

15

Conclusion: Records, Bodies and Contested Justice

Political Economy and Health

For most of the twentieth century, gold mining dominated South Africa's economy. Unlike agricultural produce, it was impervious to bad seasons or fluctuating prices. For several decades, it accounted for more than half of the country's exports. Between 1931 and 1961, it supplied between 12 and 42 per cent of government revenue. In 1941, the mines employed 410,000 workers; at their peak in 1986, 534,000. The mining corporations drew on international capital and advanced technologies to consolidate their size and success. The fixed price of gold enabled them to act in unison in pursuing a common political agenda. Through their extensive use of migrant labour, they helped shape the economies of the southern half of the continent.

Several factors drove recruitment to South Africa's gold mines. Foremost among them was the Chamber's determination to control production costs. Recruiting labour from territories to its north and paying

---

This chapter was written four years after Jock McCulloch's death from mesothelioma by his partner Pavla Miller. It summarises the arguments made in the previous chapters and outlines some of the challenges in implementing the class-action settlement.

J. McCulloch, P. Miller, *Mining Gold and Manufacturing Ignorance*,
https://doi.org/10.1007/978-981-19-8327-6_15

black miners below-subsistence wages was a central part of this process. Equally important was Britain's (and Portugal's) desire to minimise outlays on colonial administration. The labour system constructed around these twin imperatives shifted the enormous social costs from the region's most profitable industry onto the families and communities of migrant workers. As a critic of the WNLA concluded in 1948, '[t]he South African gold mining industry is permanently dependent upon being subsidised by a semi bankrupt African peasant pastoral economy in the distant reserves beyond its own national borders'.[1]

The Rand mines had a distinct relationship with the state and with science. In this way, they helped shape a particular and deeply contradictory form of what Foucauldian scholars call biopolitics.[2] They were the first in the world to be subject to a comprehensive system of state regulation, including the mandate to compensate hard rock miners who contracted silicosis (1911). The mining houses also led the industrial world in the use of science and the collection of data. By 1916 they helped set up the South African Institute of Medical Research and the Miners' Phthisis Medical Bureau. One of the first professional associations of occupational health specialists, the Mine Medical Officers' Association, was founded in 1922. Those who admired the South African legislation were unaware that in 1916 the state effectively handed over the conduct of black miners' medical examinations to employers, who thereby gained control over access to compensation. That control went unchallenged until the Leon Commission in 1994.

Under the South African Acts, a three-step occupational health regime was gradually built up for white miners, with periodic X-rays, sanatorium rehabilitation and care for those who fell ill, and education for families. The routine for dealing with black miners was considerably different. Sick and dying men whose tuberculosis was identified during entry, periodic and occasional exit medicals were repatriated to their rural homes. Early warnings about dangers of the Chamber's approach to black miners proved to be justified. From 1920, evidence about the devastating impact

[1] A. Sandiland. Review of I. Schapera, *Migrant Labour and Tribal Life. Africa. Journal of the International African Institute,* Vol. 18, No. 2, 1948, p. 145.

[2] As Foucault theorised it, biopolitics describes the way the biological health and well-being of populations is intertwined with the operations of political power.

of NRC and WNLA's recruiting mounted. It came in the form of repeated warnings from District Medical Officers about men returning from the mines with tuberculosis, and it was obvious in the high rejection rates at entry medicals of men who had served multiple contracts. There were also clear warnings from medical specialists. The spread of tuberculosis from the mines, they argued, could have been reduced by the use of X-rays at exit medicals, the segregation of infectious men in mine hospitals, notifications of repatriated miners to local health authorities, and the provision of information to miners and their families about avoiding infection. Finally, the industry could have monitored miners after they returned home.

There were many factors which, in theory at least, should have made recommendations such as these easy to implement. British governments, the administrations of the HCTs and the ILO all took a keen interest in the welfare of migrant labour. In principle, that interest was supported by the progressive South African legislation with its government-controlled system of medical surveillance designed to protect miners and their families. South Africa's gold mines were also distinguished by their remarkably high levels of bio-medical knowledge. The research clusters of the Bureau and the SAIMR in Johannesburg, which did not exist elsewhere in the colonial world, were created specifically to service the gold mines. The size and profitability of the gold mining industry meant that each of the measures for controlling an emerging epidemic of tuberculosis lay well within its grasp. Instead of implementing them, the industry chose a different path: the manufacture of ignorance.

The Manufacture of Ignorance

This book has charted the making, updating, revising, challenging and transforming the complex edifice of employment and risk management in the South African gold mining industry. A core strand of the argument has been the discrepancy between the official invisibility of occupational lung disease during most of the twentieth century and its prominence since South Africa's transition to majority rule. The best shorthand description of this discrepancy is medical involution, or more starkly the manufacture of ignorance. The histories of asbestos and tobacco litigation show that those industries have systematically suppressed evidence of hazardous work practices and the risks associated with their products. The creation of ignorance is subtler and more effective than simply hiding the evidence of risk. Suppression leaves gaps in the knowledge of risks; the fabrication of false knowledge fills such gaps with misleading information.

South African gold mining companies have long argued that the unique character of their deep mines made it inappropriate to judge working conditions there against international standards. That claim is open to question. However, another aspect of the South African gold mining industry almost certainly *was* unique: namely, the degree of collusion between the state and the industry in suppressing the knowledge of risk, the publication of misleading data and the use of publicity. The gold and asbestos industries in Southern Africa used different methods in dealing with the science of occupational health. Before the 1950s there was no state regulation of the asbestos industry, no medical monitoring and no data collection. The absence of engagement in public and scientific debates about asbestos contrasted with high levels of state regulation and the endless streams of data on disease rates, miners' health and dust levels which spilled out of the Rand gold mines. South Africa's gold industry preferred to capture the scientific debates from within and achieve state agreement with its labour practices by using massive information flows.

Several factors helped produce this outcome. The racialised labour regimes and the insidious character of pneumoconiosis were influential. So too was the nature of the research community in Johannesburg. For most of the twentieth century, there was no employment for researchers outside of the Chamber or the state, and it was the Chamber which largely set the research agenda and decided upon models of data collection. While the industry strenuously opposed gathering and sharing information on the incidence of tuberculosis, it was preoccupied with measuring dust counts. Such data was expensive to collect and to process. It also presented many technical problems, among them equipment limits and failures, the size of the particles, their silica content and vastly different dust levels at different times and places in vast mine complexes.

The free silica content on the Rand, for example, varies between 35 per cent and 80 per cent of the rock. As a result, dust counts in themselves meant little; whatever readings were recorded did not constitute a definitive proof that dust levels remained at a 'safe level'. [3] And yet, as the Johannesburg conferences of 1930, 1959 and 1969 demonstrate, questions relating to dust counts preoccupied researchers at the SAIMR and PRU for decades. In turn, the more data on silicosis released into the public domain, the less visible was the actual disease and mortality burden.

The manufacture of ignorance on South African gold mines, which arguably underpinned the industry's profitability for most of the twentieth century, was a challenging and expensive project. The Commissions of 1902, 1912 and 1914, augmented by the 1914 Gorgas Report, provided sufficient evidence of occupational disease for the South African government to ban the recruitment of labour from the Tropical north. That decision was supported by research conducted in the 1920s and 1930s, which suggested that there were high rates of silicosis and tuberculosis, and that the mines' policy of repatriating sick men was spreading tuberculosis to rural communities within South Africa and through the migrant labour system to adjoining states. These findings were endorsed by the Stratford and Lansdown Commissions of 1943, and again by the Oosthuizen Committee Enquiry of 1954. Despite all these warnings, the Chamber failed, over several decades, to inform the administrations of labour-sending states of the risk to public health posed by men with tuberculosis returning from the mines. By 1937, it succeeded in having the ban on the recruitment of tropical labour lifted. Until the 1990s, it also failed to carry out follow-up research into the health of ex-miners.

From the first decades of the twentieth century, the industry argued persuasively that it gradually succeeded in engineering silicosis out of the mines, and that tuberculosis was not an occupational disease, but rather was brought to the mines by infected African workers. The data presented in the annual reports of the Miners' Phthisis Boards offered a record of

[3] Gerald E. Markowitz and David Rosner. 'The Limits of Thresholds: Silica and the Politics of Science, 1935 to 1990'. *American Journal of Public Health*, Vol. 85, No. 2, 1995, pp. 253–262; I.G. Irvine, A. Mavrogordato and Hans Pirow. 'A Review of the History of Silicosis on the Witwatersrand Goldfields'. *Silicosis: Records of the International Conference Held at Johannesburg 13–27 August 1930*. Geneva: ILO 1930, pp. 178–208.

ceaseless improvement in which South Africa led the world in mine safety. Between 1917 and 1935, the reported annual silicosis incidence rate among the whole workforce of white miners fell from 2.195 per cent to 0.885 per cent. For black miners the rates presented were even lower, falling to 0.122 per cent by 1934–1935. The data was accepted as authoritative by the international community and South Africa's achievements featured prominently in science and policy debates in Australia, France, Belgium, Germany, Italy, Japan, the UK and the US.[4]

Migrant labour, the Chamber maintained, was episodic. Men would go to the mines to meet specific needs such as the purchase of cattle, and permanently left the industry once those needs were met. The Chamber also promoted the fiction that migrant labour had important health benefits. While at work, the miners were 'gloriously well fed'. The periodic breaks from mining associated with oscillating migration reduced dust exposures and therefore silicosis rates. By preventing urbanisation, migrant labour also protected peasant communities from the tuberculosis that was rife in urban centres. Somehow, confining men in single-sex compounds for months at a time, and preventing their wives and children from joining them, at once preserved and modernised traditional societies. During the 1940s a number of Commissions, including Lansdown (1943) and Strafford (1943), subjected that orthodoxy to scrutiny. They found that migrant labour was driven by poverty and that without such income families in labour-sending areas starved. Of necessity, men continued to work on the mines as long as they could. Typically, they left the industry when their health deteriorated to such an extent that they failed the next entry medical.

The Chamber did not initially recruit miners from outside of South Africa in order to hide occupational lung disease. Over time, however, it structured the migrant labour system so it did just that. The cursory nature of mine medical examinations, the recruitment of Tropical labour, repatriation of sick miners without compensation, the failure to conduct exit medicals or notify regional authorities of tuberculosis cases, combined to obscure the extent of pneumoconiosis produced in the gold

[4] Jock McCulloch. 'Air Hunger: The 1930 Johannesburg Conference and the Politics of Silicosis'. *History Workshop Journal,* Vol. 72, No. 1, 2011, pp. 118–37.

mines. Such manufactured ignorance was enhanced by the industry's refusal to pay pensions (which would have required monitoring the men's post-employment fate) rather than lump sum compensation, and consolidated by its failure to carry out follow-up studies of ex-miners' health.

Lies and Statistics

Many scholars rightly point out that during the nineteenth and twentieth centuries, the collection and ever-more sophisticated manipulation of statistics went hand in hand with increasing governmentality, or the capacity of state institutions to perceive, monitor and govern populations. Such process, many imply, is a bad thing. This book makes a different point. The systematic failure to collect statistics too could become a technique of rule, not least by impeding the accumulation of comparable health data and preventing the consolidation of epidemiological evidence. In particular, the deliberate obstruction of state intelligence played a significant part in hiding an epidemic of occupational lung disease—not only from the miners themselves but from the white electorate in South Africa, the administration of adjoining states, the British government, international organisations such as the ILO and UN as well as a transnational public of social justice advocates.

The lack of meaningful statistics about miners' health provoked a chorus of complaints from medical practitioners and local administrators. The regular reports published by mining industry bodies such as the WNLA did contain numerous statistical tables and graphs. However, valuable information presented for a short period was suddenly discontinued, key categories changed from year to year and important details were missing. There was, for example, no information about the miners' age, work experience, years since first exposure, the kind of work they performed or the types of mines on which they were employed.[5] For its part, during what were often protracted negotiations with the South

[5] For a valiant effort to make sense of the longitudinal data, see J-P. Leger. 'Occupational Disease in South African Mines—a Neglected Epidemic'. *South African Medical Journal*, Vol. 81, 15th February 1992, pp. 197–201.

African mining industry, the British Colonial Office never specified what it considered an acceptable annual number of deaths per thousand recruits, nor did it ask for detailed morbidity and mortality statistics.

Negotiated Diseases

The compilation (or not) of statistics overlapped with a subtler process at play, one concerned with identifying and classifying diseases, and establishing the medical and legal boundaries between health and disability. As this book has shown, these contests were not just about words; they were crafted out of a complex fabric of toil and misery, and in turn had far-reaching material consequences.

In some instances, the identification of disease involves a simple binary divide: you either have it or you do not. More frequently, it requires drawing a line at a particular point on a continuum—or indeed debates about whether there *is* a continuum, and what it looks like. Where the line is drawn is subject to contestation: it depends on scientific knowledge and medical technology, but always also includes a complex reconciliation of social norms, expert knowledge, local capacity and the views of powerful actors. At times, consensus is reached easily. More often, as in the story told in this book, it is subject to fierce and extensive contestation, played out in commissions of inquiry, legislation and government and industry regulations. The articulation of what does and does not constitute *occupational disease* is particularly fraught. By definition, occupational diseases are negotiated categories, forged in complex contests and accommodations between different powerful groups, political parties, unions and social movements, insurance companies, professionals, policy frameworks, scientific knowledges, legal systems and fractions of capital. 'Safe level' of silica dust, for example, was a rubbery negotiated category, far more arbitrary than the common-sense understanding of dust levels low enough not to cause injury. Rather, the thresholds represented a political compromise between medical uncertainty regarding the causes, symptoms and diagnosis of silicosis; technical problems encountered by the manufacturers of measuring equipment; companies' preoccupation with healthy profits, and workers' concerns with their own health.[6]

Finally, even in jurisdictions with well-funded medical facilities, there are wide disparities between different social groups in the incidence, diagnosis and access to treatment of specific diseases. In Southern Africa, gaps between the health status, diagnosis of disease and access to care between black and white populations have historically been vast, as has been their legal—and practical—access to compensation for injury.

Law, Politics and Health

There are narrow and broader definitions of politics. The narrow ones focus on state institutions, governments, political parties and elections. The broader ones widen the scope to encompass the deliberate input of powerful actors and effects of social structures into the negotiations, decisions and non-decisions affecting the life of the community. Using such broader definition, this book charted the negotiations, decisions and non-decisions which structured approaches to the mining of gold, the buying and selling of mine labour, the descriptions and classifications of risk and the manufacture of ignorance about occupational lung disease. In effect, while not employing the technical terms used by scholars writing on biopolitics, the book has traced a complex story of the interface between politics and population health.

In the first half of the twentieth century, the South African mining industry played an important role in the shaping and consolidation of apartheid. The gradual dismantling of this system in the 1990s, as well as the wave of anti-colonial struggles throughout the continent, presented the mining industry with far-reaching challenges. Certainly, as the Marikana story with which this book opened shows, majority rule did not result in a sudden transformation of South African politics. More broadly, as Francis Wilson concisely put it, 'the very process that generated wealth in the economy simultaneously produced poverty and patterns of unemployment that still hobble South Africa as it struggles to

[6] Markowitz and Rosner, 'The Limits of Thresholds'.

democratize in the twenty-first century'.[7] But neither is it possible to underestimate the extent and significance of the changes brought about by the transition to majority rule. The legal standing of citizens, whatever the colour of their skin, has been particularly important. Miners' compensation for occupational injury provides a compelling picture of both sides of the coin.

As noted in previous chapters, for most of the twentieth century the mining houses succeeded in preserving a compensation system designed to minimise their liability for occupational injury. When, in 1943, the Government Actuary costed what the Stafford Commission regarded as a fair compensation system, the mines protested that they could not afford to pay. The Commission responded that if that were true the mines must be so dangerous that they should be closed.[8] The Commission's report was shelved, and the mining houses continued as before. Half a century later, many of the racial categories and hurdles built into the system having been removed under the new post-apartheid constitution, the extent of occupational disease became clearer—and the problems of insolvency have been amplified. As one of the medical researchers working to reform the system explained it, the Miners' Compensation Fund is financed by levies on the mining industry, although at various points the state has had to supplement this fund from general taxation. However, the budget for the administration costs of the system, including medical staffing, comes from South African taxpayers via the Ministry of Health. This funding has failed to keep up with the needs of affected miners. In 2004, an actuarial study confirmed that the Fund was technically insolvent even on the basis of the then current claims, and that the levies paid by the mining industry needed to be raised 100-fold over a 15-year period to meet liabilities. In the same year the auditor-general referred to the financial controls of the Fund as a shambles.[9]

---

[7] Francis Wilson. 'Miners and Migrants: How the Mining Industry Has Shaped South Africa'. *Daedalus,* Vol. 130, No. 1, Winter 2001, pp. 99–121.

[8] *Report of the Miners' Phthisis Acts Commission, 1941–1943.* Pretoria: Government Printer 1943, pp. 9; 30–32. See also Chap. 11, Dissenting voices.

[9] Rodney Ehrlich. Inaugural Professorial Lecture: 'The body as history: on looking at the lungs of miners', 26th September 2007, University of Cape Town, p.9. See also S. Barker, R. Ehrlich, J.M. Spiegel, B. Kistnasamy, F. Riera, A. Fourie, N. Mtshali, M. Rabada, K. Lockhart and A. Yassi. 'Reforming the Workers' Compensation Process for Occupational Lung Disease among Miners in South Africa: An Efficiency Study of Claims Assessment'. *International Archives of Occupational and Environmental Health,* Vol. 95, No. 4, 2022, pp. 825–833.

At their peak in 1986, the mines employed well over half a million workers, by 1999, less than half that number, and by 2022, around 95,000. In effect, a declining industry is faced with compensating ever increasing numbers of miners suffering from occupational lung disease, which often flares up a decade or more after exposure. As a result, what had already been an imperfect system all but ground to a halt. Under-resourcing of the compensation agencies responsible for certifying occupational lung disease, unwieldy panels tasked with adjudicating compensation claims and administrative failures of the Occupational Diseases in Mines and Works Act (ODMWA) system resulted in massive backlogs in payment and verification of claims.[10] As before, the burden has fallen particularly heavily on black workers from Southern Africa's labour-sending regions.[11] One typical study of the claims experience of 90 former gold miners diagnosed with silicosis at Groote Schuur Hospital in Cape Town between 1993 and 2005 found that only one in five received compensation, and even those had to wait between one to five years for the certification and award of their claims. In the intervening months and years, many of those entitled to compensation had simply given up.[12] Despite some progress in clearing the backlog, in 2019 an average of 12,000 claims per month, excluding those not yet entered into the new electronic database, awaited final medical adjudication, with an average delay from the initial medical examination to certification of 583 days. The claims management information system has been upgraded in recent years and peripheral examination centres were opened. At the same time, occupational health specialists and medical researchers made progress in working out how to deal more expeditiously with the backlog

---

[10] Barker et al., 'Reforming the workers' compensation process'.

[11] Haeidee Williams, Rodney Ehrlich, Stephen Barker, Sophia Kisting-Cairncross, Muzimkhulu Zungu, and Annalee Yassi. 'The Utility of Length of Mining Service and Latency in Predicting Silicosis among Claimants to a Compensation Trust'. *International Journal of Environmental Research and Public Health*, Vol. 19, No. 6, 2022, pp.3562–74.

[12] Lerato Maiphetlho and Rodney Ehrlich. 'Claims Experience of Former Gold Miners with Silicosis—a Clinic Series', *Occupational Health Southern Africa*, March/April 2010 Www. Occhealth.Co.Za

of claims and speed up the processing of new ones. Nevertheless, it has been difficult to find and appoint sufficient skilled medical staff to adjudicate a large volume of claims within statutory requirements.[13]

The interface between impoverished former miners and an imperfect administrative system requiring precise information compounded the problems. Among the hurdles identified by Trapido in her 1990s research on former gold mineworkers in the Eastern Cape were incorrect transcription of addresses at the MBOD or CCOD, poor postal services in rural areas, low literacy among miners and their families and limited banking services.[14] Rodney Ehrlich, writing in 2007, painted a similar picture. Even should a former mineworker find his way into the system, he noted, a further hurdle awaited—to prove his goldmining service. Some record of mine service had always been required, but until recent years the miner's own recall, together with a fingerprint record, was accepted. In response to some incidents of fraud, the Compensation authority began insisting on some original record of service. In the experience of his team, only one in two and perhaps as little as one in three mineworkers could produce such evidence. A few, particularly those with more recent service and longer contracts, retained their personnel card and may have an extant record at the mine. Frequently, however, service was long past and consisted of a number of contracts at different mines. A surprising number of mineworkers lost the remaining scraps of documents tying them to their past in the fires which frequently ravage informal settlements. As far as the system was concerned, their mine service had been expunged from the record.[15] In order to address issues such as these, the Minerals Council undertook a massive process of digitising data relevant to compensation for occupational lung disease—a project hampered by the historic reluctance of mining houses to generate and to

[13] Barker et al., 'Reforming the Workers' Compensation rocess'.

[14] A. Trapido. An Analysis of the Burden of Occupational Lung Disease in a Random Sample of Former Gold Mineworkers in the Libode District of the Eastern Cape. PhD thesis, University of the Witwatersrand, 2000.

[15] Ehrlich, Inaugural Professorial Lecture, pp.9–10. See also Rodney Ehrlich. 'A century of miners' phthisis on the south African gold mines. Any end in sight?' Presentation at the 25th Anniversary meeting of the Collegium Ramazzini, held in Carpi, Italy on October 25–28, 2007.

keep employment records.[16] It is in the context of challenges facing ODMWA system that the settlement of the miners' class action needs to be seen.

The Miners' Class Action

Prior to majority rule, the concept of an individual miner suing an employer for occupational lung disease was foreign to South African law. The system established from 1911 under the various Miners Compensation Acts was always assumed to preclude workers' capacity to sue employers, and there is no evidence of a black miner bringing a case to court. The law also made no provision for class actions. This has now changed. In 1994, the ODMWA was amended to remove overt racial discrimination for compensation.[17] More broadly, the respective legal standing of employers and workers was transformed. Under Section 34 of the Constitution, every South African has the right of access to justice while Section 38 opens South African courts to class actions. Those clauses have altered the types of claims which can be brought before a court and the kinds of people who can seek legal redress.

The improbable legal victories by asbestos miners over the British company Cape Plc and the South African conglomerate Gencor in 2003 have pioneered such claims. The plaintiffs were impoverished communities in the Northern Cape who on appeal to the Law Lords in London successfully pursued an extra-territorial claim against a multi-national corporation. Those victories led immediately to the recent class actions by Southern African gold miners against 32 mining companies, representing virtually the entire gold mining industry in South Africa. The settlement

[16] Minerals Council South Africa. 'Joint Effort to Improve the Occupational Lung Disease Compensation System'. Fact sheet. www.mineralscouncil.org.za

[17] G.J. Churchyard and E.L. Corbett. 'Tuberculosis and Associated Diseases'. In *Handbook on Occupational Health Practice in the South African Mining Industry*. R. Guild, R.I. Ehrlich, J.R. Johnston and M.H. Ross (eds). Safety in Mines Research Advisory Committee (SIMRAC), 2001, p. 155.

of the consolidated class action, reached in May 2018, was approved by the South African High Court in July 2019 and became effective in December 2019.

The miners' class action opened a new chapter in South African history. It expanded eligibility for compensation for miners or their surviving dependants across the Southern African region. On paper at least, it relaxed the burden of proof for occupational injury and set aside funds for the tracking and tracing of potentially eligible claimants, assistance with the claims process, the processing of claims, claimants' travel expenses and the facilitation of benefit medical examinations. The court's sympathy with the mineworkers' plight and its reworking of legal precedents and regulatory frameworks to accord with the human rights principles enshrined in the South African constitution entrench this judgement as a landmark in the South African jurisprudential landscape.[18]

The Tshiamiso Trust

The Tshiamiso Trust was formally set up to implement and fulfil the terms of the class-action settlement.[19] As with previous path-breaking South African interventions into miners' occupational health, the terms under which the Trust was set up and operates represent a process of contestation and compromise. The Trust needs to conform to a long and complex court-approved trust deed and adhere to eight steps involved in processing each claim. While these requirements have delayed the compensation process, there is no formal limit to the number of people the Trust can deal with, with estimates ranging from seventeen thousand to half a million of potential claimants.[20] No one is in a position to provide more precise figures. The Trust does not have funds for research into the

[18] Tracy-Lynn Field. 'Exacting Silicosis Sustice Through the lass Action Mechanism'. *Mineral Economics*, Vol. 32, 2019, pp. 213–221.

[19] https://www.tshiamisotrust.com/about/ Tshiamiso is a Setswana word meaning 'to make good' or 'to correct'.

[20] Republic of South Africa, The High Court of South Africa, 2019, 1/pp. 35–36 #57, #58; Shula Marks. 'The Silent Scourge: Silicosis, Respiratory Disease and Goldmining in South Africa'. *Journal of Ethnic and Migration Studies*, Vol. 32, No. 4, 2006, p. 571.

matter, and previous studies only looked at small sub-sections of the mining workforce. Miners suffering from occupational lung disease can get compensation from both the ODMWA and the Tshiamiso Trust. The Trust began functioning in February 2020 and is expected to operate for 12 years after that date. It processed the first few pilot payments in December 2020.[21] In January the following year, claimants were able to begin booking appointments at 50 lodgement offices in mining centres and areas from which labour has historically been drawn in South Africa, Lesotho, Mozambique, eSwatini and Botswana. The offices began opening a month later, and sites for the Benefit Medical Examinations were established in South Africa and Lesotho. The first batch of more than 100 compensation payments was processed in August 2021. A year later, by the end of August 2022, 7418 claims were paid. At that date, 13,018 claimants were deemed medically eligible; 6753 claims were 'Medically Deferred' for reasons such as missing documentation and need for additional medical examinations, and more than half, or 25,626, were deemed Medically Ineligible.[22] Those rejected for compensation had a month in which to appeal, something which is difficult to organise for people who find bureaucracies hard to navigate.

In considering the miners' claims for compensation, the South African High Court acknowledged the overwhelming problems faced by those who became ill after working in the mines. Logistically, these problems include the remoteness of labour-sending areas with poor and expensive transport and communication systems, language and literacy barriers, particularly for those in distant labour-sending regions, and the poor state of medical and diagnostic services in rural areas. Senior citizens located in remote villages, with patchy phone coverage, few mobile phones and no access to the internet, face particular problems. The class-action settlement and the Tshiamiso Trust did make some provisions for overcoming these obstacles. Although the Trust does not have funding for the setting up of new clinics, some of its offices are located outside of South Africa, and basic information on the Trust website is available in seven languages other than English. Despite efforts to engage with

[21] https://mg.co.za/news/2021-06-04-the-gold-miners-widows-tell-of-daily-battles-for-survival/

[22] https://www.tshiamisotrust.com/information/progress-report/

communities in the old labour-sending regions, however, progress is slow. Although mine workers from Mozambique formed between tenth and a quarter of the mine workforce between 1971 and 1999, for example, as of August 2022 none so far reached the sixth part of the 8-step compensation process, the Medical Certification Panel, or received compensation. None of the claimants from Botswana, eSwatini or Malawi have reached that stage either.[23]

Many of the challenges facing the Tshiamiso Trust are similar to those confronting the ODMWA system. All have been exacerbated by the COVID pandemic and the problems it poses to administration, staffing and travel. COVID-19 also presents a serious threat to gold miners whose health is already compromised by TB and silicosis. With increased chances of dying as a result of the potential combination of COVID-19 and occupational lung diseases, claimants need to be diagnosed and compensated as soon as possible. That is not always easy.

Diagnosis

While the Trust operates with a more generous definition of eligibility than much of the previous legislation, the diagnosis of complex occupational diseases remains difficult. Few doctors are qualified to diagnose silicosis. Even before the pandemic, many African countries faced critical shortages of medical staff. This shortage is particularly acute in Mozambique and Malawi, and severe in Zimbabwe and Zambia.[24] Everywhere, it has been aggravated by the COVID epidemic. Combined with lack of appropriate training of many of those conducting the qualifying medical examinations, this has meant that many sick miners—even those found eligible for compensation under the ODMWA—have been diagnosed as healthy.[25] As in the past, some X-rays are not of sufficient

[23] Progress report—Tshiamiso Trust

[24] A. Ahmat, S.C. Okoroafor, I. Kazanga et al. 'The Health Workforce Status in the WHO African Region: Findings of a Cross-sectional Study'. *BMJ Global Health, Vol.7* (Suppl 1) 2022: e008317.

[25] See, for example, https://www.justiceforminers.org/press-releases/tshiamiso-trust-medical-examinations-found-to-be-compromised, which reported that Lung Function and Spirometry tests were administered incorrectly at the medical examination facility of the Tshiamiso Trust in the South African mining town of Welkom.

quality to determine the extent of injury, and so miners have to travel back for re-examination. The length of mining service presents other difficulties. The Benefit Medical Examinations run by the Trust are free of charge if the ex-mineworker worked for five or more years at qualifying mines during qualifying periods. Claimants with less than five years of service at qualifying mines have to bear the cost of the examination. Although this cost will be reimbursed should they be found to be eligible for compensation, raising money for transport and examination is difficult for men no longer able to work. Yet other problems confront those who lived for years in apparent good health after leaving the mines, only to discover they have silicosis after a long period of latency.

Mining Service Records

To be eligible for compensation, applicants or their surviving family members have to prove they worked at particular mines, sometimes many decades earlier. As noted above, while some miners have retained scraps of the required documentation, many others have not. In theory, such service records should also be available from the mine recruiting agency and individual mines.[26] For a number of reasons, these records are incomplete. And while the Medical Bureau for Occupational Diseases (MBOD) should have a complete record of all risk shifts worked by gold mineworkers, a substantial breakdown of its system after 1994 resulted in records either lost, destroyed or simply not kept. In cases such as these, the recent digitisation of miners' records is of limited use. As Booi Mohapi, chapter leader of Justice for Miners Campaign in Lesotho noted, most of the ex-miners and their dependants asking for assistance at their offices complain of a lot of paperwork that is involved and hard to obtain from the relevant mines. 'Some complain that even when they believe that the paperwork is complete, the Trust officers will demand the most insignificant document which they would not be aware of. They are then turned away. Some of those needed documents can be availed by the mines themselves. It is the duty of each mine to facilitate attempts to

[26] Williams et al., 'The Utility of Length of Mining Service', p. 3565.

obtain the required documents', Mohapi explained with much indignation. 'How can Tshiamiso demand mine documents from a woman whose husband died in the 1980s which the trust itself cannot find from the mine even though they have access to the mines?'[27] In a recent media release, Daniel Kotton, CEO of Tshiamiso Trust, explained some of the delays:

> Claims related to deceased mineworkers are especially difficult to process as death certificates often state that the claimant died from natural causes, making no reference to silicosis or TB … we have partnered with various government bodies and provincial health departments to access historical health data, unabridged death certificates, post-mortem reports, and medical records from clinics and hospitals. This is a mammoth task in itself, since some archives dating back to 1965 have not been digitized.[28]

Administration

In order to expedite the processing of claims, the Tshiamiso Trust negotiated a partnership agreement with TEBA Ltd. to record applicants' details and facilitate payments of compensation awards. As a result, the Trust lodgement sites are situated in existing TEBA offices. According to the communications manager of the Trust, while in the future lodgement centres will not be only limited to this organisation, 'Engaging with TEBA, which has a single management structure, is for Tshiamiso far more efficient than would be the case dealing with multiple organisations which may lack the infrastructure TEBA has'. The problem is that many former miners view TEBA with deep distrusts. As noted in Chap. 2, the Rand Native Labour Association, subsequently the Witwatersrand Native Labour Association, was formed in 1886 to eliminate competition between the mining houses and centralise the supply of migrant labour.[29]

[27] Majirata Latela. 'Ex-miners' Woes Continue'. *The Reporter*, Friday 11 June 2021, Vol. 3, No. 23, p.2.

[28] Tshiamiso Trust Media Release, 8th July 2022, https://www.tshiamisotrust.com/wp-content/uploads/2022/07/TSHIAMISO-MINEWORKERS-COMPENSATION-SURPASSES-HALF--A-BILLION.pdf

[29] *Report of the Witwatersrand Mine Native Wages Commission on the Remuneration and Conditions of Employment of Natives on the Witwatersrand Gold Mines.* Pretoria: Government Printer 1943, p. 2.

The Native Recruiting Corporation was formed in 1912 by the Chamber of Mines to carry out the same function within South Africa and the High Commission Territories. For decades, the two recruitment agencies supplied workers to the mines. In 1997, they were amalgamated to form The Employment Bureau of Africa (TEBA). Several years later, the agency diversified to assist mineworkers with programmes that included screening for TB and silicosis and processing their labour and medical claims with the authorities. In 2005 the founding president of the National Union of Mineworkers, Dr James Motlatsi, bought 75 per cent of TEBA, with 25 per cent of shares given to its employees.[30] Despite the change in name, function and ownership, many mineworkers strongly oppose the partnership between the Tshiamiso Trust and TEBA Ltd. The recruiting agency, they argue, has not served them well in the decades since it was set up. They accuse the company of many cases of fraud and see it as part of the mining companies they are fighting against. Rather than rely on its partnership with TEBA, they argue, the Trust should work with the network of community-based organisations, including paralegals and medical staff, that helped sign up miners to the class-action litigation. This is particularly important in the current social context.[31]

In a society still struggling with the legacies of apartheid, miners seeking compensation face many difficulties in organising themselves and working out who to trust. Ex-miner organisations are weak, and there is an endemic culture of non-caring. At the same time, many people are trying to defraud claimants. As the Tshiamiso Trust website explains:

> There are many people out there who are making promises to people that they can get them compensation from the Trust and taking money from them, when they have no way to make that happen. … Only the Trust's officials are authorised by the settlement agreement to help process claims and determine if a claim meets the criteria for compensation. The authorised officials will not expect any payment from you for their involvement and/or assistance. Those who dishonestly make promises of payment from

[30] https://mg.co.za/news/2021-06-04-the-gold-miners-widows-tell-of-daily-battles-for-survival/

[31] Justice for Miners Campaign Press Release, Issued by Justice for Miners Campaign Southern African Resource Watch, Johannesburg, South Africa, Civil Society endorses Justice for Miners Campaign, 18 February 2021, pp. 2–3.

the settlement fund in order to persuade people to pay them money are not authorised to do so by the settlement agreement and are not officials of the Trust.[32]

The modern and efficient methods for making appointments too present problems. Using mobile phones to send and receive text messages and download documents saves time and money—but not for those whose only access to a smart phone are neighbours' children working in a regional town. By the same token, requesting banking details from claimants via SMS is problematic as this platform is frequently used for phishing for banking details.

Justice for Miners Campaign

Many of the activists and groups who supported the miners in their long struggle for compensation have come together in an organisation called the Justice for Miners Campaign (JFM). With the assistance of funding partners such as The Open Society Institute of Southern Africa, The Ford Foundation, The Heinrich Boll Foundation, Open Society Foundation and Oxfam South Africa, the JFM has been building up a civil society interface between mining communities and the Tshiamiso Trust. The JFM organisational units, or Chapters, work on the ground within mineworker communities severely affected by TB and Silicosis. Chapters are composed of community members and groupings, such as local ex-miner organisations, church groups, women's groups and student and university collectives. Each Chapter is independent and draws upon its localised needs in contributing to the transnational JFM campaign. Currently there are JFM Chapters in Lesotho, Eastern Cape and Free State in South Africa, and in Botswana.

Looking towards mining communities, JFM activists undertake extensive training, capacity building, induction and orientation to new team members and JFM Chapters. This covers sound knowledge of what the JFM does, skills necessary for building effective campaigns and up-to-date information about the Tshiamiso Trust Deed and Legal Reform.

[32] https://www.justiceforminers.org/activities/engage-with-tshiamiso-trust-and-government [undated, JFM campaign document; 611fc9ca47da7839f446e7b1_The Tshiamiso Trust]

Looking towards the wider society, the JFM is trying to build a movement that will influence and lobby government institutions, mining houses, Tshiamiso Trust, TEBA and the Minerals Council to do the work required to ensure fast and just compensation is paid to gold miners and their dependants affected by TB and silicosis. Media campaigning, often using the award-winning documentary *Dying for Gold*, is an important part of JFM work.[33] In order to guide and strengthen the JFM campaign and link it to broader civil society, in November 2019 a group of civil society organisations and ex-miners' associations met in Johannesburg to set up a regionally representative forum.[34]

Justice for Miners Campaign activists believe that the current compensation system is broken, failing to process and deliver payments to miners whose health and lives deteriorate further as justice is delayed. From their perspective, sick miners struggle to navigate two complex and onerous systems: statutory compensation under the ODMWA and the Tshiamiso Trust. Their three key demands emphasise speed: in achieving just compensation, improved administration and legal reform. The first demand entails rapid and equitable compensation for all affected mineworkers. The second has to do with quick reform of institutions tasked with paying compensation. Simple claims and payments processes, JFM notes, must be accompanied by sufficient staffing capacity to follow through on claims within a reasonable time period. At the same time, administrative and diagnostic services must be decentralised and established in all present and former labour-sending areas. Accountability and transparency in executing these reforms must be reflected in quarterly reports to Parliament and civil society. The final demand calls for speedy legal reform to enable just compensation for all affected mineworkers. Elected representatives of the people, the Justice for Miners Campaign insist, must take up this issue and push for legal reform as a matter of the highest urgency.[35]

---

[33] http://www.dyingforgold.com/film The film tells the background story to the silicosis and TB class action that was settled out of court in June 2019. This story is a telling example of how those whose contributions to society create the foundations of wealth are rewarded with illness, poverty and suffering.

[34] https://www.justiceforminers.org/jfm-forums

[35] https://www.justiceforminers.org/what-we-do accessed 30.6.2022.

Bodies, Records and Contested Justice

For the best part of the twentieth century, South Africa held a reputation as a world leader in researching and compensating miners' occupational lung disease. During that period, the centrepiece of the mines' management of tuberculosis was repatriation of sick workers, without notification of local authorities, isolation of those with infective disease or education of family members on how to avoid infection. This approach to TB enhanced the gold mines' profitability but confounded every principle of public health policy. As previous chapters have shown, the system persisted despite detailed and repeated criticisms from community leaders, medical practitioners, local government officials, researchers and members of parliamentary commissions and committees of enquiry. At the same time, delays in the onset and difficulties in the diagnosis of silicosis helped obscure the public visibility of the disease—but did not stop retired miners from getting sick. In practice, the requirements for diagnosis and compensation for the majority of miners were such that few of those with occupational lung disease were able to prove they were ill, and fewer surmounted the hurdles in submitting claims and receiving compensation. Instead, impoverished labour-sending communities shouldered the extra burden of caring for men whom mining made ill.

Making sense of his career in treating miners for occupational lung disease, Rodney Ehrlich called his 2007 Inaugural Professorial Lecture 'The body as history'.[36] Whatever the state of politics, scientific knowledge or administration, he pointed out, the bodies of miners were indelibly marked with traces of their past employment. The vast extent of these bodily records stands in stark contrast with many decades of manufactured official ignorance regarding the extent of miners' occupational injuries. And while the individual bodies of workers carry indelible signs of past mining employment, the paucity of paper-based employment records obstructs the process of compensation. Just as the diligent collection of some statistics and studious efforts to avoid collecting others contributed to the manufacture of ignorance, so too did systematic non-compliance with repeated requests to keep records of miners'

[36] Ehrlich, Inaugural Professorial Lecture.

employment and medical histories.[37] At the same time, while the mines were subject to remarkably high levels of government surveillance and produced copious records, today the archive holdings are patchy, with what must have been voluminous correspondence files of key players missing.

This book has paid tribute to some of those who warned against looming health crises, advocated for miners and attempted to stem the tide of occupational lung disease. The Justice for Miners Campaign is among those who continue this tradition today. In assisting miners injured in South Africa's process of wealth accumulation, proposing alternative solutions to practical problems and mobilising local communities, JFM also helps heal injuries to the region's civil society.

Advocacy and practical support for miners overlap with contesting and dismantling manufactured ignorance regarding occupational lung disease. Previous chapters noted some of the work of those who contributed to this process. Jock McCulloch, who researched and wrote the bulk of this book, is among them. A prominent historian of colonialism and occupational health, Jock died from mesothelioma in 2018. His fatal exposure to asbestos (the only known cause of mesothelioma) almost certainly occurred in the 1990s, while researching the history of asbestos mining in Southern Africa. To see the mines first hand and interview mine owners and former workers, he toured Zimbabwe and the Northern Cape, where the landscape was littered with tailing dumps of fibre from abandoned workings, and where even today asbestos contaminates the environment. It was a bitter irony that he died trying to save others from being exposed to the same lethal asbestos fibres that tragically claimed his own life.

Confronted with the paucity of records which commonly accompanies company behaviour contrary to that publicly espoused, Jock believed that conventional archival research needed to be augmented by what he called 'forensic history'. This approach, which he used successfully on previous projects, involves reviewing a large volume of archival material

[37] See, for example, Memorandum of Further Evidence Submitted by the Agent for The High Commission Territories in Johannesburg to the Miners' Phthisis Commission, 5th August 1942, p.1. BNA S332/5/1 Phthisis: Miners' Compensation for B.P. Native Mine Workers. See also Chap. 8, p. 22.

to find the fragments of a story, drawing on secondary literature and field visits to identify relevant clues and triangulating accounts of one event from a number of different actors and sources. Like Jock's previous books, *Mining gold and manufacturing ignorance* is distinctive in bringing together and juxtaposing a vast range of sometimes ephemeral evidence to document what happened behind the scenes. Jock passed his enormous collection of materials on silicosis and asbestosis in Southern Africa and Australia to colleagues at Columbia University and City University of New York to form part of *Project Toxicdocs*.[38] Hopefully, this book will inspire others to chart the evolving story of money, mining, politics and health. The evidence presented in *Mining gold and manufacturing ignorance*, together with the materials in *Project Toxicdocs*, should make such contributions to social justice more manageable.

References

Ahmat A., Okoroafor S.C., Kazanga I. et al. 'The Health Workforce Status in the WHO African Region: Findings of a Cross-sectional Study'. *BMJ Global Health,* Vol. 7 (Suppl 1), May 2022: e008317.

Barker, S., R. Ehrlich, J.M. Spiegel, B. Kistnasamy, F. Riera, A. Fourie, N. Mtshali, M. Rabada, K. Lockhart and A. Yassi. 'Reforming the Workers' Compensation Process for Occupational Lung Disease Among Miners in South Africa: An Efficiency Study of Claims Assessment'. *International Archives of Occupational and Environmental Health,* Vol. 95, No. 4, 2022, pp. 825–833.

Churchyard, G.J. and E.L. Corbett. 'Tuberculosis and Associated Diseases'. In *Handbook on Occupational Health Practice in the South African Mining Industry*. R. Guild, R.I. Ehrlich, J.R. Johnston and M.H. Ross (eds). Safety in Mines Research Advisory Committee (SIMRAC), 2001.

Ehrlich, Rodney. 'A Century of Miners' Phthisis on the South African Gold Mines. Any End in Sight?' Presentation at the 25th Anniversary meeting of the Collegium Ramazzini, held in Carpi, Italy on 25th to 28th October 2007.

Ehrlich, Rodney. Inaugural Professorial Lecture: The body as history: on looking at the lungs of miners, 26th September 2007, University of Cape Town.

[38] See https://www.toxicdocs.org/

Field, Tracy-Lynn. 'Exacting Silicosis Justice Through the Class Action Mechanism'. *Mineral Economics,* Vol. 32, 2019, pp. 213–221.

Irvine, I.G., A. Mavrogordato and H. Pirow. 'A Review of the History of Silicosis on the Witwatersrand Goldfields'. *Silicosis: Records of the International Conference Held at Johannesburg 13–27 August 1930.* Geneva: ILO 1930, pp. 178-208.

Leger, J-P. 'Occupational Disease in South African Mines - a Neglected Epidemic'. *South African Medical Journal,* Vol. 81, 15th February 1992, pp. 197–201.

Maiphetlho, Lerato and Rodney Ehrlich. 'Claims Experience of Former Gold Miners with Silicosis – a Clinic Series'. *Occupational Health Southern Africa,* March/April 2010 www.Occhealth.Co.Za

Markowitz, Gerald E. and David Rosner. 'The Limits of Thresholds: Silica and the Politics of Science, 1935 to 1990'. *American Journal of Public Health,* Vol. 85, No. 2, 1995, pp. 253–262.

Marks, Shula. 'The Silent Scourge: Silicosis, Respiratory Disease and Goldmining in South Africa'. *Journal of Ethnic and Migration Studies,* Vol. 32, No. 4, 2006, pp. 569–589.

McCulloch, Jock. 'Air Hunger: The 1930 Johannesburg Conference and the Politics of Silicosis'. *History Workshop Journal,* Vol. 72, No. 1, 2011, pp. 118–37.

Sandiland, A. 'Review of I. Schapera, *Migrant Labour and Tribal Life*'. *Africa: Journal of the International African Institute,* Vol. 18, No. 2, 1948, p.145.

Trapido, A. An Analysis of the Burden of Occupational Lung Disease in a Random Sample of Former Gold Mineworkers in the Libode District of the Eastern Cape. PhD thesis, University of the Witwatersrand, 2000.

Williams, Haeidee, Rodney Ehrlich, Stephen Barker, Sophia Kisting-Cairncross, Muzimkhulu Zungu and Annalee Yassi. 'The Utility of Length of Mining Service and Latency in Predicting Silicosis among Claimants to a Compensation Trust'. *International Journal of Environmental Research and Public Health,* Vol. 19, No. 6, 2022, pp. 3562–74.

Wilson, Francis. 'Miners and Migrants: How the Mining Industry Has Shaped South Africa'. *Daedalus,* Vol. 130, No. 1, Winter 2001, pp. 99–121.

BY

Notes

Many of the key organisations and territorial units changed their names and governance over the period covered in this book.

The **Miners' Phthisis Medical Bureau**, which was established in 1916, underwent numerous transformations from the **Silicosis Medical Bureau** to the **Pneumoconiosis Research Unit**, established in 1956 for the study occupational disease and occupational hygiene, abbreviated PRU and part of the responsibility of the CSIR (Council for Scientific and Industrial Research). It was changed some years later to be part of the MRC (Medical Research Council) and was known as the **National Research Institute for Occupational Diseases** (NRIOD) It was later taken over from the MRC and became responsible to the Department of Health and Population Development. It was named the **National Centre for Occupational Health** (NCOH) and was headed by a doctor in a joint Government and University post (Wits). It was absorbed into the NHLS as the **National Institute for Occupational Diseases** – and changed its name shortly thereafter to **National Institute for Occupational Health** (NIOH).

J. McCulloch, P. Miller, *Mining Gold and Manufacturing Ignorance*,
https://doi.org/10.1007/978-981-19-8327-6

The Witwatersrand **Chamber of Mines** was founded in 1889. In 1897 it became the Chamber of Mines of the South African Republic until 1902, when it became the Transvaal Chamber of Mines. In 1953 the name was again changed to the Transvaal and Orange Free State Chamber of Mines, until 1968 when it became the Chamber of Mines of South Africa.

The International Labour Office became the International Labour Organisation.

The British currency of the time, pounds, shillings and pence (£sd) were in use in South Africa until 1961 (pegged in value to that of the United Kingdom until 1931). There were twenty shillings to the pound and twelve pennies to the shilling. The units were abbreviated as, e.g., £10. 5s. 6d. (ten pounds, five shillings and six pence). Further, two pounds, ten shillings (for example) was often further abbreviated to £2/10/- and two shillings and six pence to 2/6. The new currency, the Rand, is sub-divided into 100 cents, expressed as, e.g., R6.50.

A number of the colonial territories which supplied labour to the gold mines changed their names at independence: Bechuanaland became Botswana, Basutoland became Lesotho, Southern Rhodesia became Zimbabwe, Northern Rhodesia became Zambia and Nyasaland became Malawi, and Portuguese East Africa became Mozambique.

Basutoland, established as a Crown Colony in 1884, was renamed the Kingdom of **Lesotho** upon its independence from the United Kingdom in 1966.

Bechuanaland became **Botswana** in 1966

Britain assumed control of what became the Bechuanaland Protectorate in 1885. In 1965, the country's capital was moved from Mafeking in South Africa to Gaborone. A year later, Bechuanaland gained independence from Britain and became Botswana.

Northern Rhodesia became **Zambia** in 1964

Northern Rhodesia was formed in 1911 by amalgamating the two earlier protectorates of Barotseland- North-Western Rhodesia and North-Eastern Rhodesia. It was initially administered, as were the two earlier protectorates, by the British South Africa Company on behalf of the British government. From 1924 administration passed to the British government. In 1953, Northern and Southern Rhodesia and Nyasaland formed the Central African Federation. The federation was dissolved in 1963, and Northern Rhodesia became the Republic of Zambia in 1964.

Southern Rhodesia became **Zimbabwe** in 1980

The British South Africa Company administered the territory as a British protectorate from 1889 until 1923, when it became a self-governing British Crown colony. Between 1953 and 1963, Southern Rhodesia was part of the Central African Federation. In 1965, Rhodesia unilaterally declared itself independent under a white-dominated government. After a long civil war, Britain resumed control of the country for a brief period in 1979. A year later Rhodesia gained independence and was renamed Zimbabwe.

Nyasaland became **Malawi** in 1964.

The British Central Africa Protectorate was proclaimed in 1889 and renamed Nyasaland in 1907. Between 1953 and 1963, Nyasaland was part of the Central African Federation. In 1964, the country gained independence from Britain and was renamed Malawi.

Portuguese East Africa became **Mozambique**

The Portuguese began colonizing East Africa in the 16th century. After a war of independence between 1964 and 1974, the country gained independence from Portugal in 1975 and was renamed Mozambique. The capital, known as Lourenço Marques from 1876 to 1976, was renamed Maputo. Civil war broke out in the country in 1977. Relative peace returned to Mozambique in 1993.

Swaziland became **eSwatini** in 2018

The kingdom of Swaziland was a British protectorate from 1903 until it regained its independence in 1968. In 2018, the official name was changed from Kingdom of Swaziland to Kingdom of eSwatini.

South Africa

The South Africa Act passed by the British Parliament in 1909 combined the self-governing British colonies of the Natal, Cape, Transvaal and the Orange Free State into the Union of South Africa. The new country was formally established in May 1910. Majority rule was gained in South Africa in 1994.

Index[1]

[1] Note: Page numbers followed by 'n' refer to notes.

J. McCulloch, P. Miller, *Mining Gold and Manufacturing Ignorance*,
https://doi.org/10.1007/978-981-19-8327-6

M

Printed in the United States
by Baker & Taylor Publisher Services